# Lecture Notes in Computer Scier

Edited by G. Goos, J. Hartmanis and J. van L

T0238268

**Springer**
*Berlin*
*Heidelberg*
*New York*
*Barcelona*
*Hong Kong*
*London*
*Milan*
*Paris*
*Singapore*
*Tokyo*

Jin Akiyama   Mikio Kano
Masatsugu Urabe (Eds.)

# Discrete and Computational Geometry

Japanese Conference, JCDCG 2000
Tokyo, Japan, November 22-25, 2000
Revised Papers

 Springer

Series Editors

Gerhard Goos, Karlsruhe University, Germany
Juris Hartmanis, Cornell University, NY, USA
Jan van Leeuwen, Utrecht University, The Netherlands

Volume Editors

Jin Akiyama
Tokai University, Research Institute of Educational Development
2-28-4 Tomigaya, Shibuya-ku, Tokyo 151-0063, Japan
E-mail: fwjb5117@mb.infoweb.ne.jp

Mikio Kano
Ibaraki University, Department of Computer and Information Sciences
Hitachi 316-8511, Japan
E-mail: kano@cis.ibaraki.ac.jp

Masatsugu Urabe
Tokai University, Department of Mathematics
3-20-1 Orido Shimizu-shi, Shizuoka 424-8610, Japan
E-mail: qzg00130@scc.u-tokai.ac.jp

Cataloging-in-Publication Data applied for

Die Deutsche Bibliothek - CIP-Einheitsaufnahme

Discrete and computational geometry : Japanese conference ; revised papers /
JCDCG 2000, Tokyo, Japan, November 22 - 25, 2000. Jin Akiyama ... (ed.). -
Berlin ; Heidelberg ; New York ; Barcelona ; Hong Kong ; London ; Milan ;
Paris ; Singapore ; Tokyo : Springer, 2001
   (Lecture notes in computer science ; Vol. 2098)
   ISBN 3-540-42306-0


CR Subject Classification (1998): F.2, E.1, G.1-2, I.3.5


ISSN 0302-9743
ISBN 3-540-42306-0 Springer-Verlag Berlin Heidelberg New York

Springer-Verlag Berlin Heidelberg New York
a member of BertelsmannSpringer Science+Business Media GmbH

http://www.springer.de

© Springer-Verlag Berlin Heidelberg 2001
Printed in Germany

Typesetting: Camera-ready by author
Printed on acid-free paper      SPIN: 10839451      06/3142      5 4 3 2 1 0

# Preface

The Japan Conference on Discrete and Computational Geometry (JCDCG) has been held annually since 1997. One of the goals of this conference series is to bring together Japanese researchers from both academia and industry with researchers in these fields from abroad to share their recent results.

JCDCG 2000 was held 22–25 November 2000 at Tokai University in Tokyo in conjunction with the celebration of World Mathematics Year 2000. A total of 120 participants from 20 countries attended. This volume consists of the papers presented at JCDCG 2000, which have been refereed and revised. Some papers which appear in short form in this volume also appear in fuller expanded versions in journals dedicated to computational geometry.

The organizers of the conference thank the principal speakers for their interest and support: Imre Barany, Erik D. Demaine, Greg N. Fredrickson, Gyula Karolyi, Naoki Katoh, David Kirkpatrick, Joseph O'Rourke, Janos Pach, Jozsef Solymosi, William Steiger, Jorge Urrutia, and Allan Wilks. They thank the major sponsors for their generous contribution: The Research Institute of Educational Development of Tokai University, the Ministry of Education of Japan (for the grant-in-aid to A. Saito (A):10304008), and Tokai Education Instruments Co., Ltd.

April 2001

The Editors,
Jin Akiyama
Mikio Kano
Masatsugu Urabe

Organizing Committee
Chairs: Jin Akiyama and Mikio Kano
Members:
Tetsuo Asano, David Avis, Vasek Chvatal, Kiyoshi Hosono, Hiroshi Imai, Takako Kodate, Joseph O'Rourke, Toshinori Sakai, Xuehou Tan, Morimasa Tsuchiya, Masatsugu Urabe, and Jorge Urrutia.

Japan Conference on Discrete and Computational Geometry 2000, Tokyo, Tokai University, November 2000

# Table of Contents

## Papers

# Dudeney Dissections of
# Polygons and Polyhedrons
# – A Survey –

Jin Akiyama and Gisaku Nakamura

Research Institute of Educational Development, Tokai University
2-28-4 Tomigaya, Shibuya-ku, Tokyo 151-0063, Japan
fwjb5117@mb.infoweb.ne.jp

**Abstract.** Given two polygons (polyhedrons) $\alpha$ and $\beta$ with the same
area (volume), the problem of finding a partition of $\alpha$ into parts that
can be reassembled to form $\beta$ is a promising area of study in geometry.
We define a new type of dissection, Dudeney dissection, for polygons and
polyhedrons. The dissection imposes two restrictions, one based on the
reversal of the perimeter (surface area) and the interior (cross-section)
of the polygon (polyhedron), and the other based on the hingeability of
parts. In this paper, we survey main results on Dudeney dissections of
polygons and polyhedrons.

# Part I. Planar Dudeney Dissections

# Chapter 1. Introduction

Given an equilateral triangle $\alpha$ and a square $\beta$ of the same area, Henry E.
Dudeney introduced in [17] a partition of $\alpha$ into parts that can be reassembled
in some way, without turning over the surfaces, to form $\beta$ (Figure 1.1). An
examination of Dudeney's method of partition motivated us to introduce the
notion of Dudeney dissection of a polygon.

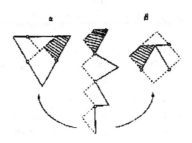

Figure 1.1

J. Akiyama et al. (Eds.): JCDCG 2000, LNCS 2098, pp. 1-30, 2001.
© Springer-Verlag Berlin Heidelberg 2001

**Definition 1.1** *Let $\alpha$ and $\beta$ be convex polygons with the same area. A* **Dudeney dissection** *of $\alpha$ to $\beta$ is a partition of $\alpha$ into a finite number of parts which can be reassembled to produce $\beta$ as follows: Hinge the parts of $\alpha$ like a tree along the perimeter of $\alpha$, then fix one of the parts and rotate the remaining parts about the fixed part to form $\beta$ (this property is called* **transformation***) with the perimeter of $\alpha$ in its interior, and with its perimeter consisting of the dissection lines in the interior of $\alpha$ (this property is called* **reversal***), without turning the surfaces over.*

Analogously we define Dudeney dissections for solids. Many important and beautiful results have been discovered during these two millennia [18,23]. Among many results on planar dissections, the following result obtained independently by Wallace[25], Bolyai[13] and Gerwein[19] is important: An arbitrary polygon can be transformed to any other polygon of the same area by partitioning it into a finite number of pieces and reassembling the pieces in some suitable way.

On the other hand, David Hilbert[21] posed a related problem for solids as one of the problems in his famous list: Give two tetrahedra which cannot be decomposed into congruent tetrahedra directly or by adjoining congruent tetrahedra.

In the same year, Max Dehn[16] solved the problem by introducing the notion of Dehn's invariant showing that a cube cannot be decomposed into pieces which are reassembled to form a regular tetrahedron even if they have the same volume. Later, Jean-Pierre Sydler[24] completed Dehn's work by demonstrating the sufficiency of Dehn's invariant.

This survey article consists of two parts: Planar Dudeney Dissections (Part I) and Dudeney Dissections of Solids (Part II). The problem of planar dissections is completely solved by the judicious use of the theory of planar tessellations. For Dudeney dissections of solids, we have succeeded in giving many distinct families consisting of infinitely many solids which are Dudeney dissectable based on some common principles; however, it is not clear whether these families exhaust all polyhedra which are Dudeney dissectable. Methods employed in obtaining planar Dudeney dissections in Part I and those used for Dudeney dissections of solids in Part II are different in general, but as you will see that for Dudeney dissections of prisms (Chapter 7) methods of planar dissections are utilized a great deal. Furthermore, the idea of "superimposition (Chapter 2)", useful for planar Dudeney dissection also plays a role in Dudeney dissections of space filling type solids in Chapter 6.

# Chapter 2. Notation and Basic Strategy (Superimposition of tilings)

The dissected and hinged version of $\beta$ is called the *Dudeney partner of the dissected and hinged version of $\alpha$*. $\beta$ will denote the Dudeney partner of $\alpha$ throughout. In the figures, the sides of polygon $\alpha$ will be drawn using solids lines while the lines of the dissection will be drawn using dotted lines (Figure 2.1(a)). We refer to the dotted lines as *dissection lines* and the resulting parts of the polygons as *components of the dissection*.

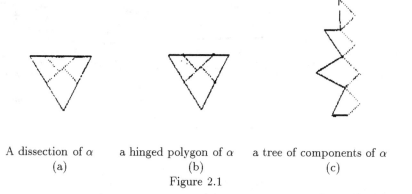

A dissection of α     a hinged polygon of α     a tree of components of α
(a)                        (b)                          (c)

Figure 2.1

Attach hinges to all the vertices of α from which at least one dissection line emanates and call the resulting polygon the *hinged polygon of α*. The hinge points are denoted by small circles (Figure 2.1(b)). All the vertices of polygon α which are not hinge points are referred to as nonhinge points. Hinge points can be suppressed from a hinged polygon, as appropriate, to obtain a *tree components of α* (Figure 2.1(c)).

Suppose β is a Dudeney partner of α. The following are immediate consequences of the fact that both α and β are convex.

**Prop. 1.1** *In a Dudeney dissection, every component of a polygon is convex.*

**Prop. 1.2** *In a Dudeney dissection, every component is bounded by both sides of α and dissection lines.*

The basic strategy used to obtain the main results relates dissections to plane tilings. By a *tiling of the plane with polygon α* is meant an exact covering of the plane with congruent copies of α such that there are no gaps or overlaps. α is referred to as a *tile*. It is well known that every triangle and every quadrilateral tiles a plane and only some special types of pentagons and three different types of hexagons tile a plane. None of polygons with more than six sides tile a plane [15,20].

Denote by α′ the polygon obtained from α by a 180rotation. α′ is called the *half turn* of α (Figure 2.2(a)). A *concatenation* of α and α′ is a polygon obtained by putting α and α′ together along a common side (Figure 2.2(b)). Note that there are infinitely many ways of concatenations of α and α′ depending on how α and α′ are put together. A convex polygon α is said to satisfy the *condition $P_2$* if a concatenation of α and α′ tiles the plane when it is repeated indefinitely in two directions (Figure 2.2(c)). A polygon α satisfying the condition $P_2$ is called a *$P_2$-tiler*.

Let α be a dissected and hinged polygon and β be its Dudeney partner. We discuss the relationship between the Dudeney dissection of α to β and two tilings of the plane, one by α and the other by β. Consider a specific hexagon α which

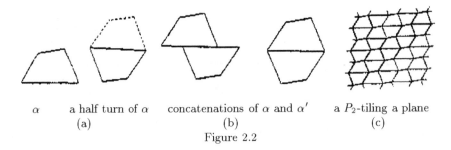

$\alpha$      a half turn of $\alpha$      concatenations of $\alpha$ and $\alpha'$      a $P_2$-tiling a plane
(a)                                    (b)                                    (c)

Figure 2.2

has a Dudeney dissection to $\beta$. Suppose that $\alpha$ can be dissected into four parts $\alpha_1$, $\alpha_2$, $\alpha_3$ and $\alpha_4$ along the dotted lines shown in (Figure 2.3(a)). Since $\beta$ is a Dudeney partner of $\alpha$, $\beta$ (bounded by dotted lines in Figure 2.3(b)) shares a common part $\alpha_1$ of $\alpha$ if we fix $\alpha_1$ and rotate the remaining parts $\alpha_2$, $\alpha_3$ and $\alpha_4$ about $\alpha_1$ (Figure 2.3(b)). Note that the three points $S$, $T$ and $U$ are collinear. Fix a part $\alpha_4$ of $\alpha$, then rotate the remaining parts about $\alpha_4$. This results in a hexagon $\beta$ (bounded by dotted lines) on the righthand side of $\alpha$, sharing $\alpha_4$ in common (Figure 2.3(c)). Note that the three points $S$, $T$ and $U$ are collinear and the straight segment $SU$ in Figure 2.3 (c) coincides with the one in Figure 2.3(b). This implies that the two $\beta$'s, the one in Figure 2.3(b) and the one in Figure 2.3(c) do not overlap and have no gaps (Figure 2.3(d)) between them. Note that the three points $S$, $V$ and $W$ are collinear (Figure 2.3(d)). Fix a part $\alpha_3$ of $\alpha$ and rotate the remaining parts about $\alpha_3$, then we again have $\beta$ (bounded by dotted lines) just under $\alpha_3$, sharing $\alpha_3$ in common without overlaps or gaps (Figure 2.3(e)). Finally fix a part $\alpha_4$ of $\alpha$ and rotate the remaining parts about $\alpha_4$. We obtain a part of a plane tiling by $\beta$, drawn with dotted lines (Figure 2.3(f)).

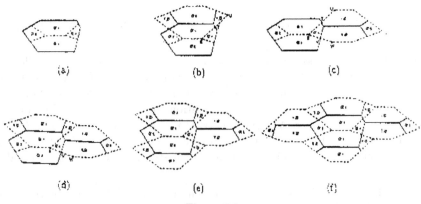

(a)                    (b)                    (c)

(d)                    (e)                    (f)

Figure 2.3

By applying the same operation on each hexagon $\beta$ of the four hexagons which enclose $\alpha$, an extended part of a plane tiling by $\alpha$ is obtained since $\alpha$ is a Dudeney partner of $\beta$. In the same manner, repeat the operation for $\alpha$ and for $\beta$ alternately. The domains tiled doubly by $\alpha$ and by $\beta$ are expanding gradually. Moreover, it follows from the manner of tiling that both tilings, by $\alpha$ and by $\beta$, satisfy the condition $P_2$. It is guaranteed that the operation mentioned above can be applied on an arbitrary $\alpha$ and its Dudeney partner $\beta$, since at every stage of this procedure every part of $\alpha$ has a side which faces to the exterior region of $\alpha$ and the same thing can be said for every part of $\beta$. Thus we have the following result for a general polygon $\alpha$ and its Dudeney partner $\beta$.

**Theorem 2.1[4]** *Let $\alpha$ be a polygon with a Dudeney dissection $\beta$, then both $\alpha$ and $\beta$ satisfy the condition $P_2$.*

**Superimposition of the two tilings**

Theorem 2.1 implies the following method to find the Dudeney dissection between $\alpha$ and $\beta$, where $\alpha$ and $\beta$ have the same area. Let T(A) be a plane tiling by A. Chose an arbitrary vertex of the tile A and color it with red, then the red vertices are distributed throughout the whole plane in T(A). Superimpose T(B), a plane tiling by B, on T(A) appropriately. If every red vertex locates in the same position in some B of T(B), it gives a Dudeney dissection between A and B.

The superimpositions drawn in Figure 2.4(a) and 2.4(c) do not give congruent Dudeney dissections, but they do in Figures 2.4(b) and 2.4(d).

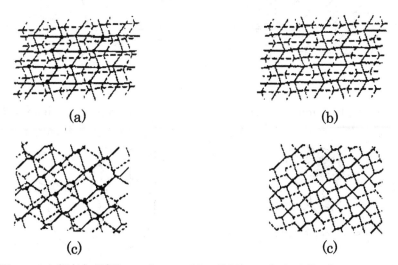

(a)                                    (b)

(c)                                    (c)

Figure 2.4 (T(A), T(B) are drawn with solid lines, dashed lined, respectively)

## Chapter 3. Verious Results of General Dudeney Dissections of Polygons

The results can be classified into two, namely general Dudeney dissections and congruent Dudeney dissections. A general Dudeney dissection of a polygon $\alpha$ produces a polygon $\beta$, which is not congruent to $\alpha$. A *congruent Dudeney dissection* of a polygon $\alpha$ produces $\alpha$ itself. In this section, we introduced various results on general Dudeney dissections which are discussed in the paper [2]. It provides procedures for constructing the dissections shown below.

Every quadrilateral has a Dudeney dissection to a parallelogram (Figure 3.1).

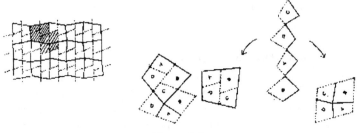

Figure 3.1

Every triangle has infinitely many Dudeney dissections to parallelograms (Figure 3.2).

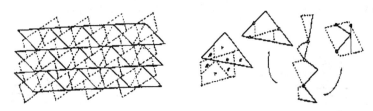

Figure 3.2

Every parallel hexagon has a Dudeney dissection to a trapezoid (Figure 3.3).

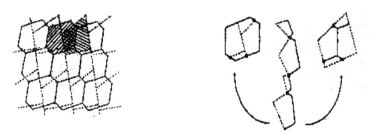

Figure 3.3

Every parallel hexagon has a Dudeney dissection to a triangle (Figure 3.4).

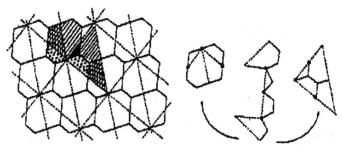

Figure 3.4

# Chapter 4. Congruent Dudeney Dissections of Polygons

A congruent Dudeney dissection of a polygon $\alpha$ produces $\alpha$ itself. In the discussion of congruent Dudeney dissections of polygons, three cases are considered depending on the location of hinge points:

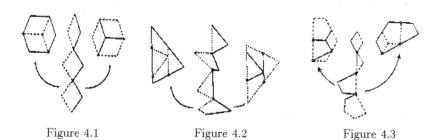

Figure 4.1              Figure 4.2              Figure 4.3

**Case 1.** All hinge points are interior points of the sides of $\alpha$ and so also of $\beta$ (Figure 4.1).

**Case 2.** All hinge points are on the vertices of polygon $\alpha$ and so also of $\beta$ (Figure 4.2).

**Case 3.** Some hinge points are vertices of either $\alpha$ or $\beta$ while others are interior points of their sides (Figure 4.3).

Cases 1,2 and 3 are discussed in papers [3], [4] and [5], respectively. The following are some of the results for cases 1, 2 and 3.

**Case 1. All hinge points are interior points of the sides of $\alpha$ and so also $\beta$.**

We state here only some of the main results in this category by illustrating examples.

Every triangle has a Dudeney dissection to a triangle (Figure 4.4(a)) and also to itself (Figure 4.4(b)).

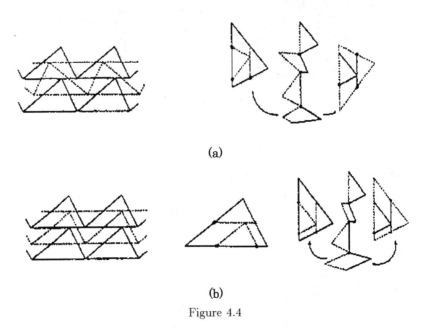

(a)

(b)

Figure 4.4

Every quadrilateral has a Dudeney dissection to another quadrilateral (Figure 4.5(a)) and also to itself (Figure 4.5(b)).

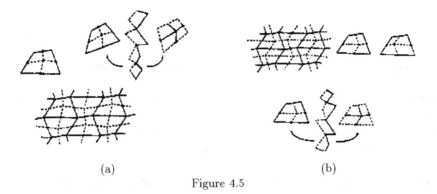

(a)                                    (b)

Figure 4.5

A parallelogram ABCD has a congruent Dudeney dissection if and only if
BD=BC (Figure 4.6(a),(b)).

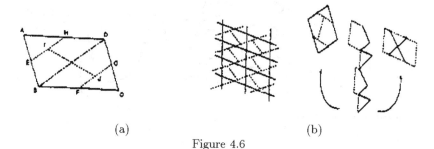

(a)                                              (b)

Figure 4.6

A $1 \times n$ rectangle has a congruent Dudeney dissection, where $n$ is any integer
greater than one (Figure 4.7).

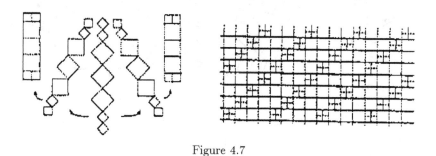

Figure 4.7

Every parallel pentagon (a pentagon with a pair of parallel sides (Figure 4.8(a)))
has a Dudeney dissection to another parallel pentagon (Figure 4.8(b)). No pentagon has a congruent Dudeney dissection.

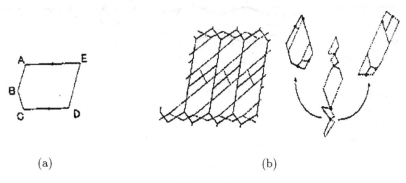

(a)                                              (b)

Figure 4.8

Every quasi parallel hexagon (Figure 4.9(a)) has a congruent Dudeney dissection (Figure 4.9(b)).

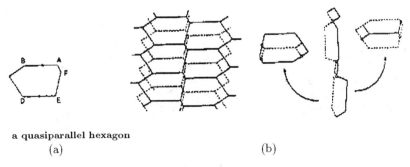

a quasiparallel hexagon
(a)

(b)

Figure 4.9

Every pentagonal hexagon (Figure 4.10(a)) has a congruent Dudeney dissection (Figure 4.10(b)).

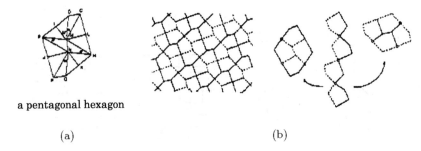

a pentagonal hexagon

(a)

(b)

Figure 4.10

Every symmetric parallel hexagon (Figure 4.11(a)) has a congruent Dudeney dissection (Figure 4.11(b)).

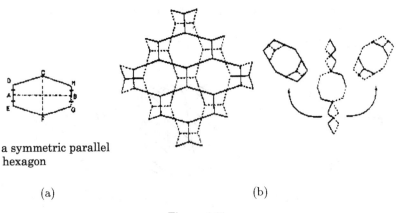

a symmetric parallel
hexagon

(a)

(b)

Figure 4.11

**Case 2. All hinge points are on the vertices of polygon $\alpha$ and so also of $\beta$.**

A hexagon $\alpha$ =ABCDEF has a Dudeney dissection if and only if $\alpha$ satisfies the following conditions (Figure 4.12(a),(b)) :

$$FA = AB, \ BC = CD, \ DE = EF,$$
$$\angle ABC + \angle CDE + \angle EFA = 2\pi,$$

where A, C, E are hinge points and B, D, F are not.

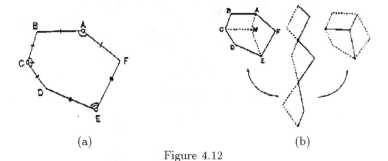

| (a) | (b) |
|---|---|

Figure 4.12

A hexagon $\alpha$ =ABCDEF with an interior point M, has a congruent Dudeney dissection if and only if either of the two conditions are satisfied (Figure 4.13):

$$\left\{ \begin{array}{l} \text{CDEM is a parallelogram,} \\ \text{ABCM=AMEF,} \\ \text{AB=AM=AF,} \end{array} \right. \qquad \text{(a)}$$

$\alpha$ is an equilateral parallel hexagon with MA=MC=ME. \hfill (b)

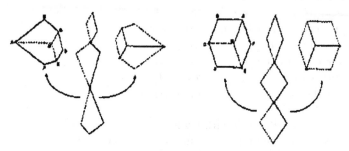

Figure 4.13

**Case 3. Some hinge points are vertices of either α or β while others are interior points of their sides.**

We show here only a few examples which are either triangles or quadrilaterals. Analysis for this case is much more complicated than the first two. Only two different types of triangles, both are isosceles triangles, have congruent Dudeney dissections. The associated dissections are shown in Figures 4.14 and 4.15.

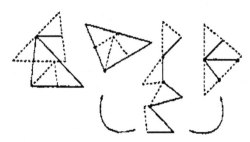

Dudeney dissection to another triangle
Figure 4.14

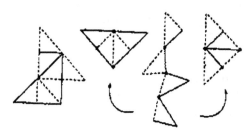

Congruent Dudeney dissection
Figure 4.15

The analysis of quadrilaterals belonging to Case 3 considers how many vertices of α become vertices of β.

[i]    All four vertices of α become vertices of β.
[ii]   Three vertices of α become vertices of β.
[iii]  Two vertices of α become vertices of β.
[iv]   One vertex of α become a vertex of β.
[v]    None of the vertices of α become vertices of β.

There are no quadrilaterals with Dudeney dissections satisfying [i] and [ii]. On the other hand, there are many quadrilaterals with either Dudeney dissections or congruent Dudeney dissections satisfying [iii], [iv] or [v]. Examples corresponding to these cases are shown in Figures 4.16, 4.17 and 4.18, respectively.

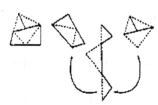
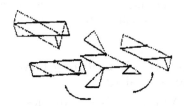

Dudeney dissection to
another quadrilateral
(a)

Congruent Dudeney dissection

(b)

Figure 4.16

Dudeney dissection to
another quadrilateral
(a)

Congruent Dudeney dissection

(b)

Figure 4.17

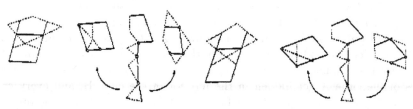

Dudeney dissection to
another quadrilateral
(a)

Congruent Dudeney dissection

(b)

Figure 4.18

# Part II. Dudeney Dissection of Solids

By extending the idea of Dudeney dissection of one polygon into another in the plane, one can consider Dudeney dissection of a polyhedron into another. In the following chapters, let us explain Dudeney dissections of a variety of only interesting solids. Refer to [6,7,8,9,10,11] for the details.

# Chapter 5. Definition and various types of Dudeney dissection of solids

Although it is easy to guess what the definition of Dudeney dissection of a solid into another should be in analogy with planar Dudeney dissection, we will give a precise definition in order to pinpoint the problems involved.

**Definition 5.1** *Suppose convex polyhedrons A and B have the same volume. We say that the polyhedron A has a* **Dudeney dissection** *into B if A has a dissection into a finite number of pieces, which are connected to form a tree, and if the pieces can be rearranged without losing the connectivity to form the solid B in such a way that all of the exterior surfaces of A get into the interior of B, and all of the faces of the dissection of A will appear as the exterior surfaces of B (call this property* **reversal***).*

*Remark.* Pieces of dissection are put together to form a connected chain by gluing together sides of two pieces by gummed tapes (this kind of hinges are called *piano hinges*). This is the same as saying that two adjoining edges of the two pieces are attached together by hinges. You can picture this by imagining a door being opened and closed.

Though this is the definition of Dudeney dissection of solids, it may not be easy to visualize it concretely, so let us give a simple example to illustrate it. Figure 5.1 below shows a dissection of a big cube into 8 smaller cubes. When you put together these little cubes in various ways, using piano hinges, you discover the way to put them together as in Figure 5.2. This way of putting together the pieces yields a Dudeney dissection of the big cube A into another cube B, if you rearrange smaller cubes in the way indicated in Figure 5.2. Namely, exterior faces of the cube A get hidden in the interior of the cube B, and every cross section of A appears as an exterior face of B, that is, this satisfies the property reversal. It is easy to construct this model, so construct one and try to convince yourself by experimenting with it.

In search for new types of Dudeney dissections of polyhedrons you encounter more and newer varieties of Dudeney dissections of solids. However, if you want to discover new types of such dissections, you need a disciplined approach. A haphazard way of searching will not yield good results usually. Then what is the strategy you should employ in your search? You should figure out mathematical principles lurking behind Dudeney dissections of solids. Through meticulous investigations, we succeeded in discovering that there are at least four different

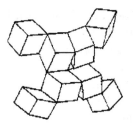

Figure 5.1                                    Figure 5.2

mathematical principles which govern Dudeney dissections of polyhedrons. If we follow each of these mathematical principles, then characteristically, you will catch all at once a large family of Dudeney polyhedrons of solids based on that principle. It may be possible that there are still other mathematical principles besides these four, but at present we do not know whether such principles exist. Thus, we have decided to classify Dudeney dissections of solids into the following six types.

1. Space filling type [6]

2. Layered type    (a) 2-Layered type [9]    (b) 3-Layered type [7]
                   (c) Multi-layered type [10]

3. Mirror image type [8]

4. Clasping type [11]

The reasons for naming each of these types as such will become clear as we proceed to investigate the mechanism underlying Dudeney dissections of solids.

# Chapter 6. Space filling type Dudeney dissections and superimposition of two tessellations

A solid is called a *space filling solid* or a *space filler* if a space can be filled without having any gaps or overlaps by using copies of the solid. There are many such solids such as cubes, rectangular blocks, a special kind of tetrahedrons, triangular prisms, semiregular 14-hedrons (truncated octahedrons) (Fig.6.1(a)), rhombic dodecahedrons (Fig6.2(a)) and so on. Refer to [1,23] for details.

By space filling type we refer to Dudeney dissections of solids constructed by a clever use of properties of space filling solids, which is a generalization of the superimposition of two tilings applied for planar cases. Let us explain this method (we call this method superimposition of two tessellation) more concretely by taking two examples of semiregular 14-hedrons and rhombic dodecahedrons.

## (a) Dudeney dissection between a semiregular 14-hedron and a rectangular block

As we described above you can place semiregular 14-hedrons of same size as in figure 1 to honeycomb the entire space. Suppose you take out just two of these 14-hedrons and stack them as in Figure 6.1(a). By connecting four vertices lying on the lower portion of each of these 14-hedrons by dotted lines you form two squares and connecting these squares with four vertical dotted lines, you see that a rectangular block will emerge from the two 14-hedrons(Figure 6.1(b)). You discover that by repeating the same process for each vertically adjacent pair of 14-hedrons you end up with the honeycombing of the space by rectangular blocks. From the way that each rectangular block was constructed, it is clear that the relative positions of a semiregular 14-hedron and a rectangular block remain constant in space, and that the volume of a 14-hedron and of a rectangular block is the same.

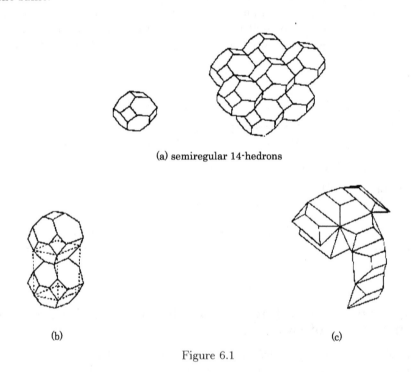

(a) semiregular 14-hedrons

(b)                                                          (c)

Figure 6.1

Space filling type Dudeney dissection of solids are constructed based on this property of space filling solids. Suppose you draw a tessellation of the space by means of semiregular 14-hedrons and draw another by means of rectangular blocks and superimpose these drawings. You then dissect a semiregular 14-hedron into 6 pieces by means of the 6 planes forming the exterior faces of a rectangular block interecting the 14-hedron, then you obtain a Dudeney dissection of the 14-hedron into the rectangular block. Figure 6.1(c) illustrates this fact. If you roll in the pieces of dissection you get the semiregular 14-hedron, while if you roll out the pieces you get the rectangular block. The fact that this is indeed what happens is more or less obvious from the way semiregular 14-hedrons and the rectangular blocks honeycomb the space.

### Superimposition of two tessellations of the space

If you superimpose a tessellation of the space by one of space filling solids onto a tessellation of the space by another such solid in an adroit way thereby keeping relative positions of these solids constant, and then dissect one of the solids of one kind by the planes forming exterior faces of the solid of the other kind that intersects the former, you will obtain a space filling type Dudeney dissection between these 2 solids. We emphasize here that the superposition of 2 tessellations has to be done adroitly as it is important that everything must fit together when corresponding vertices and sides of solids are put together by gummed tapes and rotated.

### (b) Dudeney dissections between a rhombic dodecahedron and a rectangular block

As another example of space filling type Dudeney dissection of solids, let us explain Dudeney dissections between a rhombic dodecahedron and a rectangular block, by applying for superimposition of two tessellations of the space. A rhombic dodecahedron is a space filling solid. We can fill up the space with no gaps and no overlaps, as in Figure 6.2(a), by using copies of this solid. By picking points located at the same position on each of the corresponding edges of the surfaces of these dodecahedrons, and connecting them by dotted lines as shown in Figure 6.2(b), we obtain rectangular blocks. These rectangular blocks also honeycomb the space as in Figure 6.2(c), and thus, by dissecting one of the rhombic dodecahedrons into 9 pieces by means of the planes comprising the exterior surfaces of the rectangular block intersecting the rhombic dodecahedron, we obtain, as in Figure 6.2(d), a Dudeney dissection between a rhombic dodecahedron and a rectangular block. By means of superimposition of tessellations we can obtain surprisingly many Dudeney dissections between two kinds of space filling solids.

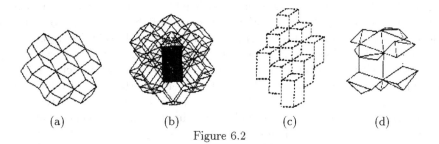

(a)              (b)              (c)              (d)

Figure 6.2

# Chapter 7. Layered type Dudeney dissections of solids
## –Reduction to the case of planar Dudeney dissections–

Solids, such as triangular or rectangular prisms, which have the shape of pillars, are called *prisms*. When lateral sides of a prism are perpendicular to the top and the bottom, it is called a *right prism*. By layered type Dudeney dissections of solids, we refer to a certain class of Dudeney dissections of right prisms. There are two essentially different Dudeney dissections of right prisms, one is 3-layered type and another is 2-layered type. On a basis of there two cases, we can produce Dudeney dissections of certain type of prisms which can be obtained by slicing them into $n$ layers for an arbitrary $n$. We start with 3-layered type.

### (a) 3-layered type Dudeney Dissections of Solids

This class of dissections shares a characteristic of having two planes parallel to the top (and bottom) of the prism slicing the prism into three layers. Here, we emphasize that we are referring to one particular class of Dudeney dissections of right prisms, as there exists another class, called 2-layered type dissections, of right prisms.

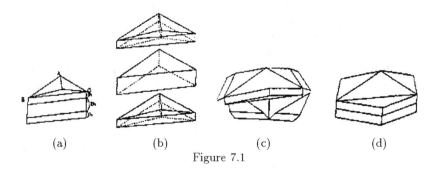

(a)            (b)            (c)            (d)

Figure 7.1

### Method of layered type Dudeney dissections of solids

Let us first take the simplest case of a triangular right prism, and explain a Dudeney dissection between a triangular and a hexagonal right prisms. In Figure 7.1 (a) a triangular right prism of height $4h$, which is partitioned into three layers of triangular right prisms by means of the two planes parallel to the top (and bottom) of the prism. Here, the first and the third layers have height $h$, while the second layer has the height $2h$. The number $h$ can be an arbitrary positive number. We leave the second layer prism intact, while the first and third layer prisms are dissected further into three smaller triangular prisms each, by means of three planes perpendicular to the top of the initial prism and intersecting the top triangle along the three line segments connecting a point P arbitrarily chosen from the interior of the triangle and the vertices A, B and C of that triangle. The three layers shown in Figure 7.1 (b) illustrate this manner of dissection. We next attach the three pieces of the first layer and those of the third layer to the second layer prism by gluing them by gummed tapes along the common edges as shown in Figure 7.1(c), and then flipping these smaller triangular prisms overs the glued edges, we obtain a hexagonal right prism of height $2h$, as shown in Figure 7.1(d). We notice that by this procedure, we end up with having the top and the bottom surfaces and the lateral surfaces of the resulting hexagonal right prism consist of the cross sections of the dissection of the original triangular right prism, that is, it satisfies the reversal property, and therefore, we have a Dudeney dissection between a triangular right prism of height $4h$ and a hexagonal right prism of height $2h$. This provides the simplest example of layered type Dudeney dissections of solids. Figure 7.2 illustrates a Dudeney dissection between a triangular right prism of height $4h$ and a pentagonal right prism of height $2h$, while Figure 7.3 shows a Dudeney dissection of the same type between a rectangular right prism of height $4h$ and a quadrilateral right prism of height $2h$.

Figure 7.2                              Figure 7.3

## Reduction to the case of planar Dudeney dissection

With some more manipulation, we can extend the layered type Dudeney dissections discussed above to more general layered type Dudeney dissections of solids. In order to explain the ideas involved, let us take a very simple example of a triangular right prism, whose top (and bottom) is a right triangle, as shown in Figure 7.4(a). The height of this prism is $4h$, and it is partitioned into three layers of triangular prisms with height $h$, $2h$ and $h$, respectively. We connect the first and the third layer to the second layer by gluing them by gummed tapes along the lines of dissection appearing on the same lateral side of the original prism, and then flip the first and the third layer prisms over the glued edges, as shown in Figure 7.4(b). Going through the stage shown in Figure 7.4(c), we end up with a triangular right prism of height $2h$, as shown in Figure 7.4(d).

The top and bottom of the resulting prism are isosceles triangles and they are made up of the cross sections of the dissection of the original prism. However, each of the three lateral sides of the resulting prism still is made up of lateral surfaces of the original triangular prism. So, let us figure out a way to convert these lateral sides to those consisting of cross sections of the dissection of the original prism. In order to achieve this, it suffices to consider a planar congruent Dudeney dissection of the isosceles triangle, as shown in Figure 7.5. In this way,

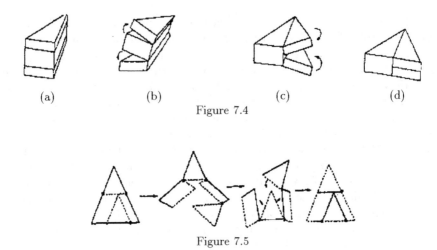

(a)                    (b)                    (c)                    (d)

Figure 7.4

Figure 7.5

we obtain a Dudeney dissection of a triangular right prism shown in Figure 7.5, which is obtained by different manner from the one discussed earlier.

**Generalization of the 3-layered method**

For general layered type Dudeney dissections of solids discussed above, the top and bottom faces of the prism were flipped over one of the edges on the same lateral side of the original prism to form polygons which are symmetric with respect to these edges. Therefore, if we can achieve a planar Dudeney dissection of this polygon, then we can obtain a Dudeney dissection of the original solid. Polygons shown in Figure 7.6, including the isosceles triangle of Figure 7.4(d), all of which are symmetric with respect to a line segment, have such planar Dudeney dissections. Dashed lines in these pictures show that these polygons are obtained by flipping pieces over these lines. By applying this method, we can obtain many different kinds of layered type Dudeney dissections of solids. Furthermore, it is known that you can apply this basic idea, with some clever manipulations, to obtain Dudeney dissections of prisms, which are not right prisms but slanted somewhat.

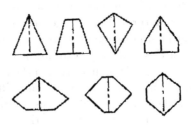

Figure 7.6

## (b) 2-layered type Dudeney Dissections of Solids

2-layered type Dudeney dissections of solids refer to Dudeney dissections of right prisms, as it was the case for 3-layered type Dudeney dissections; however, 2-layered type dissections have a characteristic of dissecting prisms into two layers by means of a single plane parallel to the top (and the bottom) of the prism, whereas 3-layered type dissections discussed earlier dissected prisms into three layers. In order to explain concretely the mechanism involved, let us take an example of a cube, which is easy to visualize. Suppose we take a cube A, and consider a way to convert it to a cube B, congruent to the cube A, by means of a Dudeney dissection of 2-layered type. We partition, as in Figure 7.7(a), the cube A by slicing it by means of a plane parallel to the top of the cube A to obtain two congruent square blocks. We then put together these square blocks by gluing along one of the edges of the cross section by a gummed tape, as in Figure 7.7(b), and by flipping the upper block over the glued edge, we get a rectangular block of Figure 7.7(c). This rectangular block has width, length and height of 1, 2 and $\frac{1}{2}$, if the length of each edge of the initial cube equals 1.

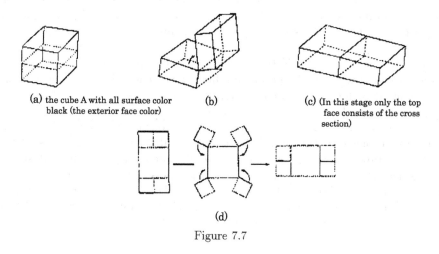

(a) the cube A with all surface color     (b)                (c) (In this stage only the top
black (the exterior face color)                                   face consists of the cross
                                                                  section)

(d)

Figure 7.7

If we observe the exterior faces of this rectangular block, we notice that the top face consists of the cross sections of the dissection of the cube A, but the lateral surfaces and the bottom face remain as exterior faces of the cube A. So, let us figure out a way to convert the lateral surfaces of the rectangular block into surfaces consisting of the cross sections of the dissection of the cube A. For this purpose, we can use a planar Dudeney dissection of the top rectangle of the block. Figure 7.7(d) illustrates this procedure. What remains now is to take care of the bottom face, but this can be done easily by going through the procedures indicated in Figures 7.8(a),(b),(c). In this way a Dudeney dissection of the cube A into the cube B is accomplished.

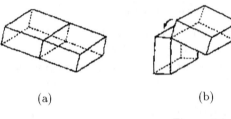

(a)                          (b)                          (c) the cube B with all surface
                                                              color white (interior color)

Figure 7.8

**Generalization for the 2-layered method**

We note that Dudeney dissection of solids of this type discussed above consists of slicing the cube into two layers to get a rectangular block with half the height, following it with a planar Dudeney dissection of the top face of the block, and finally stacking one-half of the rectangular block on top of the other half to restore the two layers. The process of slicing the cube and that of stacking pieces to restore two layers are both quite simple, and therefore, the heart of 2-layered type Dudeney dissections lies in performing a planar Dudeney dissection of the top face of the rectangular block. A characteristic property of this planar Dudeney dissection is that it deals with two polygons symmetric with respect to an axis. Since the problem of finding Dudeney dissections of solids is thus reduced to finding planar Dudeney dissections, it becomes much simpler. As further examples of this type of Dudeney dissections of solids, we give two more examples of Dudeney dissections between two congruent solids. Figure 7.9 illustrates a congruent Dudeney dissection of a double thickness right prism having the top (and bottom) face of an equilateral triangle, while Figure 7.10 shows a congruent Dudeney dissection of a double thickness right prism having the top (and bottom) face of an isosceles trapezoid. At a glance, they look rather simple, but they are actually quite intricate. We can construct many other examples of 2-layered type Dudeney dissections of similar solids.

**(c) Multi-layered Type Dudeney Dissections of Solids**

Taking hints from the procedures involved in two layered or three layered type Dudeney dissections of solids discussed earlier, we can show that Dudeney dissections of certain type of prisms can be obtained by slicing them into $n$ layers for an arbitrary $n$. We have to have prisms with their top (and bottom) consisting of polygons of rather special kind in order to consider $n$ layered dissection for an arbitrary $n$. The simplest example is given by the case of a rectangular prism. Since this case illustrates well the essential feature of multi-layered Dudeney dissections, and it can be explained quite simply, we will consider a rectangular prism in the sequel.

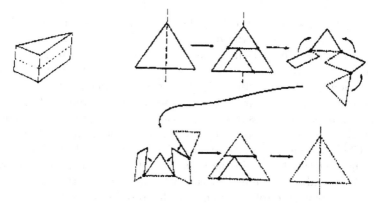

Figure 7.9

(Start and end with a double thickness right prism having the top face of an equilateral triangle.)

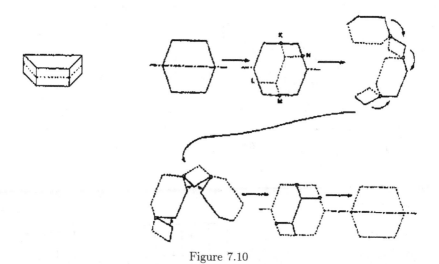

Figure 7.10

(Start and end with a double thickness right prism having the top face of an isosceles trapezoid.)

**An example of a multi-layered Dudeney dissection of a rectangular prism**

Let us consider a prism having rectangular top and bottom faces as in Figure 7.11(a). This prism, however, is not a right prism but has a slight slant. This slant will play a significant role in the $n$-layered type Dudeney dissections of solids. The reason for this comes from the fact that it is necessary to go through an operation to change the top and bottom faces of the prism to surfaces consisting of cross-sections of the dissection before we perform slicing into $n$ layers. This operation can be executed quite easily, however. For example, if we want to change the top face to the surface consisting of cross-sections, we divide the top rectangle into two equal parts by cutting along the dotted line indicated in Figure 7.11(b), and cut out a triangular wedge beneath one half of the rectangle and flip this over the dotted line and glue it to the other half of the rectangle. Figures 7.11(c) and 7.11(d) illustrate this procedure. We see that the dotted line on the top face plays the role of a hinge in this process. If we perform the same operation on the bottom face, we obtain a rectangular prism with both top and bottom consisting of cross-sections of the dissection. If the triangular wedge used in the procedure above has a right triangle on its side face, then, the rectangular prism obtained after this operation will become a right prism. We started off with a rectangular prism with a slant in order to achieve this goal, namely to end up with a right rectangular prism. In this way, we get a rectangular right prism shown in Figure 7.11(e) with top and bottom faces consisting of surfaces of cross-section. We now dissect this right prism, for instance, into five layers as

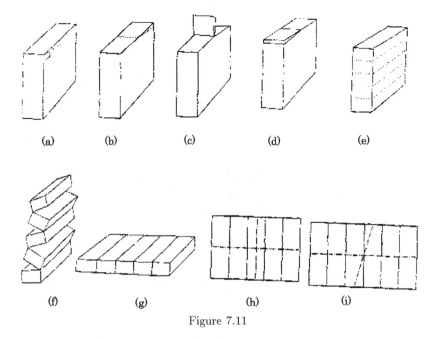

Figure 7.11

in Figure 7.11(e) and spread out by unfolding dissected pieces over the lines of dissection appearing on the lateral sides of the prism. Figure 7.11(f) illustrates this process, and Figure 7.11(g) shows the resulting rectangular block. This rectangular block has the width five times the width of the original rectangular prism and the height one fifth of the prism. It is easy to check that the top and the bottom faces of this rectangular block consist of the surfaces of cross-section of dissection. Exterior surfaces of the original rectangular prism remain on this rectangular block only on the four lateral faces. Therefore, a Dudeney dissection of the original rectangular prism will be obtained if we perform a planar Dudeney dissection of the top face of the rectangular block. In performing such a planar dissection, we should be careful not to use the original lines of dissection into the five layers appearing on the top face of the block. Because of this restriction, there is a subtle distinction in choosing a correct planar Dudeney dissection depending on the number of layers being even or odd. Figure 7.11(h) gives one example of the planar dissection for a case of odd number of layers, and Figure 7.11(i) shows one example for an even numbered layer case. Lines of planar dissections are indicated by dashed lines with dots in these figures.

Although we discussed only the case of a prism with rectangular top and bottom faces above, there are number of cases of the prisms with different type of polygons on top and bottom, to which multi-layered type Dudeney dissection can be performed. However, when we unfold the prism after slicing it into layers, we may have to unfold not only in one direction, as was the case in the example above, but to two directions (from left to right, and up and down), or as in the case of unfolding a triangle, we may have to unfold zigzaggedly.

# Chapter 8. Mirror image type Dudeney dissections of solids

By mirror image type Dudeney dissections of solids, we mean Dudeney dissections of solids obtained by a clever use of the properties of reflection by mirrors. In order to explain more concretely what this type of Dudeney dissections is like, let us take an example of a cube that can be visualized clearly.

### Method of mirror image type Dudeney dissections of cubes

Suppose we consider a plane passing through the center and one of the edges of a cube. This plane contains another edge of the cube situated symmetrically across the center from the first edge. Therefore, if we consider all the planes containing the center and one of the twelve edges of the cube, you obtain six different planes. If we dissect the cube by means of these six planes, we get, as in Figure 8.1, six congruent right square pyramids.

Next, we consider an image of each of these pyramids obtained by reflecting it by a mirror situated on its base (which is one of the faces of the cube). Then we obtain six octahedrons formed by each combination of a pyramid and its mirror image (sandwiching the mirror from above and below). Figure 8.2 (a) illustrates the way two square pyramids (one being the mirror image of the other) approach each other from above and below, and Figure 8.2(b) shows the octahedron obtained when these two pyramids come together. As a result we

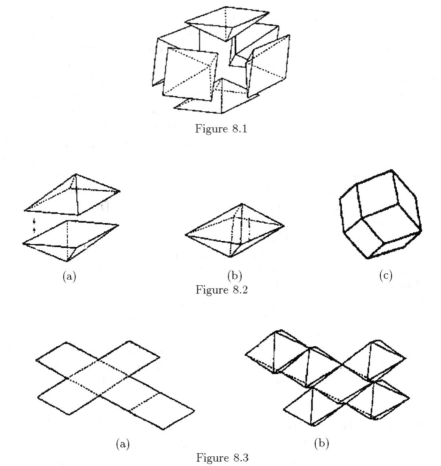

Figure 8.1

(a)                    (b)                    (c)

Figure 8.2

(a)                    (b)

Figure 8.3

obtain a solid having a right square pyramid attached to each of the six faces of the cube, and this solid is a rhombic dodecahedron having as vertices the six top vertices of the attached square pyramids. Figure 8.2(c) illustrates the rhombic dodecahedron obtained in this way.

If we use this procedure, you obtain a mirror image type congruent Dudeney dissection of a rhombic dodecahedron (hinged inversion of a rhombic dodecahedron). Suppose we draw a development of a cube, as in Figure 8.3(a),(b), and we attach to each of the six squares two right square pyramids of Figure 8.2 (a), one on the top side and the other on the bottom side of the square to get six octahedrons as in Figure 8.2 (b). We then glue together, as shown in Figure 8.3(b), the adjacent octahedrons by gummed tapes along the edges (indicated by dotted lines) of adjacent squares of the development. By folding these octahedrons by flipping them inwards we get a rhombic dodecahedron, and by folding them by flipping them outwards we also get a rhombic dodecahedron of the same shape. This means that we have a congruent Dudeney dissection of a rhombic dodecahedron. If we use solids different from square pyramids, we can obtain Dudeney dissections of solids such as decahedron as well.

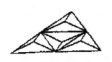   

Figure 8.4                    Figure 8.5

The characteristic feature of Dudeney dissections of solids of this type is the fact that a basic solid, cube in the case above, is given to start with, and by reflecting each of the six square pyramids lying inside the cube by means of a mirror located on its bottom surface (which is one of the faces of the cube) we get six square pyramids, which are mirror images of the interior pyramids, attached to the cube from outside. The reason why the type of Dudeney dissections obtained in this way is called "mirror image" type is clear. We may use also, in place of cubes, tetrahedrons and triangular prisms as the basic solid to start the operation. The reason why these solids can be used as the basic solids lies in the fact that any of the angles formed by two abutting exterior faces of such solids is always less than 90 degrees. If we go through the same procedure as above by starting with a tetrahedron or with a triangular prism, for example, we obtain a Dudeney dissection of a dodecahedron of Figure 8.4 or of Figure 8.5, respectively. Investigating carefully the mechanism involved in mirror image type Dudeney dissections of solids, we may also obtain exotic Dudeney dissections of solids which involve no reflected images.

# Chapter 9. Clasping Type Dudeney Dissections of Solids

By clasping type Dudeney dissections of solids, we mean those Dudeney dissections which involve at least one application of the "clasping" operations described below. Because of this definition, this class of Dudeney dissections is not disjoint from the classes of Dudeney dissections of solids discussed earlier: space filling type, two layered type, three layered type and mirror image type. However, since there exist Dudeney dissections of solids, which belong to the clasping type, but to none of the other types, it is useful to regard this as constituting one class.

Let us now explain what this clasping operation is. Take a solid (polyhedron) on which we want to perform a Dudeney dissection and pick out one of the edges. If the two faces (which are polygons) of the polyhedron that abut this edge are symmetric with respect to this edge, then we can clasp these two faces together by slicing the polyhedron by means of a plane through this edge, joining the two faces with a hinge along the edge, and then turning over one sliced half of the polyhedron over the edge. Let us all operation described be *the clasping operation*.

*Examples*

## (a) Cubes

Let us look at a very simple example. This involves a congruent Dudeney dissection of a cube of Figure 9.1(a). Let us designate the top face (a square) of the cube ABCD and the bottom square EFGH, and pick out the edge AE. Two square faces AEFB and AEHD that abut the edge AE are symmetric with respect to the edge AE, and therefore, the clasping operation can be performed on these square faces. We slice the cube into two parts, as indicated in Figure 9.1(b), by means of the plane going through the three points A, E and C. Then put a hinge on the edge AE, and turn one of the triangular prisms obtained by the slicing over the edge AE so as to clasp together the two square faces. Figure 9.1(c) illustrates the outcome of this operation: it gives a triangular prism with right isosceles triangles on the top and bottom faces. If we apply a three layered type dissection to the lateral face containing the edge CI, we get a rectangular block, as in Figure 9.1(d), with the height half as that of the original cube and the area of the top rectangle twice that of the square face of the cube. Note that the top and bottom faces of the resulting rectangular block are made up with the surfaces of the cross-section obtained by the slicing of the cube. Therefore, if we apply a planar Dudeney dissection to the top rectangle of the rectangular block, then we accomplish a congruent Dudeney dissection of the cube.

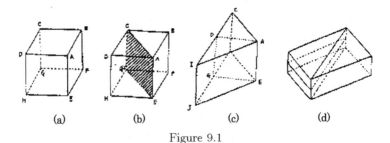

(a)        (b)        (c)        (d)

Figure 9.1

## (b) Hexahedrons

Although the example above indicated in Figures 9.1(a)–(d) is almost trivial, the next example gives a spectacular Dudeney dissection of a hexahedron into another hexahedron. For this example, we start off with a special tetrahedron ABCD as in Figure 9.2(a). On this tetrahedron, four triangular faces consist of two pairs, triangles ABC and ABD, and triangles CDA and CDB. The triangles ABC and ABD in the first pair are symmetric with respect to the edge AB, while CDA and CDB in the second pair are symmetric with respect to the edge CD. Consequently, AC = AD = BC = BD holds, and the paired triangles are congruent and they are isosceles triangles. If we now choose an arbitrary point E on the edge CD of the tetrahedron, as indicated in Figure 9.2(b) and dissect the tetrahedron by means of the plane passing through the three points A, B and E, then we can clasp together the two triangular faces ABC and ABD by turning one of the two sliced portions over the hinged edge AB. Figure 9.2(c) illustrates the hexahedron obtained by this procedure. In the same manner, we can clasp together the two triangular faces CDA and CDB by taking an arbitrary point G on the edge AB and dissecting the tetrahedron by means of the plane going through the three points C, D and G and turning one of the sliced portions over the hinged edge CD. Figure 9.2(e) shows the resulting hexahedron. If we suppose that we were given the hexahedron of Figure 9.2(c) to start with and transform it to the hexahedron of Figure 9.2(e) via the procedures illustrated in Figures 9.2(b) and 9.2(d), then we obtain a Dudeney dissection of a hexahedron into another one. This is a typical example of clasping type Dudeney dissection of solids, and this does not belong to any of the types, (space filling, two layered, three layered, or mirror image) discussed earlier.

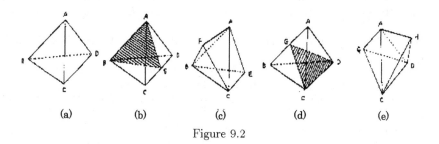

(a)        (b)        (c)        (d)        (e)

Figure 9.2

**Acknowledgments:** The authors thank Greg N. Frederickson for his invaluable suggestions, Mari-Jo Ruiz and Yuji Ito for their assistance throughout the preparation of this paper.

# References

1. Akiyama, J., Nakamura, G.: A lesson on double packable solids. *Teaching Mathematics and Its Applications* **18** No.1, Oxford Univ. Press (1999) 30–33
2. Akiyama, J., Nakamura, G.: Dudeney dissection of polygons. *Discrete and Computational Geometry* (J. Akiyama et al. (eds.), Lecture Notes in Computer Science **1763**), Springer-Verlag (2000) 14–29,
3. Akiyama, J., Nakamura, G.: Congruent Dudeney dissections of polygons I, –All hinge points are vertices of the polygon–. submitted to *Proc. the $9^{th}$ Quadrennnial International Conference on Graph Theory and Combinatorics, Algorithms and Applications, Discrete Math.*
4. Akiyama, J., Nakamura, G.: Congruent Dudeney dissections of polygons II,–All hinge points are on the sides of the polygon–. *Proc. Mathematics and Art Conference*, Bond University (2000) 115–145

5. Akiyama, J., Nakamura, G.: Congruent Dudeney dissections of polygons III, –Hinge points are on both vertices and sides of a polygon–. (In Japanese) *Tech. Reports, Res. Inst. Edu. Develop. Tokai Univ.* (June, 2000)

6. Akiyama, J., Nakamura, G.: Dudeney dissections of polyhedrons I, –Space filings solids–. (In Japanese) *The Reports of the Res. Inst. of Educ., Tokai Univ.* **8** (2000) 1–12

7. Akiyama, J., Nakamura, G.: Dudeney dissections of polyhedrons II, –Three layered type–. (In Japanese) *The Reports of the Res. Inst. of Educ., Tokai Univ.* **8** (2000) 13–26

8. Akiyama, J., Nakamura, G.: Dudeney dissections of polyhedrons III, –Mirror image type–. (In Japanese) *The Reports of the Res. Inst. of Educ., Tokai Univ.* **8** (2000) 27–35

9. Akiyama, J., Nakamura, G.: Dudeney dissections of polyhedrons IV, –Two layered type–. (In Japanese) *The Reports of the Res. Inst. of Educ., Tokai Univ.* **8** (2000) 36–57

10. Akiyama, J., Nakamura, G.: Dudeney dissections of polyhedrons V, –Multi layered type–. (In Japanese) *The Reports of the Res. Inst. of Educ., Tokai Univ.* **8** (2000) 58–62

11. Akiyama, J., Nakamura, G.: Dudeney dissections of polyhedrons VI, –Clasping type–. (In Japanese) *The Reports of the Res. Inst. of Educ., Tokai Univ.* **8** (2000) 63–66

12. Boltyanski, V. G.: *Hilbert's Third Problem.* V. H. Winston & Sons. Translated by R. A. Silverman (1978)

13. Bolyai, F.: *Tentamen juventutem.* Maros Vasarhelyini: Typis Collegii Reformatorum per Josephum et Simeonem Kali (In Hungarian) (1832)

14. Busschop, P.: Problèmes de géométrie. *Nouvelle Correoindance Mathématique* **2** (1876) 83–84

15. Coxeter, H. S. M.: *Introductions to Geometry.* Wiley & Sons (1965)

16. Dehn, M.: Über den Rauminhalt. *Nachrichten von der Gesellschft der Wissenschaften zu Göttingen, Mathematisch-Physikalische Klasse* (1900) 345–354. Subsequently in *Mathematische Annalen* **55** (1902) 465–478

17. Dudeney, H. E.: *The Canterbury Puzzles.* Thomas Nelson & Sons (1907)

18. Frederickson, G. N.: *Dissections: Plane & Fancy.* Cambridge University Press (1997)

19. Gerwien, P.: Zershneidung jeder beliebigen Anzahl von gleichen geradlinigen Figuren in dieselben Stücke. *Jouranl für die reine und angewandte Mathematik (Crelle's Journal)* **10** (1833) 228–234 and Taf.III

20. Grünbaum, B., Shephard, G. C.: Tiling with congruent tiles. *Bull. AMS* **3** (1980) 951–973

21. Hilbert, D.: Mathematische Probleme. *Nachrichten von der Gesellschaft der Wissenschaften zu Göttingen, Mathematisch-Physikalische Klasse* (1900). Subsequently in *Bulletin of the American Mathematical Society* **8** (1901-1902) 437–479

22. Klarner, D. A.: *The Mathematical Gardner.* Wadsworth International (1981)

23. Lindgren, H.: *Geometric Dissections.* D. Van Nostarand Company (1964)

24. Sydler, J.-P.: Conditions nécessaries et suffisantes pour l'équivalence des polyèdres de l'espace euclidien à trios dimensions. *Commentarll Mathematici Helvetica* **40** (1965) 43–80

25. Wallace, W.: (Ed.) *Elements of Geometry* (8$^{th}$ ed.). Bell & Bradfute. First six books of Euclid, with a supplement by John Playfair (1831)

# Universal Measuring Devices Without Gradations

Jin Akiyama[1], Hiroshi Fukuda[2], Gisaku Nakamura[1], Toshinori Sakai[1],
Jorge Urrutia[3]*, and Carlos Zamora-Cura[3] *

[1] Research Institute of Educational Development, Tokai University, 2-28-4 Tomigaya
Shibuya-ku Tokio, JAPAN `fwjb5117@mb.infoweb.ne.jp`
`tsakai@yoyogi.ycc.u-tokai.ac.jp`
[2] School of Administration and Informatics, University of Shizuoka, 52-1 Yada
Shizuoka, 422-8526 JAPAN `fukuda@u-shizuoka-ken.ac.jp`
[3] Instituto de Matemáticas, Universidad Nacional Autónoma de México, Area de la
Inv. Científica, Circuito Exterior, Ciudad Universitaria, México D.F., C.P. 04510
MÉXICO `czamora@matem.unam.mx`, `urrutia@matem.unam.mx`

**Abstract.** Measuring cups are everyday instruments used to measure
a required amount of liquid for many common tasks such as cooking,
...etc. A measuring cup usually has gradations marked on its sides. In
this paper we study measuring devices without gradations but which
nevertheless can measure any integral amount, say liters, of liquid up
to their full capacity. These devices will be called *universal measuring
devices*. We determine the largest volume of a measuring device with
triangular base and obtain one with rectangular base which can measure
up to 691 liters.

## 1   Introduction

A common device used to measure liquid in many Japanese stores some years
ago was a measuring box with a square base of area 6 and height 1. By dipping
liquid only once from the container and then tilting the box and using its edges
and vertices as markers, it is possible to keep 6, 3, and 1 liters in the box, as
shown in Figure 1. The box would have no extra gradations to measure 2, 4, and
5 liters. A store would keep a container holding large amounts of the liquid.

If a customer wanted to buy a certain integral amount of liquid between 1
and 6, the store owner would proceed as follows:

1. The store clerk would dip his measuring box into the store's container, filling
   it only once.
2. He would then empty certain amounts of liquid alternately into the store's
   container and into the customer's.

For example, to measure 5 liters, the clerk would first fill the measuring box
to the rim and then pour into the customer's container until 1 liter of liquid was

---

* Partially supported by CONACyT–REDII projects.

J. Akiyama, M. Kano, and M. Urabe (Eds.): JCDCG 2000, LNCS 2098, pp. 31–40, 2001.

left in the measuring box. The remaining liquid would go back into the store's container. To measure 4 liters, he would first fill the box. Next he would pour liquid into the customer's container until 3 liters were left in the box. Then he would pour two liters into the store's container, and the remaining liter into the customer's container.

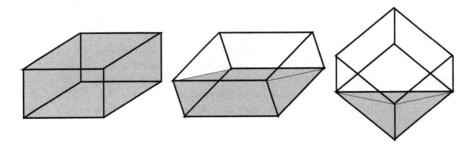

**Fig. 1.** Measuring 6, 3 and 1 liters

In this paper we are interested in studying measuring devices without gradations which nevertheless can be used for measuring any integral amount of liquid up to their full capacity by using the procedure described above. We call these *universal measuring devices*. We determine the largest volume of universal measuring devices with *triangular* base. More precisely, we determine the dimensions of a universal measuring device of maximum volume obtained by cutting a triangular cylinder perpendicular to the $x$–$y$ plane by the $x$–$y$ plane, and another plane that cuts the cylinder above its base. In the case of rectangular cylinders we consider devices similar to those shown in Figure 5. We will assume that the lengths of the edges of the device which are contained in the edges of the original triangular or rectangular cylinder are $h_1 < h_2 < h_3$ or $h_1 < h_2 < h_3 < h_4$ respectively. These will be called the *heights* of the device. We show that the largest possible volume of a device with triangular base is 41. We have also obtained a device with a rectangular base that measures up to 691 liters. The areas of the bases of these devices are 3 and 6, while their heights are 12, 13, 16, and 1, 32, 83, 691 (or 2, 64, 166, 691) respectively.

## 2    Universal measuring devices with triangular bases

In this section and the next, we prove that the maximum volume of a universal measuring device with triangular base of area 3 is 41. The assumption that the area of the base is 3 can be removed easily. The assumption will, however, simplify our analysis, as the reader will soon see. Under this restriction, the heights of a universal measuring device with triangular base are 12,13, and 16.

## 2.1    An interesting relationship

We begin by determining the volumes that can be measured using corner points of the device. We first prove the following theorem.

**Theorem 1.** *Let $\mathcal{B}$ be a device with base area 3, and heights $h_1 < h_2 < h_3$. Then the following amounts of liquid can be measured: $h_1$, $h_2$, $h_3$, $h_1+h_2$, $h_1+h_3$, $h_2+h_3$, $h_1+h_2+h_3$.*

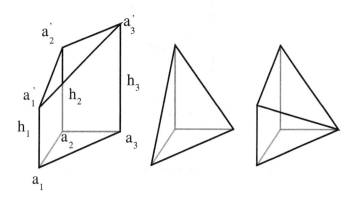

**Fig. 2.** The volumes of the polyhedron shown here are $h_1 + h_2 + h_3$, $h_2$ and $h_1 + h_2$ respectively.

**Proof:** Assume that the vertices of $\mathcal{B}$ are labelled $a_1$, $a_2$, $a_3$, $a_1'$, $a_2'$, $a_3'$ as shown in Figure 2. Assume that the distance between $a_i$ and $a_i'$ is $h_i$, $i = 1, 2, 3$. It is well known that the volume of a tetrahedron with base area $A$ and height $h$ is $\frac{Ah}{3}$. Since the area of the base of the device, i.e. the area of the triangle with vertices $a_1$, $a_2$, $a_3$ is 3, it follows immediately that the volume of the tetrahedron with vertices $a_1$, $a_2$, $a_3$, $a_i'$ is $h_i$, $i = 1, 2, 3$.

We now show that the volume of the polyhedron with vertices $a_1$, $a_2$, $a_3$, $a_i'$, and $a_j'$ is precisely $h_i + h_j$, $i \neq j$, $i$, $j \in \{1, 2, 3\}$. We make the following observation:

Let $\mathcal{T}$ be a triangular cylinder, and assume that the area of the triangle obtained by cutting $\mathcal{T}$ along any plane perpendicular to its edges is 3. Then if we choose two points $p$, $q$ on one of its edges, and any two points $r$ and $s$, one in each of the remaining edges of $\mathcal{T}$, then the volume of the tetrahedron with vertices $p$, $q$, $r$, $s$ is equal to the distance between $p$ and $q$; see Figure 3.

Suppose w.l.o.g. that $i = 1$, $j = 2$. Dissect the polyhedron with vertices $a_1$, $a_2$, $a_3$, $a_1'$, $a_2'$ into two tetrahedra with vertices $a_1$, $a_2$, $a_3$, $a_1'$ and $a_2$, $a_3$, $a_1'$, $a_2'$ respectively, see Figure 4. The volume of the first tetrahedron is $h_1$, and by our

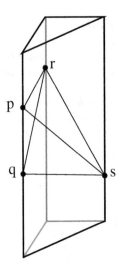

**Fig. 3.** Once $p$ and $q$ are fixed, the volumes of the tetrahedra with vertices $p$, $q$, $r$, and $s$ remains constant regardless of the positions of $r$ and $s$.

previous observation, the area of the second one is the distance from $a_2$ to $a_2'$ which is $h_2$.

Using similar arguments, we can now show that the volume of the measuring device is $h_1 + h_2 + h_3$, and the result is proved.                     $\square$

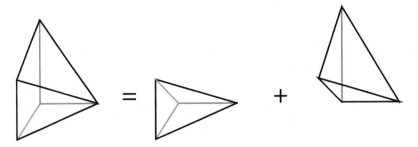

**Fig. 4.** Dissecting the polyhedron with vertices $\{a_1,\, a_2,\, a_3,\, a_1',\, a_2'\}$ into two tetrahedra with volumes $h_1$ and $h_2$ respectively.

## 3    The second reduction

Using the previous result, it follows that if $h_1 = 12$, $h_2 = 13$, and $h_3 = 16$ we can measure the following amounts, sorted in increasing order: $l_1 = h_1 = 12$,

$l_2 = h_2 = 13$, $l_3 = h_3 = 16$, $l_4 = h_1 + h_2 = 25$, $l_5 = h_1 + h_3 = 28$, $l_6 = h_2 + h_3 = 29$, and $l_7 = h_1 + h_2 + h_3 = 41$.

Now consider any subsequence $S = (l_{i_1} < \cdots < l_{i_k})$ of the sequence

$$(12, 13, 16, 25, 28, 29, 41).$$

We can associate to this subsequence an amount in liters $V_S$ that can be measured with a container with heights $12, 13$, and $16$ as follows:

$$V_S = (l_{i_k} - l_{i_{k-1}}) + (l_{i_{k-2}} - l_{i_{k-3}}) + \ldots + l_{i_1}, \quad \text{for } k \text{ odd}$$

or

$$V_S = (l_{i_k} - l_{i_{k-1}}) + (l_{i_{k-2}} - l_{i_{k-3}}) + \ldots + (l_{i_2} - l_{i_1}), \quad \text{for } k \text{ even.}$$

For example if $S = (16, 29, 41)$, $V_S = (41 - 29) + 16 = 28$. If $S = (12, 16, 25, 28)$ then $V_S = (28 - 25) + (16 - 12) = 7$. From this, it is straightforward to develop a method to measure $V_S$ liters using the device. Two examples will suffice to illustrate this:

If $S = (12, 16, 25, 28)$, then to measure $V_S = 7$ liters we proceed as follows. First fill the measuring device, and pour back $41 - 28$ liters into the store's container. Now empty $3 = 28 - 25$ liters into the customer's container until $25$ liters are left in the measuring device. Then pour $9 = (25 - 16)$ liters into the store's container. Next pour $4 = 16 - 12$ liters into the customer's container and finally empty the remaining $12$ liters into the store's container.

If $S = (16, 29, 41)$, to measure $V_S = 28$ liters, fill the device, then empty $12 = 41 - 29$ liters into the customer's container. Then pour $13 = 29 - 16$ liters into the store's container, and finally pour the remaining $16$ liters into the customer's container. The reader may now easily verify that for any integer $i$, $1 \le i \le 41$ there is a subset $S$ of $(12, 13, 16, 25, 28, 29, 41)$ such that $i = V_S$. In the next section we develop an easy test to verify this fact.

Observe that since there are exactly $2^7$ subsets of $\{h_1, h_2, h_3, h_1 + h_2, h_1 + h_3, h_2 + h_3, h_1 + h_2 + h_3\}$, it follows that the maximum number of liters that a universal measuring device with triangular base can measure is at most $2^7 - 1 = 128 - 1$ (the empty set corresponds to $0$ liters).

In the next section we show that the largest volume a universal measuring device with triangular base can have is $41$, thus proving that $12, 13, 16$ are the heights of a universal measuring device with maximal volume and triangular base of area $3$.

## 3.1  Proof of maximality

We now prove that the maximum volume of a universal measuring device with triangular base is $41$. Note first that there are only two possible orderings for the measurable quantities of such a device. These are:

$$0 < h_1 < h_2 < h_3 < h_1 + h_2 < h_1 + h_3 < h_2 + h_3 < h_1 + h_2 + h_3, \qquad (1)$$

and

$$0 < h_1 < h_2 < h_1 + h_2 < h_3 < h_1 + h_3 < h_2 + h_3 < h_1 + h_2 + h_3. \qquad (2)$$

Consider the second ordering and take the sequence formed by the differences of consecutive elements for this ordering: $s_1 = h_1 - 0$, $s_2 = h_2 - h_1$, $s_3 = h_1$, $s_4 = h_3 - (h_1 + h_2)$, $s_5 = h_1$, $s_6 = h_2 - h_1$, and $s_7 = h_1$.

Observe that for any subsequence $S$ of

$$(h_1, h_2, h_1 + h_2, h_3, h_1 + h_3, h_2 + h_3, h_1 + h_2 + h_3) \qquad (3)$$

we can express $V_S$ as a sum of some subset of $(s_1, \ldots, s_7)$, e.g. if

$$S = (h_1, h_1 + h_2, h_1 + h_3, h_1 + h_2 + h_3)$$

then

$$V_S = ((h_1 + h_2 + h_3) - (h_1 + h_3)) + ((h_1 + h_2) - h_1)$$

which equals

$$s_7 + s_6 + s_3 + s_2.$$

It is easy to see that to any subset of $\{s_i; \ i = 1, \ldots, 7\}$ we can associate a unique subsequence of

$$(h_1, h_2, h_1 + h_2, h_3, h_1 + h_3, h_2 + h_3, h_1 + h_2 + h_3).$$

Thus the number of integers $V_S$ that can be formed with subsequences of

$$(h_1, h_2, h_1 + h_2, h_3, h_1 + h_3, h_2 + h_3, h_1 + h_2 + h_3)$$

equals the number of elements that can be obtained as the sum of the elements of subsets of $\{s_i; \ i = 1, \ldots, 7\}$.

Denote by $a = h_1$, $b = h_2 - h_1$ and $c = h_3 - (h_1 + h_2)$. Note that any number obtained as the sum of the elements of a subset of $\{s_i; i = 1, \ldots, 7\}$ must be of the form $ai + bj + ck$, where $i \in \{0, 1, 2, 3, 4\}$, $j \in \{0, 1, 2\}$, and $k \in \{0, 1\}$. This is because $h_1, h_2 - h_1$ and $h_3 - (h_1 + h_2)$ repeat themselves in the set of differences 4, 2 and 1 times respectively. It follows now that the maximum capacity of a device in this case is at most $5 \times 3 \times 2 - 1 = 29$.

Now consider the sequence

$$h_1, h_2 - h_1, h_3 - h_2, h_1 + h_2 - h_3, h_3 - h_2, h_2 - h_1, h_1$$

of differences of consecutive elements for (1).

As in the previous paragraph, let $a = h_1$, $b = h_2 - h_1$, $c = h_3 - h_2$ and $d = (h_1 + h_2) - h_3$. In this case, any number generated by (1) is of the form $ai + bj + ck + d\ell$, where $i, j, k \in \{0, 1, 2\}$ and $\ell \in \{0, 1\}$. Again this is because $h_1$,

$h_2 - h_1$, $h_3 - h_2$ and $(h_1 + h_2) - h_3$ repeat themselves in the sequence of differences 2, 2, 2 and 1 times respectively. Thus an upper bound for the number of distinct numbers generated when using a sequence like (1) is $3 \times 3 \times 3 \times 2 - 1 = 53$.

We now prove that 12 of these quantities are repeated. To this end, note that $a - c = d$. Thus it is true that

$$ai + bj + ck + d\ell = a(i + 1) + bj + c(k - 1) + d(\ell - 1)$$

whenever $i \in \{0, 1\}$, $k \in \{1, 2\}$, $\ell = 1$ and $j \in \{0, 1, 2\}$. Therefore the number of repetitions is 12, and thus in this case at most $53 - 12 = 41$ numbers can be generated. Our result now follows since using the device with triangular base of area 3 and heights 12, 13, and 16 we can measure from 1 to 41 liters.

## 4    Generating sequences

Let $\Sigma = (n_1, \ldots, n_k)$ be a sequence of numbers such that $n_i < n_{i+1}$, $i = 1, \ldots, k - 1$. For each subsequence $S = (n_{i_1}, \ldots, n_{i_j})$ of $\Sigma$ we associate the number $V_S = (n_{i_j} - n_{i_{j-1}}) + (n_{i_{j-2}} - n_{i_{j-3}}) + \ldots + n_{i_1}$, $j$ odd, or $V_S = (n_{i_j} - n_{i_{j-1}}) + (n_{i_{j-2}} - n_{i_{j-3}}) + \ldots + (n_{i_2} - n_{i_1})$, $j$ even. For example if $\Sigma = (4, 12, 23, 24, 69, 71, 81, 213, 225, 282)$ and $S = (12, 69, 81, 282)$ then $V_S = (282 - 81) + (69 - 12)$.

Call $\Sigma$ a *generating sequence* if for each integer $i$, $1 \le i \le n_k$ there is a subsequence $S$ of $\Sigma$ such that $i = V_S$. The best known generating sequences are the ones corresponding to powers of 2, i.e. $(1, 2, 4, \ldots, 2^k)$. In this section we develop a linear-time test to decide if a sequence of numbers is a generating sequence.

Let $\Sigma' = (0, n_1, \ldots, n_k)$, and if $s_0 = 0$, let $\Psi = (s_0, s_1, \ldots, s_k)$ be the sequence obtained by sorting the sequence $(n_i - n_{i-1}, i = 1, \ldots, k)$, taking $n_0 = 0$. For example if $\Sigma$ is the sequence

$$(4, 12, 23, 24, 69, 71, 81, 213, 225, 282),$$

then

$$\Psi = (0, 1, 2, 4, 8, 10, 11, 12, 45, 57, 132).$$

We now prove:

**Theorem 2.** *The sequence $\Sigma$ is a generating sequence if and only if $s_i - 1 \le s_0 + \cdots + s_{i-1}$, $i = 1, \ldots, k$.*

Using the previous result, it is easy to verify that

$$\Sigma = (4, 12, 23, 24, 69, 71, 81, 213, 225, 282)$$

is a generating sequence since

$$2 - 1 = 1$$
$$4 - 1 = 1 + 2$$
$$8 - 1 = 1 + 2 + 4$$
$$10 - 1 < 1 + 2 + 4 + 8$$
$$11 - 1 < 1 + 2 + 4 + 8 + 10$$
$$12 - 1 < 1 + 2 + 4 + 8 + 10 + 11$$
$$45 - 1 < 1 + 2 + 4 + 8 + 10 + 11 + 12$$
$$57 - 1 < 1 + 2 + 4 + 8 + 10 + 11 + 12 + 45$$
$$132 - 1 < 1 + 2 + 4 + 8 + 10 + 12 + 45 + 57$$

We now proceed to prove the result. Let $\Sigma = (n_1, \ldots, n_k)$. Observe that $n_i - n_j = (n_i - n_{i-1}) + (n_{i-1} - n_{i-2}) + \cdots + (n_{j+1} - n_j)$. Thus for any subsequence $S = (n_{i_1}, \ldots, n_{i_j})$ of $\Sigma$, $V_S$ can be written as a sum of elements of $\Psi$. For instance, in the previous example if

$$S = (23, 24, 69, 81, 282),$$

$$N_S = (282 - 81) + (69 - 24) + 23,$$

which equals

$$(282 - 225) + (225 - 213) + (213 - 81) + (69 - 24) + (23 - 12) + (12 - 4) + (4 - 0),$$

i.e.,

$$V_S = 57 + 12 + 132 + 45 + 11 + 8 + 4.$$

This implies that $\Sigma$ is a generating sequence if any integer from 1 to $n_k$ can be written as the sum of some elements of $\Psi$.

Suppose next that $s_i - 1 \leq s_1 + \ldots + s_{i-1}$, $i = 1, \ldots, k$. We prove that $\Sigma$ is a generating sequence. Note that $s_1 = 1$. By induction, assume that any integer between 1 and $s_1 + \ldots + s_{i-1}$ can be expressed as the sum of some subset of $\{s_1, \ldots, s_{i-1}\}$. Observe that any integer of the form $m + s_i$, $0 \leq m \leq s_1 + \cdots + s_{i-1}$ can be expressed as the sum of the elements of a subset of $\{s_1, \ldots, s_{i-1}, s_i\}$. Since by hypothesis $s_i - 1 \leq s_1 + \ldots + s_{i-1}$, it follows that any integer less than or equal to $s_1 + \ldots + s_i$ can be expressed as the sum of the elements of some subset of $\{s_1, \ldots, s_i\}$.

Conversely if for some $i$ we have $s_i - 1 > s_1 + \ldots + s_{i-1}$ then clearly $\Sigma'$ is not a generating sequence since $s_1 + \ldots + s_{i-1} + 1$ can not be generated. The result follows.

## 5   Devices with rectangular base

In this section we find a universal measuring device that can measure up to 691. It has square base of area 6 and heights 1, 32, 83, and 691. We give a brief description of how it was obtained. Since the ideas used are very similar to those used previously, we will only sketch the main arguments.

As in the case of triangular base devices, we first find some of the amounts of liquid that can be measured directly with a device of heights $h_1 \leq h_2 \leq h_3 \leq h_4$. Additional problems arise since there are a number of different shapes for the devices, resulting in different sets of measurable quantities. As in the case of a triangular base device we assume that the vertices of these devices are labeled $a_i, a_i'$ for $i = 1, 2, 3, 4$, where the distance between $a_i$ and $a_i'$ is $h_i$; see Figure 5. Consider any device of heights $h_1 \leq h_2 \leq h_3 \leq h_4$.

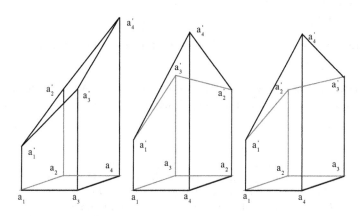

**Fig. 5.** Shapes of measuring devices with square base.

If we begin listing the heights of a device starting with $h_1$ in the clockwise order, then there are at most 6 different shapes that the devices can have, namely: $(h_1, h_4, h_3, h_2)$, $(h_1, h_4, h_2, h_3)$, $(h_1, h_3, h_4, h_2)$, $(h_1, h_3, h_2, h_4)$, $(h_1, h_2, h_4, h_3)$, and $(h_1, h_2, h_3, h_4)$. Notice that first and sixth, second and fourth, and third and fifth devices have the same set of measurable quantities. Thus up to symetry there are only 3 types that are relevent, those are shown in Figure 5. It is easy, but tedious work to check that the device $\mathcal{B}$ with heights $(h_1, h_3, h_4, h_2)$ can measure the following amounts: $h_1$, $3h_1$, $6h_1$, $h_2$, $3h_2$, $3(h_1 + h_2)$, $h_3$, $3h_3$, $3(h_1 + h_3)$, and $h_4$. If in addition $h_1 + h_4 \geq h_2 + h_3$ we can also measure $3(h_2 + h_3)$. For example, $h_1$ can be measured by the points $a_2, a_3$ and $a_1'$, $3h_1$ by $a_2, a_4$ and $a_1'$, $6h_1$ by the plane parallel to the base passing through $a_1'$, and so on. For some of the devices shown in Figure 5, there are some volumes that can be measured that might be integral, and it could be productive to increase the capacity of a rectangular base universal measuring device. Further work along these lines is in progress. We can summarize all the above observations in the next theorem.

**Theorem 3.** *There is a universal measuring device $\mathcal{B}$ of rectangular base with area 6 such that $h_1 + h_4 \geq h_2 + h_3$ that can measure at least the following ammounts:* $h_1$, $3h_1$, $6h_1$, $h_2$, $3h_2$, $3(h_1 + h_2)$, $h_3$, $3h_3$, $3(h_1 + h_3)$, $3(h_2 + h_3)$, $h_4$.

Having these results at hand, we were able to find a device of of capacity 691 by brute force. We did this by writing a C program that did the following: for any given set of heights $1 \leq h_1 < h_2 < h_3 < h_4 \leq 2^{11}$ compute its set $\{h_1, 3h_1, 6h_1, h_2, 3h_2, 3(h_1 + h_2), h_3, 3h_3, 3(h_1 + h_3), 3(h_2 + h_3), h_4\}$ of measurable quantities. We then applied the criterion from theorem 2 to them, by which we found two universal measuring devices with heights 1, 32, 83, 691 and 2, 64, 166, 691 that can measure all integer quantities from 1 to 691. For any other set of heights, the sets of measurable amounts obtained were not generating, or had smaller capacities.

Thus we have:

**Theorem 4.** *There is a universal measuring device with rectangular base which can measure up to 691 liters.*

**Acknowledgements**: We would like to thank Prof. Y. Watanabe for his interest in our work, and Prof. Mari-Jo Ruiz for editing our paper.

# References

1. Akiyama, J., Hirano, Y., Kawamura, K., Matsunaga, K., Nakamura, G., Watanabe, Y.: *Models and Experiments of Mathematics for Junior High School and High School Students Vol.2.* Suken Shuppan (in Japanese) (1999)
2. Nakamura, G.: *Why Are Manhole Covers Round?* Nihon Keizai Shimbun Publishers (in Japanese) (1984)

# A Note on the Purely Recursive Dissection
# for a Sequentially $n$-Divisible Square

Jin Akiyama[1], Gisaku Nakamura[1], Akihiro Nozaki[2], and Ken'ichi Ozawa[3]

[1] Research Institute of Educational Development, Tokai University
2-28-4 Tomigaya, Shibuya-ku, Tokyo 151-0063, Japan
fwjb5117@mb.infoweb.ne.jp
[2] School of Social Information Studies, Otsuma Women's University
2-7-1 Karakida, Tama-shi, Tokyo 206-8540, Japan
nozaki@otsuma.ac.jp
[3] Higashino Highschool
112-1 Nihongi, Iruma-shi, Saitama 358-8558, Japan
ozawa314@ba2.so-net.ne.jp

**Abstract.** A dissection for a sequentially $n$-divisible square is a partition of a square into a number of polygons, not necessarily squares, which can be rearranged to form two squares, three squares, and so on, up to $n$ squares successively. A dissection is called *type-k* iff $k$ more pieces needed to increase the maximum number $n$ of composed squares by one. Ozawa found a general dissection of type-3, while Akiyama and Nakamura found a particular, "purely recursive" dissection of type-2. Nozaki has given a mixed procedure for a dissection of type-1.

In this note, we shall show that there is no type-1 purely recursive dissection for a sequentially $n$-divisible square. Therefore Akiyama and Nakamura's dissection is optimal with respect to the type, among the purely recursive dissections.

## 1 Introduction

It is an interesting puzzle to decompose a square into a number of squares. Fig.1 shows a simple way to decompose a square into two subsquares, and Fig.2 shows a dissection of a square into nine pieces, which can be rearranged into two squares, three squares, and four squares. We call it "dissection for a sequentially 4-divisible square". Busschop ([2]) found a dissection for a sequentially 3-divisible square. Duijverspein([3]) found that a square can be divided into 21 pieces, all of which are squares of different sizes.

Ozawa([5]) has found an interesting way of dissection, shown in Fig.3, which can be applied repeatedly several times. The result is a "dissection for a sequentially $n$-divisible square": we can rearrange the pieces into two squares, three squares, and so on, up to $n$ squares successively.

The number of pieces in the first dissection is five. But, in the second dissection, the subsquare to be dissected contains some cutting lines already (see Fig.4), and therefore the total number of pieces increases only by three in every

J. Akiyama, M. Kano, and M. Urabe (Eds.): JCDCG 2000, LNCS 2098, pp. 41–52, 2001.

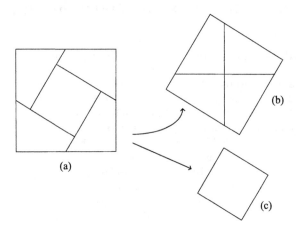

**Fig.1.** A square (a) is divided into five pieces, from which two squares (b) and (c) are constructed.

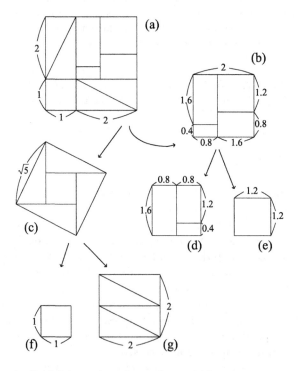

**Fig.2.** A square is divided into nine pieces, from which we can construct two squares (b) and (c), three squares (c), (d) and (e), and four squares (d), (e), (f) and (g).

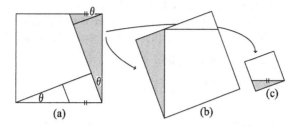

**Fig.3.** The symbol $\theta$ stands for an arbitrary angle between $0°$ and $90°$.

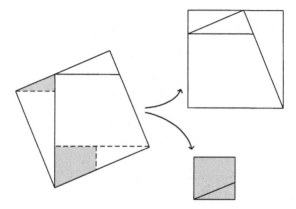

**Fig.4.** The square (b) in Fig.3 is dissected by applying the same "dissecting pattern" (a) in Fig.3, proportionally contracted and turned over. Dotted lines show the line segments to be newly dissected.

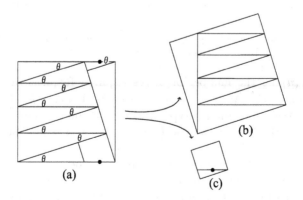

**Fig.5.** The symbol $\theta$ stands for a deliberately chosen angle ($\theta \approx 17.48°$).

repetition of the dissection, provided that the angle $\theta$ has been chosen carefully, depending on the desired maximum number $n$ of subsquares. A sufficient condition for $\theta$ to obtain $n\,(> 1)$ subsquares economically is as follows.

For an even integer $2p$ not less than $n$,

$$\cos^{2p}\theta + \tan\theta \geq 1. \tag{1}$$

Table 1(Ozawa[6]) shows the values of $\theta$ satisfying the equality in the above condition for each $p$ greater than two. By these values $\alpha$ and $\beta$, the condition (1) can be represented more explicitly as follows.

$$0° < \theta \leq \alpha \quad \text{or} \quad \beta \leq \theta < 90°. \tag{2}$$

**Table 1.** (Ozawa[6])

| $2p$ | $\alpha(°)$ | $\beta(°)$ |
|------|-------------|------------|
| 6 | 29.68 | 34.31 |
| 8 | 17.48 | 42.32 |
| 10 | 13.17 | 43.90 |
| 12 | 10.64 | 44.50 |
| 14 | 8.94 | 44.76 |
| 16 | 7.72 | 44.88 |

Akiyama and Nakamura([1]) found independently the essentially same dissection, shown in Fig.5, and noticed the following facts.

(i) By taking $\theta = \alpha$ or $\beta$, the total number of pieces increases only by two for each application of the dissection, which can be applied recursively for any times.

(ii) When $p = 3$, some subsquares have the same size. But when $p = 4$, all subsquares have different sizes.

Now let $f(n)$ be the number of pieces required for constructing up to $n$ subsquares sequentially. We say a dissection method is type-$k$, iff $\frac{f(n)}{n}$ tends to $k$ for large $n$. Ozawa's method is type-3, while Akiyama-Nakamura's method is type-2. Nozaki([4]) has given a type-1 procedure for a sequentially $n$-divisible square.

Nozaki's procedure is a mixture of the three methods: the dissection in Fig.2, Duijverspein's partition, and Ozawa's dissection. It is optimal with respect to the type, but not so simple and elegant as Ozawa's method and Akiyama-Nakamura's method. In what follows, we define the notion of "purely recursive dissection", and prove that Akiyama-Nakamura's method is optimal with respect to the type among the purely recursive dissections.

## 2    Basic Notions

**Definition 1.** *A dissection for sequentially n-divisible square is a partition of a square into several polygons which can be rearranged into two squares, three squares, and so on, up to n squares successively.*

**Definition 2.** *Let $f(m)$ be the number of pieces required to construct $m$ subsquares. If $\frac{f(m)}{m}$ tends to $k$ for large $m$, the dissection is said to be type-$k$.*

**Definition 3.** *A dissection pattern is a figure representing a way of dissecting a square. A composition pattern is a figure showing how to rearrange some pieces into a square.*

The pattern in Fig.1(a) is a dissection pattern, and the patterns in (b), (c) are composition patterns.

**Definition 4.** *A dissection for a sequentially n-divisible square is said to be recursive iff it satisfies the following conditions.*

*(I) It is obtained by recursive application of a fixed dissection pattern D.*
*(II) The pattern D is applied to squares under proportional adjustment of the size, possibly with turning over.*

*In other words, a similar figure of D is overlayed to the square to be dissected.*

*(III) By the first application of D, two subsquares $S'$ and $S^*$ can be constructed.*
*(IV) The dissection pattern is recursively applied to one of the subsquares, say $S^*$.*

So, after applying $D$ four times recursively, we have five subsquares as follows.

$$S', \quad S^{*\prime}, \quad S^{**\prime}, \quad S^{***\prime}, \text{ and } S^{****}.$$

By repeating the same process $(n-1)$ times, we obtain a dissection for sequentially $n$-divisible square.

*It is said to be purely recursive iff the following condition is also satisfied.*

*(V) The composition pattern of $S^{**}$ is similar to that of $S^*$.*

*Example.* Akiyama-Nakamura's dissection is purely recursive: its dissection pattern is the partition in Fig.5(a).

*Remark.* Here we don't require that the sizes of subsquares are different.

Now we can state our main theorem.

**Theorem 1.** *There is no purely recursive dissection for sequentially n-divisible square, whose type is less than two.*                                                            □

**Corollary 1.** *Akiyama-Nakamura's dissection is optimal with respect to the type, among the purely recursive dissections.*                                                     □

## 3    Proof of Theorem 1

### 3.1    Preparatory Consideration

Suppose that there is a purely recursive dissection for sequentially $n$-divisible square, whose type is less than two. Let $D$ be the dissection pattern, and $C$ the composition pattern of $S^*$.

Suppose that $S^*$ contains $s$ pieces

$$P_0, P_1, \ldots, \text{ and } P_{s-1},$$

in the decreasing order of their sizes. By applying the dissection pattern $D$ to $S^*$, we have two subsquares $S^{*\prime}$ and $S^{**}$. Since the composition pattern of $S^{**}$ is similar to that of $S^*$, the number of pieces of $S^{**}$ is equal to that of $S^*$, and therefore the total number of pieces is increased by the number $k$ of pieces in $S^{*\prime}$. In every further application of $D$, the total number of pieces is always increased by $k$. Since the type of the dissection is asssumed to be less than two, $k$ should be equal to one: $S^{*\prime}$ is a one-piece square, and the dissection is type-1.

Now let

$$Q_0, Q_1, \ldots, \text{ and } Q_{s-1}$$

be the pieces of $S^{**}$ in the decreasing order of sizes. Since its composition pattern is similar to that of $S^*$, the ratio of the sizes of $Q_i$ and $Q_j$ is equal to the ratio of the sizes of $P_i$ and $P_j$. Since $k = 1$, only one of original pieces $P_i$'s is divided into two. The piece divided should be $P_0$,: if $P_j$ is divided for some $j > 0$, then $P_i = Q_i$ for all $i < j$, and $Q_j$ is smaller than $P_j$. Therefore the ratio of the sizes of $P_0$ and $P_j$ is not equal to the ratio of the sizes of $Q_0$ and $Q_j$. By a similar argument, it is shown that $P_0$ is divided into a square $S^{*\prime}$ and $Q_{s-1}$, and $Q_i = P_{i+1}$ for $i < s - 1$. Moreover, $P_i$'s are similar polygons, which are shown to be rectangles. Of course, the ratio of the length of their two sides is equal to the golden ratio $\tau$:

$$\tau^2 + \tau - 1 = 0, \quad \tau = \frac{-1 + \sqrt{5}}{2}.$$

Taking a suitable scale, we can assume without loss of generality that the lengths of the longer and shorter edges of $P_0$ are 1 and $\tau$, respectively. Then the lengths of the longer edges form a geometrical sequence $\{\lambda_j\}$, where $\lambda$ denotes the $s$-th root of $\tau$.

Since the pieces are rectangulars, it is obvious that s is greater than two. The area $M$ of the square $S^*$ is equal to

$$\tau\left(1 + \lambda^2 + \lambda^4 + \ldots + \lambda^{2(s-1)}\right).$$

On the other hand, the length of an edge $AB$ of $S^*$ is the sum of the lengths of edges of some $P_i$'s. So we have the following equality:

$$\tau\left(1 + \lambda^2 + \lambda^4 + \ldots + \lambda^{2(s-1)}\right) = \left(\sum_{i=0}^{s-1} a_i \lambda^i\right)^2 \lambda^k, \tag{3}$$

where $a_i = 1$ or $\tau$ if the longer or shorter edge of the $i$-th piece is contained in the edge $AB$, respectively, and $a_i = 0$ otherwise.

We shall show that this equality (3) will *never* hold.

## 3.2 Technical Part of the Proof

**Lemma 1.** *The following polynomial in $\lambda$ over the field $\mathbf{Q}[\tau]$ is irreducible.*

$$P(\lambda) = \lambda^s - \tau. \qquad \square$$

An elementary proof will be given in the appendix.

Now, suppose that the equality (3) holds:

$$\tau \left(1 + \lambda^2 + \lambda^4 + \ldots + \lambda^{2(s-1)}\right) = \left(\sum_{i=0}^{s-1} a_i \lambda^i\right)^2 \qquad (3)$$

$$= \sum_{k=0}^{s-1} \left(\sum_{i+j=k} a_i a_j + \tau \sum_{i+j=s+k} a_i a_j\right), \qquad (4)$$

where $s > 2$, and $a_i = 0$, $1$, or $\tau$.

**Case 1.** The number $s$ is odd.

In this case, we have the following equality:

$$\tau \left(1 + \lambda^2 + \lambda^4 + \ldots + \lambda^{2(s-1)}\right)$$
$$= \tau \left(1 + \lambda^2 + \lambda^4 + \ldots + \lambda^{s-1}\right) + \tau^2 \lambda \left(1 + \lambda^2 + \lambda^4 + \ldots + \lambda^{s-3}\right).$$

Therefore, the constant term of the lefthand side is $\tau$. On the other hand, the constant term of the righthand side is

$$a_0^2 + 2\tau \sum_{i+j=s, i<j} a_i a_j$$

which should be, by Lemma 1, equal to the constant $\tau$ in the lefthand side. But actually, it can be written in the form:

$$2(c\tau + d), \ 1 + 2(c\tau + d), \ \text{or} \ (1 - \tau) + 2(c\tau + d)$$

according to the value of $a_0$, for some integers $c$ and $d$. But none of them can be equal to $\tau$. So the equality (3) can't hold in this case.

**Case 2.** The number $s$ is even.

In this case, $s \geq 4$ and

$$\tau \left(1 + \lambda^2 + \lambda^4 + \ldots + \lambda^{2(s-1)}\right)$$
$$= \tau \left(1 + \lambda^2 + \lambda^4 + \ldots + \lambda^{s-2}\right) + \tau^2 \left(1 + \lambda^2 + \lambda^4 + \ldots + \lambda^{s-2}\right)$$
$$= \left(\tau + \tau^2\right) \left(1 + \lambda^2 + \lambda^4 + \ldots + \lambda^{s-2}\right)$$
$$= 1 + \lambda^2 + \lambda^4 + \ldots + \lambda^{s-2}.$$

So by Lemma 1, the following lemma is immediate.

**Lemma 2.** *Suppose that the equality (3) holds, and s is even.*

(1) $\displaystyle\sum_{i+j=k} a_i a_j + \tau \sum_{i+j=s+k} a_i a_j = 1$ *for even* $k$, *and*

(2) $\displaystyle\sum_{i+j=k} a_i a_j + \tau \sum_{i+j=s+k} a_i a_j = 0$ *for odd* $k$.      □

**Lemma 3.** *Suppose that the condition (2) in Lemma 2 is valid. Then*

$$a_i a_j = 0$$

*if* $i + j$ *is odd.*

*Proof.* Let us denote the numbers of particular terms, $a_i a_j$'s, in the first sigma (i.e., $i + j < s$), in the following manner.

$$p = \text{the number of terms } a_i a_j\text{'s such that } a_i a_j = \tau^2,$$
$$q = \text{the number of terms } a_i a_j\text{'s such that } a_i a_j = \tau,$$
$$r = \text{the number of terms } a_i a_j\text{'s such that } a_i a_j = 1.$$

We denote by $p'$, $q'$, and $r'$ the numbers of corresponding terms in the second sigma (i.e., $i + j \geq s$). Then

$$\sum a_i a_j + \tau \sum a_i a_j = p\tau^2 + q\tau + r + \tau(p'\tau^2 + q'\tau + r')$$
$$= (-p + q + 2p' - q' + r')\tau + (p + r - p' + q') = 0$$

Therefore

$$(-p + q + 2p' - q' + r) = 0 \quad \text{and} \quad (p + r - p' + q') = 0,$$
$$-p + q + 2p' - q' + r = -p + q + 2(p + r + q') - q' + r$$
$$= p + q + 3r + q' = 0.$$

Since all these values, $p$, $q$, etc., are non-negative,

$$p = q = r = q' = 0.$$

It follows immediately that $p' = r' = 0$. So $a_i a_j = 0$ for $i + j = k$ or $i + j = s + k$, when $k$ is odd.

Now if $i + j$ is odd, then both $k = i + j$ and $k = i + j - s$ are odd, since $s$ is even, and therefore $a_i a_j = 0$.      □

**Corollary 2.** *If* $a_i \neq 0$ *for some even* $i$, *then* $a_j = 0$ *for every odd* $j$.      □

Let us denote:

$$c_k = \left( \sum_{i+j=k} a_i a_j + \tau \sum_{i+j=s+k} a_i a_j \right).$$

Then we have:

$$c_{2j} = a_j^2 + \tau a_{\frac{s}{2}+j}^2 + 2 \left( \sum_{i+j=k,\, i<j} a_i a_j + \tau \sum_{i+j=s+k,\, i<j} a_i a_j \right).$$

**Lemma 4.** *Suppose that* $c_{2j} = 1$.

(1) *If* $a_j = 0$, *then* $a_{\frac{s}{2}+j} = \tau$.

(2) *If* $a_{\frac{s}{2}+j} = 0$, *then* $a_j = 1$.

*Proof.* The expression

$$a_j^2 + \tau a_{\frac{s}{2}+j}^2 - 1$$

is equal to

$$2 \left( \sum a_i a_j + \tau \sum a_i a_j \right),$$

which can be represented in the form $2(s\tau + t)$ for some integers $s$ and $t$.

(1) If $a_j = 0$, then the value of the above expression is equal to $-1$, $\tau - 1$, or $2\tau - 2$, according to the case: $a_{\frac{s}{2}+j} = 0$, 1, or $\tau$, respectively. So only the case when $a_{\frac{s}{2}+j} = \tau$ is possible.

The property (2) is verified by a similar argument.  □

**Case 2.1.** $a_i \neq 0$ for some even $i$.

In this case, $a_j = 0$ for every odd $j$. Therefore by (1) in Lemma 4, $a_{\frac{s}{2}+j} = \tau$ for odd $j$. This means that $\frac{s}{2}$ is odd (otherwise $a_{\frac{s}{2}+j} = 0$), $a_{\frac{s}{2}+j} = 0$ for an even integer $j$, and $a_j = 1$ for an even integer $j$ less than $\frac{s}{2}$.

After all, $a_j = 1$ for an even integer $j$ less than $\frac{s}{2}$, and $a_j = \tau$ for other even $j$. Therefore

$$c_0 = 1 + 0 + 2\tau \sum_{i+j=s,\, i<j} a_i a_i = 1 + 2\tau \times \frac{\frac{s}{2} - 1}{2} \times (1 \times \tau) = 1 + \left( \frac{s}{2} - 1 \right) \tau^2 \neq 1,$$

since $s > 2$. This contradicts the property (1) in Lemma 2.

**Case 2.2.** $a_i = 0$ for every even integer $i$.

By the property (1) in Lemma 4, $\frac{s}{2} + i = \tau$ for every even $i$, and $\frac{s}{2}$ is odd (otherwise $a_{\frac{s}{2}+i} = 0$ in this case). This means $a_{\frac{s}{2}+i} = 0$ for odd $j$, and $a_j = 1$ for every odd $j$ less than $\frac{s}{2}$. But if so, we have

$$c_2 = a_1^2 + \tau a_{\frac{s}{2}+1}^2 + 2 \left( a_0 a_2 + \tau \sum_{i+j=s+1,\, i<j} a_i a_j \right)$$

$$= 1 + 2\tau \times \frac{\frac{s}{2} - 1}{2} \times (1 \times \tau) \neq 1.$$

Once more, the property (1) in Lemma 2 is violated.

Thus we can conclude that the equality (3) can't hold in any case.

**Acknowledgments:** The authors are very grateful to Professor G. N. Frederickson for giving them valuable information relating to this study, and to Professor T. Sakai of Tokai University for his help in completing this paper in this form.

# References

1. Akiyama, J., Nakamura, G.: An efficient dissection for a sequentially $n$-divisible square. Proc. of Discrete and Computational Geometry Workshop, Tokai University (1997), 80–89
2. Busschop, P.: Problèmes de géométrie. Nouvelle Correspondance Mathématique **2** (1876) 83–84
3. Hitotumatu, S.: Tile Problem (in Japanese), in "One Hundred Mathematical Problems". Nihon-Hyouron-sya, (1999) 23–26
4. Nozaki, A.: On the dissection of a square into squares (in Japanese). Suugaku-Seminar No.12 (1999) 52–56
5. Ozawa, K.: Entertainer in a classroom (in Japanese). Suugaku-Seminar No.10 (1988) cover page
6. Ozawa, K.: private communication (1999)

# Appendix

*The following polynomial $P(x)$ over the field $\mathbf{Q}[\tau]$ is irreducible.*

$$P(x) = x^s - \tau,$$

*where $s > 1$ and $\tau^2 + \tau - 1 = 0$.*

First, we shall show the following fact.

**Fact 1.** *Let $m$ be an integer greater than 1. Then, the $m$-th root $\lambda$ of $\tau$ is not in $\mathbf{Q}[\tau]$.*

*Proof.* Suppose that $\lambda \in \mathbf{Q}[\tau]$, that is,

$$\lambda = p(\tau + c)$$

for some $p$ and $c$ in $\mathbf{Q}$. For an arbitrary integer $n$, we put:

$$(\tau + c)^n = A_n \tau + B_n.$$

Obviously, both $A_n$ and $B_n$ are polynomials in $c$ with integer coefficients. For instance:

$$A_1 = 1 \quad \text{and} \quad B_1 = c.$$

Since

$$(\tau + c)^2 = \tau^2 + 2c\tau + c^2 = (1 - \tau) + 2c\tau + c^2 = (2c - 1)\tau + (c^2 + 1),$$

we have

$$A_2 = 2c - 1 \quad \text{and} \quad B_2 = c^2 + 1.$$

In general,

$$A_{n+1}\tau + B_{n+1} = (A_n\tau + B_n)(\tau + c) = (-A_n + cA_n + B_n)\tau + (A_n + cB_n)$$

and therefore

$$A_{n+1} = -A_n + cA_n + B_n \quad \text{and} \quad B_{n+1} = A_n + cB_n.$$

Now since $B_n$ is a monic polynomial with integer coefficients, any rational solution c of the equation $B_n(c) = 0$ should be an integer. So, we shall show that $B_n(c) \neq 0$ for any integer c and any $n > 1$.

**Case 1.**  $c \geq 1$.
Let $H_n = \frac{B_n}{A_n}$. In this case, $H_1 = \frac{c}{1} \geq 1$. Besides, if $H_n \geq 1$, then

$$\begin{aligned}
H_{n+1} &= \frac{A_n + cB_n}{-A_n + cA_n + B_n} \\
&= \frac{1 + cH_n}{-1 + c + H_n} \\
&= 1 + \frac{1 + (c-1)(H_n - 1)}{-1 + c + H_n} > 1
\end{aligned}$$

So, $H_n > 1$ for any $n > 1$ in this case.

**Case 2.**  $c \leq 0$.
In this case, $H_1 = \frac{c}{1} \leq 0$. If $H_n \leq 0$, then

$$H_{n+1} = \frac{1 + cH_n}{-1 + c + H_n} < 0,$$

since the numerator is positive and the denominator is negative. Thus $H_n < 0$ for any $n > 1$ in this case.

*Remark.* If $c = 0$, then $H_1 = B_1 = 0$, but $H_n < 0$ and $B_n \neq 0$ for any $n > 1$.

After all, $H_n \neq 0$ and therefore $B_n \neq 0$ for any $n > 1$. So $\lambda^m = p^m(A_m\tau + B_m) \neq \tau$, although $\lambda$ was defined to be the m-th root of $\tau$. Contradiction.    □

**Fact 2.**  *Let s be an integer greater than 1, and $\lambda$ the s-th root of $\tau$. Let t be a positive integer less than s $(0 < t < s)$. Then $\lambda^t$ is not in $\mathbf{Q}[\tau]$.*

*Proof.* Let d be the greatest common divisor of t and s. If $\lambda^t$ is in $\mathbf{Q}[\tau]$, then so is $\lambda^d$, since $\lambda^s = \tau \in \mathbf{Q}[\tau]$. But this is impossible, since $\lambda^d$ is the $\frac{s}{d}$-th root of $\tau$.    □

Now the polynomial $P(x)$ is factorized in $\mathbf{C}$ as follows:

$$P(x) = \prod_{j=0}^{s-1} \left(x - \lambda\zeta^j\right),$$

where $\zeta$ is a primitive $s$-th root of 1. If $P(x)$ is reducible in $\mathbf{Q}[\tau]$, then it is divisible by a monic polynomial $R(x)$ over the field $\mathbf{Q}[\tau]$, which can be written as follows.

$$R(x) = \prod_{j\in D} \left(x - \lambda\zeta^j\right).$$

Its constant term $c \in \mathbf{Q}$ is the product of $\left(-\lambda\zeta^j\right)$'s, and is written in the following form:

$$c = (-1)^t \lambda^t \left(\prod \zeta^j\right),$$

where $t$ is the degree of $R$ (i.e., the number of elements in $D$). Since the absolute value of $\zeta$ is 1 and $c$ is supposed to be in $\mathbf{Q}$, the product $\left(\prod \zeta^j\right)$ is equal to 1 or $-1$. Consequently, $\lambda^t$ is in $\mathbf{Q}[\tau]$, against the property stated in Fact 2.

Thus the polynomial $P(x)$ is irreducible in $\mathbf{Q}[\tau]$.

# Sequentially Divisible Dissections
## of Simple Polygons

Jin Akiyama[1], Toshinori Sakai[2], and Jorge Urrutia[3]

[1] Research Institute of Educational Development, Tokai University,
2-28-4 Tomigaya, Shibuya-ku, Tokyo 151-0063, Japan,
fwjb5117@mb.infoweb.ne.jp,
[2] Research Institute of Education, Tokai University,
2-28-4 Tomigaya, Shibuya-ku, Tokyo 151-0063, Japan,
tsakai@yoyogi.ycc.u-tokai.ac.jp,
[3] Instituto de Matemáticas, Ciudad Universitaria,
Universidad Nacional Autónoma de México, México D.F., México,
urrutia@math.unam.mx

**Abstract.** A $k$-dissection $\mathcal{D}$ of a polygon $\mathcal{P}$, is a partition of $\mathcal{P}$ into a set of subpolygons $\{\mathcal{Q}_1, \ldots, \mathcal{Q}_m\}$ with disjoint interiors such that these can be reassembled to form $k$ polygons $\mathcal{P}_1, \ldots, \mathcal{P}_k$ all similar to $\mathcal{P}$. If for every $j$, $1 \leq j \leq k$, the pieces of $\mathcal{D}$ can be assembled into $j$ polygons, all similar to $\mathcal{P}$, then $\mathcal{D}$ is called a sequentially $k$-divisible dissection. In this paper we show that any convex $n$-gon, $n \leq 5$, has a sequentially $k$-divisible dissection with $(k-1)n+1$ pieces. We give sequentially $k$-divisible dissections for some regular polygons with $n \geq 6$ vertices. Furthermore, we show that any simple polygon $\mathcal{P}$ with $n$ vertices has a $(3k+4)$-dissection with $(2n-2) + k(2n + \lfloor \frac{n}{3} \rfloor - 4)$ pieces, $k \geq 0$, that can be reassembled to form 4, 7, ..., or $3k+4$ polygons similar to $\mathcal{P}$. We give similar results for star shaped polygons.

## 1 Introduction

Dissections of polygons is a truly classical field of study in the mathematical sciences. A classical result of the 19-th century by Lowry, Wallace, Bolyai, and Gerwien, asserts that given two simple polygons $\mathcal{P}$ and $\mathcal{Q}$ of the same area, we can dissect $\mathcal{P}$ into a finite number of polygons which can be reassembled to form $\mathcal{Q}$.

Books mentioning dissections of polygons appear from time to time in the literature, each bringing new advances and interesting puzzles to the topic we study here, e.g. Fourrey [6](1907), Kraitchik [9](1942), Gardner [8](1961), .... Books by Lindgren [10](1964) and Frederickson [7](1997) devoted solely to the study of dissections.

Let $\mathcal{P}$ be a polygon on the plane. A *dissection* $\mathcal{D}$ of $\mathcal{P}$ is a partitioning of $\mathcal{P}$ into $m$ subpolygons $\mathcal{P}_1, \ldots, \mathcal{P}_m$ such that $int\mathcal{P}_i \cap int\mathcal{P}_j = \phi$, $1 \leq i < j \leq m$, where $int\mathcal{P}$ denotes the interior of $\mathcal{P}$. Each $\mathcal{P}_i$ is called a *piece* of $\mathcal{D}$. If none of the pieces of $\mathcal{D}$ is similar to $\mathcal{P}$, $\mathcal{D}$ will be called a *non-trivial* dissection of $\mathcal{P}$.

A *k-dissection* $\mathcal{D}$ of $\mathcal{P}$ is a partitioning of $\mathcal{P}$ into subpolygons $\{\mathcal{P}_1, \ldots, \mathcal{P}_m\}$ with disjoint interiors such that they can be reassembled to form $k$ polygons all similar to $\mathcal{P}$. A dissection of $\mathcal{P}$ is called *sequentially k-divisible* if for every $j$, $1 \leq j \leq k$, its pieces can be assembled so as to form $j$ polygons similar to $\mathcal{P}$. In Fig.1(a), we show a 2-dissection of a triangle. Fig.1(b) and (c) show a non-trivial 2-dissection and a sequentially 4-divisible dissection of the same triangle.

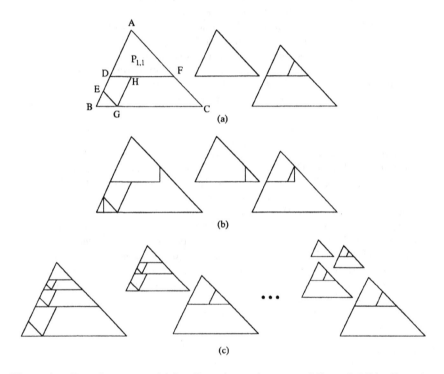

**Fig. 1.** A 2-dissection, a nontrivial 2-dissection and a sequentially 4-divisible dissection of a triangle.

Collison([5], see also page 97 of [7]) gave 2-dissections of regular $n$-gons with $n + 1$ or $n$ pieces for $n$ odd or even, respectively. These dissections allows us to construct sequentially $k$-divisible dissections with $(k - 1)n + 1$ or $(k - 1)n - k + 2$ pieces, respectively (Theorem 4 in this paper). Sequentially $k$-divisible dissections of squares have been studied in $[1\text{-}4, 11, 12]$. In this paper we present sequentially $k$-divisible dissections of triangles, convex quadrilaterals, and convex pentagons with $3k - 2$, $4k - 3$, and $5k - 4$ pieces respectively. For triangles we present non-trivial sequentially $k$-divisible dissections with $3k - 1$ pieces. For regular polygons with $n = 4m$ vertices, we present 2-dissections with $n$ pieces which allow us to construct sequentially $k$-divisible dissections with $(k-1)n-k+2$

pieces. For *simple polygons*, not necessarily convex, with $n$ vertices we present a 4-dissection with $2n - 2$ pieces. This allows us to construct $(3k + 4)$-dissections with at most $(2n - 2) + k(2n + \lfloor \frac{n}{3} \rfloor - 4)$ pieces. Finally for *star shaped polygons*, we show similar results.

The following notation will be useful throughout the paper: Given two points $P$ and $Q$ on the plane, $PQ$ will denote the line segment joining them. The point $(1 - \lambda)P + \lambda Q$ will be denoted as $\lambda(PQ)$. Note that $\lambda(PQ)$ is different from $\lambda(QP)$. For example when $\lambda = 0$, we obtain $P$, when $\lambda = 1$ we get $Q$, and when $\lambda = \frac{1}{2}$ we obtain the mid point of the segment $PQ$. Similarly, let $\mathcal{Q}$ be a polygonal (resp. a polygon) with vertices $Q_1, \ldots, Q_n$, then $\lambda(P, \mathcal{Q})$ will denote the polygonal (resp. the polygon) with vertices $\lambda(PQ_1), \ldots, \lambda(PQ_n)$. If line segments $PQ$ and $RS$ are parallel, we will write $PQ \parallel RS$. If a polygon has vertices $V_1, \ldots, V_n$ we will often refer to it as the polygon with vertices $\{V_1, \ldots, V_n\}$, or simply the polygon $\{V_1, \ldots, V_n\}$.

Two polygons $\mathcal{P}$ and $\mathcal{Q}$ are called *congruent* if there is a translation $T$, a rotation $R$, and perhaps a reflection that maps $\mathcal{P}$ onto $\mathcal{Q}$. $\mathcal{P}$ and $\mathcal{Q}$ are called *similar* if there is mapping $f : \mathbf{R}^2 \to \mathbf{R}^2$ such that $f(x) = p_0 + \lambda x$, where $p_0$ is a point in $\mathbf{R}^2$, and such that $f(\mathcal{P})$ and $\mathcal{Q}$ are congruent. Observe that if two polygons $\mathcal{P}$ and $\mathcal{Q}$ are similar, any dissection $\mathcal{D} = \{\mathcal{P}_1, \ldots, \mathcal{P}_m\}$ of $\mathcal{P}$ induces, in a natural way, a dissection $\mathcal{D}'$ of $\mathcal{Q}$ such that the pieces of $\mathcal{D}'$ are the sets $f(\mathcal{P}_i)$, $i = 1, \ldots, m$. Dissection $\mathcal{D}'$ will be referred to as the *dissection induced* in $\mathcal{Q}$ by $\mathcal{D}$.

## 2  Sequentially divisible dissections of triangles

As for triangles, we show the following theorem:

**Theorem 1.** *Any triangle has a sequentially $k$-divisible dissection (resp. non-trivial sequentially $k$-divisible dissection) with $3k - 2$ pieces (resp. $3k - 1$).*

*Proof.* Let $\mathcal{P}_{1,1}$ be a triangle with vertices $\{A, B, C\}$. Let $\mathcal{D}_1$ be the dissection of $\mathcal{P}_{1,1}$ obtained as follows: Let $D$ and $E$ be the points on AB such that $D = \frac{3}{5}(AB)$ and $E = \frac{4}{5}(AB)$, and let $F \in AC$, $G \in BC$ and $H \in DF$ be the points such that $DF \parallel BC$, $EG \parallel AC$ and $GH \parallel AB$ (Fig.1(a)). Then triangle $\mathcal{P}_{2,1}$ with vertices $\{A, D, F\}$ is similar to $\mathcal{P}_{1,1}$ and their ratio of similitude is $\frac{3}{5}$. Note that triangle $\mathcal{P}_{2,2}$ with vertices $\{B, E, G\}$ and trapezoids $\mathcal{P}_{2,3}$ and $\mathcal{P}_{2,4}$ with vertices $\{D, E, G, H\}$, and $\{C, F, H, G\}$, respectively, can be assembled into a triangle similar to $\mathcal{P}_{1,1}$ with ratio of similitude equal to $\frac{4}{5}$, see Fig.1(a).

In a recursive way, let $\mathcal{D}_j$ be the dissection induced in $\mathcal{P}_{j,1}$ by $\mathcal{D}_1$, where $\mathcal{P}_{j,1}$ is the triangle of $\mathcal{D}_{j-1}$ containing vertex $A$, $j \geq 2$. For any fixed integer $k \geq 3$, $(\mathcal{D}_1 - \{\mathcal{P}_{2,1}\}) \cup (\mathcal{D}_2 - \{\mathcal{P}_{3,1}\}) \cup \ldots \cup (\mathcal{D}_{k-2} - \{\mathcal{P}_{k-1,1}\}) \cup \mathcal{D}_{k-1}$ defines a dissection $\mathcal{S}_k$ of $\mathcal{P}_{1,1}$ with exactly $3(k - 2) + 4 = 3k - 2$ pieces. See Fig.1(c). Clearly $\mathcal{S}_k$ is a sequentially $k$-divisible dissection of $\mathcal{P}_{1,1}$.

Finding non-trivial sequentially $k$-divisible dissections of triangles is more challenging. First we start by modifying $\mathcal{S}_k$ to obtain a sequentially $k$-divisible

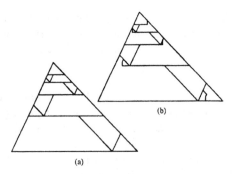

**Fig. 2.** Finding a non-trivial sequentially 5-dissection of a triangle.

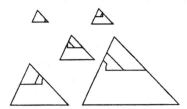

**Fig. 3.** Reassembling the dissection in Fig 2(b) into five triangles.

dissection $\mathcal{D}$ of triangle $\{A, B, C\}$ as shown in Fig.2(a). The main objective of our modification, is to make sure that every second triangle of $\mathcal{S}_k$ from top to bottom touches $AB$, and the others touch $AC$ (with the exception the triangle containing $A$, which touches $AB$, and $AC$). The details of this modification are straightforward, and are left to the reader. We now proceed to show how we can modify this construct to obtain a non-trivial sequentially $k$-divisible dissection of our triangle.

Suppose that we relabel the triangles of $\mathcal{D}$ from top to bottom by $\mathcal{T}_1, \ldots, \mathcal{T}_k$. Split $\mathcal{T}_2$ into two triangles $\mathcal{L}_2$ and $\mathcal{R}_2$ by drawing a segment through its top vertex such that resulting triangles are not similar to triangle $ABC$. Join triangle $\mathcal{L}_2$, the left piece of $\mathcal{T}_2$, to the element of $\mathcal{D}$ below it (throughout this paragraph, a left (resp. right) piece means a piece having (resp. not having) an edge contained within $AB$). Next split $\mathcal{T}_1$ into two pieces, one of which, the right piece, is similar to $\mathcal{R}_2$, and join the right piece to the element of $\mathcal{D}$ below it, as shown in Fig.2(b). In a recursive way, we now split $\mathcal{T}_i$ into a right piece $\mathcal{R}_i$ and a left piece $\mathcal{L}_i$ such that if $i$ is odd, then $\mathcal{L}_i$ is congruent with $\mathcal{L}_{i-1}$, and if $i$ is even then $\mathcal{R}_i$ is congruent to $\mathcal{R}_{i-1}$, $i = 2, \ldots, k$. Next if $i$ is odd, join $\mathcal{R}_i$ to the piece of $\mathcal{D}$ below it, else if $i$ is even join $\mathcal{L}_i$ to the piece of $\mathcal{D}$ below it, $i < k$, see Fig.2(b). It is now easy to see that the pieces of the dissection thus obtained form a sequentially $k$-divisible dissection. In Fig.3 we show how to assemble the pieces of the dissection in Fig.2(b) into five triangles. $\qquad \square$

*Remark.* We can modify a 2-dissection of triangles due to Collison (see Figure 9.8 in [7]) and shown in Fig.4(a) (solid lines) to obtain different $k$-divisible dissections also with $3k - 1$ pieces, see Fig.4(b).

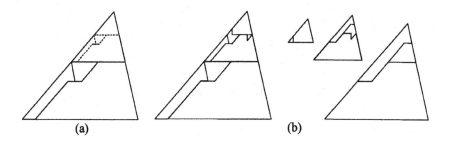

**Fig. 4.** Finding a sequentially 3-divisible dissection of a triangle.

# 3 Quadrilaterals

We now show:

**Theorem 2.** *Any convex quadrilateral has a sequentially $k$-divisible dissection with $4k - 3$ pieces.*

*Proof.* Let $\mathcal{P} = \{A, B, C, D\}$ be the convex quadrilateral to be dissected. We may assume that $\angle A + \angle B \leq \pi$ and $\angle B + \angle C \leq \pi$.

We first give a sequentially 2-divisible dissection $\mathcal{D}$ of $\mathcal{P}$ consisting of five pieces. Let $E \in AB$, $F \in AD$ be the points such that $E = \frac{3}{5}(AB)$, $F = \frac{3}{5}(AD)$ and let $G$ be the point on the diagonal $AC$ such that $EG \parallel BC$ (so $FG \parallel DC$). Let $H \in BC$, $I \in DC$ be the points such that $H = \frac{1}{5}(BC)$, $I = \frac{1}{5}(DC)$, let $J \in EG$, $K \in FG$ be the points such that $JH \parallel AB$ and $KI \parallel AD$ and let $L$ be the mid-point of $JH$. Let $M$ be the intersection point of the line passing through $E$ and parallel to $DC$ and the line passing through $L$ and parallel to $AD$. Since $\angle A + \angle B \leq \pi$ and $\angle B + \angle C \leq \pi$, $M$ is a point in the parallelogram $\{B, E, J, H\}$. This produces the dissection of $\mathcal{P}$ with pieces $\{\mathcal{P}_{1,1}, \ldots, \mathcal{P}_{1,5}\}$ as shown in Fig.5(a). It is easy to see that this is a sequentially 2-divisible dissection of $\mathcal{P}$ (Fig.5(b)).

We will now take the dissection induced by $\mathcal{D}$ in $\mathcal{P}_{1,1}$ to obtain a 3-dissection of $\mathcal{P}$. By iterating this process, we get a sequence of sequentially $k$-divisible dissections of $\mathcal{P}$ with $4k - 3$ pieces. $\qquad\square$

# 4 Pentagons

In this section we prove the following result:

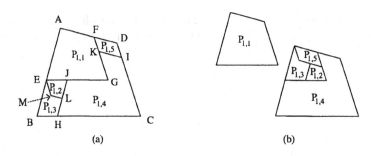

**Fig. 5.** A 2-dissection of a convex quadrilateral.

**Theorem 3.** *Any convex pentagon has a sequentially k-divisible dissection with $5k - 4$ pieces.* □

We note that using Collison's 2-dissection of a regular polygon with six pieces (see Figure 9.17 in [7]) we can obtain our result for regular pentagons. We now present two 2-dissections for non-regular convex pentagons that will prove our result for general convex pentagons. Some extra results will be now proved.

Let $\mathcal{P}$ be a convex pentagon. We use the following lemma without proof:

**Lemma 1.** *Let $\mathcal{P}$ be a pentagon. Then we can label its vertices A, B, C, D, and F in the clockwise or counter-clockwise direction such that:*

1. *$\angle A + \angle B > \pi$, $\angle B + \angle BCE \geq \pi$,*
2. *at least one of $\angle ACD + \angle D \geq \pi$ or $\angle D + \angle E > \pi$ holds.*

Let $\mathcal{P}$ be a pentagon with its vertices labelled $A$, $B$, $C$, $D$, $E$ as in Lemma 1. Let $A_1 = A$, $B_1 = \frac{3}{5}(AB)$, ..., $E_1 = \frac{3}{5}(AE)$, $A_2 = \frac{2}{5}(DA)$, ..., $C_2 = \frac{2}{5}(DC)$, $D_2 = D$, and $E_2 = \frac{2}{5}(DE)$. Let $F_1 = \frac{1}{5}(BC)$, and $G_1 = \frac{1}{3}(B_1C_1)$.

Since $\angle B + \angle C > \pi$ by the first condition 1 in Lemma 1, pentagons $\{A_1, B_1, C_1, D_1, E_1\}$ and $\{A_2, B_2, C_2, D_2, E_2\}$ have no common interior point. Note that $BB_1$ and $B_2A_2$ have the same length. Since $\angle A + \angle B > \pi$ and $\angle B + \angle C > \pi$ the translation that maps $BB_1$ to $B_2A_2$ maps the parallelogram $\{B, F_1, G_1, B_1\}$ to a subset of the pentagon $\{A_2, B_2, C_2, D_2, E_2\}$.

Consider next the pentagon with vertices $C_3 = C_2$, $A_3 = \frac{1}{2}(C_2A_2)$, $B_3 = \frac{1}{2}(C_2B_2)$, $D_3 = \frac{1}{2}(C_2D_2)$, $E_3 = \frac{1}{2}(C_2E_2)$. Rotate pentagon $\{A_3, B_3, C_3, D_3, E_3\}$ 180 degrees around $A_3$ as shown in Fig.6(a) to obtain the pentagon $\mathcal{P}_4$ with vertices $\{A_4, B_4, C_4, D_4, E_4\}$. Two cases arise: $E_4$ belongs to the interior of pentagon $\{A, B, C, D, E\}$ as in Fig.6(a), or $E_4$ does not lie in the interior of the same pentagon.

In the first case, the reader can verify that the dissection shown in Fig.7(a) is realizable. In the case when $E_4$ lies outside the original pentagon (this situation can arise if we move point $E$ in Fig.6(a) far enough to the right, for aesthetic reasons we don't show a picture for this case), we will show that the pentagon $\mathcal{P}'_4 = \{A_5, \ldots, E_5\}$, obtained by translating $\{A_4, \ldots, E_4\}$ in such a way that

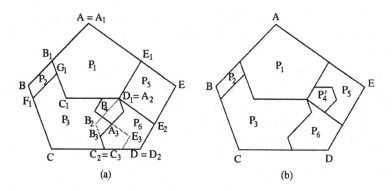

**Fig. 6.** Finding a dissection of a convex pentagon.

$E_4$ lies on $D_1$, is contained in the parallelogram $\{D_1, E_2, E, E_1\}$ (Fig.6(b)). The dissection shown in Fig.7(b) will now be realizable.

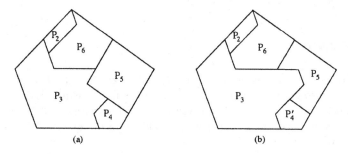

**Fig. 7.** Showing the regrouping of the pieces of the dissection from Fig.6. $P_1$ is not shown here, but it is a part of each 2-dissection illustrated here.

Given two points $P$ and $Q$, $\overrightarrow{PQ}$ will denote the vector $Q-P$. Let $a = \overrightarrow{AC}$, $b = \overrightarrow{AE}$ and consider the unique real numbers $\alpha$ and $\beta$ such that $\overrightarrow{AD} = \alpha a + \beta b$. Clearly $\alpha > 0$, $\beta > 0$ and $\alpha + \beta > 1$. We also have $\beta \leq 1$ or $\alpha < 1$, depending on whether $\angle ACD + \angle D \geq \pi$ or $\angle D + \angle E > \pi$. Two cases arise:

**Case 1.** $\beta > \max\{\frac{1}{3}, \alpha - 1\}$:

In this case, we show that $E_4$ belongs to the interior of $\{F_1, C, C_2, B_3, B_4, C_4, C_1, G_1\}$ to prove that the dissection shown in Fig.6(a) is realizable. Since

$$\overrightarrow{C_4E_4} = \tfrac{1}{5}(a - b) = \tfrac{1}{3\beta}\overrightarrow{C_4C_1} + \tfrac{\alpha+\beta-1}{2\beta}\overrightarrow{C_4C_2}$$

and since $0 < \frac{1}{3\beta} < 1$ and $0 < \frac{\alpha+\beta-1}{2\beta} < 1$, $E_4$ is an interior point of the parallelogram $\{C, C_2, C_4, C_1\}$. Since $\angle B_4 A_4 E_4 < \pi$, the desired conclusion now follows.

**Case 2.** $\beta \leq \max\{\frac{1}{3}, \alpha - 1\}$:

We will show that in this case the dissection shown in Fig.6(b) is realizable. To see this we show that $B_5$ and $C_5$ are interior points in the parallelogram $\{E_1, E_5, E_2, E\}$. Note that $\alpha > \frac{2}{3}$ and $\beta \leq 1$ in this case.

Let $\gamma$ and $\delta$ be unique real numbers such that $\overrightarrow{EB} = \gamma \overrightarrow{EA} + \delta \overrightarrow{EC} \, (= \gamma(-b) + \delta(a - b))$. Since $\angle A + \angle B > \pi$ and $\angle B + \angle BCE \geq \pi$, $0 < \delta < 1$ and $0 < \gamma \leq 1$.

On the other hand, we have

$$\overrightarrow{E_5 B_5} = -\frac{1}{5}[\gamma(-b) + \delta(a - b)] = \frac{\delta}{3\alpha} \overrightarrow{E_5 E_1} + \frac{(\gamma+\delta)\alpha+\delta(\beta-1)}{2\alpha} \overrightarrow{E_5 E_2} \quad \text{and}$$

$$\overrightarrow{E_5 C_5} = -\frac{1}{5}(a - b) = \frac{1}{3\alpha} \overrightarrow{E_5 E_1} + \frac{\alpha+(\beta-1)}{2\alpha} \overrightarrow{E_5 E_2}.$$

Then since $0 < \frac{\delta}{3\alpha} < \frac{1}{3\alpha} < \frac{1}{2}$, $0 < \frac{(\gamma+\delta)\alpha+\delta(\beta-1)}{2\alpha} < 1$ and $0 < \frac{\alpha+(\beta-1)}{2\alpha} \leq \frac{1}{2}$, the desired conclusion follows.

In a recursive way, we get a sequence of sequentially $k$-divisible dissections of $\mathcal{P}$ with $5k - 4$ pieces.

# 5 Hexagons

J. Schmerl found a 2-dissection of the regular hexagon with five pieces, see Fig.8(a) (see also Figure 9.18 in [7]). By recursively using Schmerl's dissection it follows that there are sequentially $k$-divisible dissections with $4k-3$ pieces. Using the 2-dissection shown in Fig.8(b) we can also obtain sequentially $k$-divisible dissections with $6k - 5$ pieces.

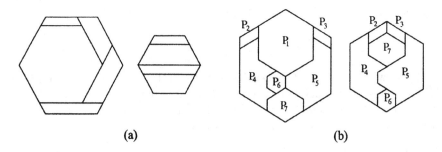

(a)         (b)

**Fig. 8.** 2-dissections of regular hexagons. We show how to construct one of two regular hexagons.

# 6 Sequentially divisible dissections of regular 4k-gons

Using the iteration process used in Sections 2–5, the following result now follows from Collison's result mentioned in Section 1:

**Theorem 4.** *Let $\mathcal{P}$ be a regular polygon with $n$ vertices. Then if $n$ is odd there is a sequentially $k$-divisible dissection of $\mathcal{P}$ with $(k-1)n+1$ pieces. If $n$ is even a sequentially $k$-divisible dissection with $(k-1)n-k+2$ pieces exists.* $\qquad\square$

For regular polygons with $n = 4m$ vertices, we now give a new 2-dissection with $n$ pieces, which allows us to cinstruct sequentially $k$-divisible dissections with $(k-1)n-k+2$ pieces. Let $\mathcal{P}$ be a regular polygon with $4m$ vertices labeled $A_0, \ldots, A_{4m-1}$ in the counterclockwise direction, with $A_0$ being the topmost vertex of $\mathcal{P}$.

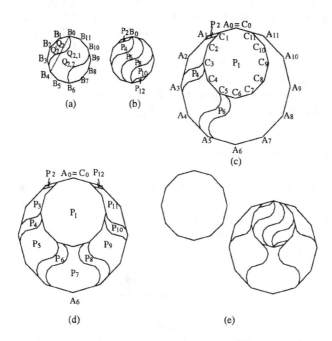

**Fig. 9.** Dissecting regular $4m$ polygons. Notice that pieces $\mathcal{P}_8$, $\mathcal{P}_{10}$ and $\mathcal{P}_{12}$ in (d) have been rotated and flipped.

Consider a second regular $4m$ polygon $\mathcal{B}$ with vertices $\{B_0, \ldots, B_{4m-1}\}$ of size $\frac{2}{5}$ that of $\mathcal{P}$. We present first a dissection of $\mathcal{B}$ with $2m$ pieces obtained as follows:

For each $i$, $1 \leq i \leq m$, let $\mathcal{Q}_i$ be the polygonal with vertices $\{B_0, \ldots, B_{2i}\}$, and let $\mathcal{Q}'_i$ be the polygonal $\frac{1}{2}(B_{2i}, \mathcal{Q}_i)$. Let $\mathcal{Q}_{i,1}$ be the polygonal obtained by joining a copy of $Q_i'$ with the polygonal obtained by rotating $Q_i'$ 180 degrees around the point $\frac{1}{2}(B_{2i}B_0)$ as shown in Fig.9(a) for the case $m = 3$. Finally let $Q_{i,2}$ be the polygonal obtained from $Q_{i,1}$ by rotating it 180 degrees around the center of $B$. The set of $Q_{i,1}$, $Q_{i,2}$, $i = 1, \ldots, m$ induces a partitioning of $B$ into $2m$ pieces. Let us label the regions of this partitioning $P_{2i}$, $i = 1, \ldots, 2m$, as shown in Fig.9(b). With this labeling $P_{2i}$, $i = 1, \ldots, m$, will contain vertex $B_{2i-1}$ of $B$.

We now show a dissection of $P$ that will contain pieces similar to $P_{2i}$, $i = 1, \ldots, 2m$. Let $P_1$ be the polygon $\frac{3}{5}(A_0, P)$. Let us label the vertices of $P_1$ by $C_i$, $i = 0, \ldots, 4m - 1$, where $C_0 = A_0$.

For each $1 \leq i \leq m$ let us translate a copy of $P_{2i}$ so that vertex $B_{2i-1}$ is mapped to vertex $A_{2i-1}$ of $P$. Since the length of segment $A_{2i-1}C_{2i-1}$ is $\frac{2}{5}$ the length of $A_{2i-1}A_0$ the point $B_0$ of $P_{2i}$ maps to vertex $C_{2i-1}$ of $P_1$. See Fig.9(c). We now flip our current construction along the line passing through $A_0$ and $A_{2m}$, to obtain a dissection of $P$ with $4m$ pieces as shown in Fig.9(d). Label the images of $P_{2i}$ under our flipping along the line determined by $A_0$ and $A_{2m}$ by $P_{4m-2i+2}$, $i = 1, \ldots, m$, as shown in the same figure. Clearly when we reflect $P_{4m-2i+2}$, $i = 1, \ldots, m$, the resulting pieces together with $P_{2i}$, $i = 1, \ldots, m$, can be reassembled to form $B$. It is now easy to verify that the remaining pieces $P_3, \ldots, P_{4m-1}$ can be assembled to form a polygon similar to $P$ of size $\frac{4}{5}$ the size of $P$ minus a polygon congruent to $B$. It now follows that the resulting partitioning of $P$ is a 2-dissection of $P$. See Fig.9(e).

## 7 Dissecting simple polygons

Consider a simple polygon $P$ with $n$ vertices. We now present a 4-dissection of $P$ that uses exactly $2n - 2$ pieces. A triangulation $T$ of $P$ is a partition of $P$ into $n - 2$ triangles $\{t_1, \ldots, t_{n-2}\}$ with disjoint interiors obtained by cutting $P$ along $n - 3$ diagonals joining pairs of vertices of $P$, see Fig.10(a). We observe now that if we dissect each $t_i \in T$ into 4 similar triangles $\{t_{i,1}, t_{i,2}, t_{i,3}, t_{i,4}\}$ by cutting it along the line segments joining the mid points of its edges we obtain a dissection $D'$ of $P$ with $4(n - 2)$ triangles, see Fig.10(b). Clearly for each $j$, the set of triangles $\{t_{1,j}, \ldots, t_{n-2,j}\}$ can be reassembled to obtain a polygon similar to $P$, $j = 1, \ldots, 4$.

We now show how to modify $D'$ to obtain a 4-dissection $D$ of $P$ with $2n - 2$ pieces. First we color the vertices of $P$ with 3 colors 1, 2, and 3 in such a way that if two vertices of $P$ are adjacent in $T$, (i.e. that are connected by a diagonal of $T$ or an edge of $P$) they receive different colors [13]. See Fig.11(a). Our new dissection $D$ is now obtained from $D'$ by eliminating the cuts made in $P$ along the diagonals used to obtain $T$, see Fig.11(b). The pieces of $D$ are $n$ polygons each of which contains exactly one vertex of $P$, plus a set of triangles, one for each $t_i \in T$. Since $T$ contains $n - 2$ triangles, it now follows that the number of pieces of $D$ is exactly $2n - 2$.

We now show how to assemble the pieces of $D$ into four polygons similar to $P$. Consider the triangulation $T$ together with the 3-vertex coloring defined before.

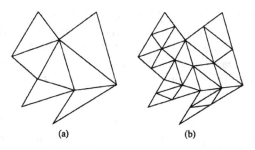

**Fig. 10.** Triangulating and dissecting $\mathcal{P}$.

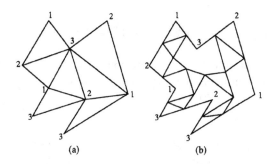

**Fig. 11.** Coloring and obtaining our final dissection $\mathcal{D}$.

Notice that each triangle in $\mathcal{T}$ has exactly one vertex of each color. For each such vertex $v_i$ of $\mathcal{P}$ let $\mathcal{P}_i$ be polygon obtained by joining the set of triangles in $\mathcal{T}$ having $v_i$ as one of its vertices. We observe now that each of the sets $\mathcal{S}_j = \{\mathcal{P}_i : v_i \text{ has color } j\}$, $j = 1,\ 2,\ 3$, induces a dissection of $\mathcal{P}$. Furthermore, observe that for each $v_i$ the polygon of $\mathcal{D}$ containing it, denoted by $\mathcal{P}'_i$ is similar to $\mathcal{P}_i$. It now follows that the sets of polygons $\mathcal{S}'_j = \{\mathcal{P}'_i : v_i \text{ has color } j\}$ can be reassembled to form polygons similar to $\mathcal{P}$, $j = 1,\ 2,\ 3$. Observe now that the remaining triangles of $\mathcal{D}$ can also be reassembled to form a fourth polygon similar to $\mathcal{P}$. Summarizing we have:

**Theorem 5.** *Every simple polygon with $n$ vertices, has a 4-dissection with $2n-2$ pieces, $n \geq 3$.* □

We now show how to obtain $(3k + 4)$-dissections of $\mathcal{P}$ with at most $(2n - 2) + k(2n + \lfloor \frac{n}{3} \rfloor - 6)$ pieces such that the pieces can be reassembled to form $4,\ 7,\ \ldots$, or $3k + 4$ polygons similar to $\mathcal{P}$. Since the coloring of the vertices of $\mathcal{T}$ induces a partition on its vertices, there is a chromatic class with at most $\lfloor \frac{n}{3} \rfloor$ vertices. Suppose then that the chromatic class containing the vertices with color 1 has at most $\lfloor \frac{n}{3} \rfloor$ elements. Let $\mathcal{D}''$ be the dissection of $\mathcal{P}$ obtained from $\mathcal{D}$ by adding cuts along the diagonals of $\mathcal{T}$ joining pairs of vertices colored with colors 2 and 3, see Fig.12(a) and (b). Observe that the number of diagonals of $\mathcal{T}$ with endpoints colored 2 and 3 is exactly the number of vertices of color 1 minus 1, and that when we cut the pieces of $\mathcal{D}$ along each of these diagonals, the number of pieces increases by 2, see Fig.12. Since there are at most $\lfloor\frac{n}{3}\rfloor$ vertices with color 1, we have that the number of pieces of $\mathcal{D}''$ is at most:

$$2n - 2 + 2\left(\left\lfloor\frac{n}{3}\right\rfloor - 1\right).$$

Observe now that the pieces of $\mathcal{S}_1' = \{\mathcal{P}_i' : v_i \text{ has color } 1\}$ when assembled properly form a polygon $\mathcal{Q}$ similar to $\mathcal{P}$, dissected along its diagonals corresponding to those of $\mathcal{P}$ joining pairs of vertices colored 2 and 3. Let $\mathcal{D}_1$ be the dissection induced in $\mathcal{Q}$ by $\mathcal{D}''$. Combining $\mathcal{D}''$ with $\mathcal{D}_1$ we obtain a 7-dissection of $\mathcal{P}$ with at most

$$(2n - 2) - \left\lfloor\frac{n}{3}\right\rfloor + 2n - 2 + 2\left(\left\lfloor\frac{n}{3}\right\rfloor - 1\right) = (2n - 2) + 2n + \left\lfloor\frac{n}{3}\right\rfloor - 4$$

pieces. Clearly we can now iterate our previous procedure on the pieces of our last dissection of $\mathcal{P}$ containing the vertices of $\mathcal{P}$ with color 1 to obtain $(3k + 4)$-dissections of $\mathcal{P}$ with $(2n - 2) + k(2n + \lfloor\frac{n}{3}\rfloor - 4)$ pieces.

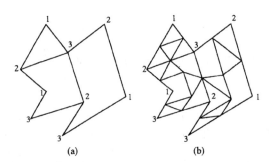

(a)          (b)

**Fig. 12.** Obtaining $\mathcal{D}''$.

Thus we have proved:

**Theorem 6.** *Every simple polygon $\mathcal{P}$ with $n$ vertices has a dissection with $(2n - 2) + k(2n + \lfloor\frac{n}{3}\rfloor - 4)$ pieces such that its pieces can be reassembled to form 4, 7, ..., or $3k + 4$ polygons similar to $\mathcal{P}$.*  □

## 8  Star shaped polygons

A polygon $\mathcal{P}$ is called star shaped if there is a point $p$ in $\mathcal{P}$ such that the line segment connecting it to any other point in $\mathcal{P}$ is contained in $\mathcal{P}$. We show how to obtain dissections of *star shaped* polygons with $n$ vertices having $2kn + 1$ pieces such that the pieces of these dissections can be reassembled to form 4, 7, ..., or $3k + 1$ polygons similar to $\mathcal{P}$. We start by proving:

**Theorem 7.** *Any star shaped polygon $\mathcal{P}$ has a 4-dissection $\mathcal{D}$ using $2n+1$ pieces, one of which is a star shaped polygon similar to $\mathcal{P}$.* □

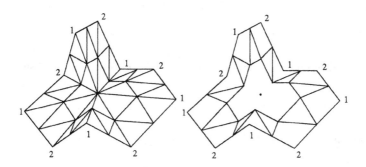

**Fig. 13.** A 4-dissection of a star shaped polygon with $n$ vertices, $n$ even, using $2n + 1$ pieces.

Let $\mathcal{P}$ be a star shaped polygon, and let $p$ be a point in the interior of $\mathcal{P}$ such that the line segment connecting $p$ to any point $q$ in $\mathcal{P}$ is totally contained in $\mathcal{P}$. Suppose first that $\mathcal{P}$ has an even number of vertices. Color the vertices of $\mathcal{P}$ with colors 1 and 2 in such a way that adjacent colors receive different colors. Connect $p$ to all the vertices of $\mathcal{P}$ to obtain a set of $n$ triangles as shown in Fig.13(a). Subdivide the triangles obtained into 4 subtriangles using the mid-points of their edges as shown in the same figure. Next delete the edges connecting $p$ to the vertices of $\mathcal{P}$, as in Fig.13(b) to obtain a dissection $\mathcal{D}$ of $\mathcal{P}$ with $2n+1$ pieces, one of which is similar to $\mathcal{P}$. Observe that all the pieces containing a vertex of color 1 (resp. 2) can be regrouped to form a star shaped polygon similar to $\mathcal{P}$. The remaining $n$ triangles can also be regrouped to form a fourth polygon similar to $\mathcal{P}$. The case when $\mathcal{P}$ has an odd number of vertices can be done in a similar way, except that we color exactly one vertex of $\mathcal{P}$ with both colors 1 and 2. The details are left to the reader. An example for this case is shown in Fig.14.

By using the piece of $\mathcal{D}$ similar to $\mathcal{P}$, we can obtain $(3k + 1)$-dissections of $\mathcal{P}$ with $2kn + 1$ pieces. Thus we have proved:

**Theorem 8.** *Any star shaped polygon with $n$ vertices has a dissection with $2kn + 1$ pieces such that its pieces can be reassembled to form 4, 7, ..., or $3k+1$ polygons similar to $\mathcal{P}$.* □

**Acknowledgments:** The authors would like to thank an anonymous referee for his helpful comments and valuable suggestions.

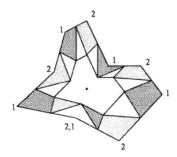

**Fig. 14.** A 4-dissection of a star shaped polygon with $n$ vertices, $n$ odd, using $2n + 1$ pieces.

# References

1. Akiyama, J., Nakamura, G.: An efficient dissection for a sequentially $n$-divisible square. Proc. of Discrete and Computational Geometry Workshop, Tokai University (1997), 80–89

2. Akiyama, J., Nakamura, G., Nozaki, A., Ozawa, K.: A note on the purely recursive dissection for a sequentially $n$-divisible square. in this proceedings

3. Akiyama, J., Nakamura, G., Nozaki, A., Ozawa, K., Sakai, T.: The optimality of a certain purely recursive dissection for a sequentially $n$-divisible square. To appear in Computational Geometry; Theory and Applications

4. Busschop, P.: Problèmes de géométrie. Nouvelle Correspondance Mathématique **2** (1876) 83–84

5. Collison, D. M.: Rational geometric dissections of convex polygons. Journal of Recreational Mathematics **12**(2) (1979–1980) 95–103.

6. Fourrey, E.: Curiosités Géométriques. Paris: Vuibert et Nony (1907)

7. Frederickson, G. N.: Dissections: Plane & Fancy. Cambridge University Press (1997)

8. Gardener, M.: The 2nd. Scientific American Book of Mathematical Puzzles and Diversions. New York: Simon and Schuster (1961)

9. Kraitchik, M.: Mathematical Recreations. New York: Northon (1942)

10. Lindgren, H.: Geometric Dissections. Princeton, N.J.: D. Van Nostrand Company (1964)

11. Nozaki, A.: On the dissection of a square into squares (in Japanese). Suugaku-Seminar No.12 (1999) 52–56

12. Ozawa, K.: Entertainer in a classroom (in Japanese). Suugaku-Seminar No.10 (1988) cover page

13. Urrutia, J.: Art Gallery and Illumination Problems in Handbook on Computational Geometry. Sack, J.R. and Urrutia, J. eds. Elsevier Science Publishers (2000)

# Packing Convex Polygons into Rectangular Boxes

Helmut Alt[1] * and Ferran Hurtado[2] **

[1] Institut für Informatik, Freie Universität Berlin
Takustr. 9, D-14195 Berlin
alt@inf.fu-berlin.de
[2] Departament de Matemàtica Aplicada II
Universitat Politècnica de Catalunya, Pau Gargallo 5
08028-Barcelona, España

**Abstract** We consider the problem of packing several convex polygons into minimum size rectangles. For this purpose the polygons may be moved either by translations only, or by combinations of translations and rotations. We investigate both cases, that the polygons may overlap when being packed or that they must be disjoint. The *size* of a rectangle to be minimized can either be its area or its perimeter. In the case of overlapping packing very efficient algorithms whose runtime is close to linear or $O(n \log n)$ can be found even for an arbitrary number of polygons. Disjoint optimal packing is known to be NP-hard for arbitrary numbers of polygons. Here, efficient algorithms are given for disjoint packing of two polygons with a runtime close to linear for translations and $O(n^3)$ for general isometries.

## 1 Introduction

An important geometric optimization problem is to find for a given object a *smallest enclosing* object from a certain class, like a rectangle, a triangle, a 3d-box, etc. (see [4, 7, 6]).

Here, we will consider the problem of enclosing not only one but several objects into a minimum size container. More precisely, we will consider packing convex polygons into rectangles. For this purpose the polygons may be moved either by translations only, or by combinations of translations and rotations. We also consider two cases: either the polygons may overlap when being packed or they must be disjoint. The *size* of a rectangle to be minimized can either be its area or its perimeter.

For *disjoint* packing under translations considerable work has been done previously by Milenkovic (see e.g. [5]) motivated by industrial applications, for

---

* Partially supported by the German Research Association (DFG), projects Al 253/4-2 and Al 253/4-3.
** Partially supported by CUR Gen. Cat. 1999SGR00356 and Proyecto DGES-MEC PB98-0933.

J. Akiyama, M. Kano, and M. Urabe (Eds.): JCDCG 2000, LNCS 2098, pp. 67–80, 2001.
© Springer-Verlag Berlin Heidelberg 2001

example cutting parts of a certain shape out of rectangle shaped stock material like cloth or sheet metal and trying to minimize waste. For packing $k$ convex $m$-gons into a minimum area isothetic rectangle he gives an $O(m^{k-1}\log m)$ algorithm using linear programming techniques. The most general problem for disjoint packing under translations was solved by Avnaim and Boissonnat [2] who found an $O(m^2 + mn)^{2k}\log n)$ algorithm for packing $k$ arbitrary simple $n$-gons into a simple $m$-gon.

Here, we will, for a start, in the disjoint case only consider two convex polygons to be packed. In fact, if we allow an arbitrary non-constant number of polygons, the problem is known to be NP-hard even if the polygons are rectangles ([3]) and only translations are allowed. We will give a linear-time algorithm for packing two convex polygons into a minimum size isothetic rectangle and for arbitrary rectangles we obtain an $O(n\log n)$ algorithm where $n$ is the total number of vertices. We also give an $O(n^3)$ algorithm for the case of allowing arbitrary *isometries* to pack the polygons.

In addition, we also consider packing *with overlap* (which could be motivated by, for example, finding the smallest rectangular area where a set of given flat objects can be *stacked*). We will develop efficient algorithms even for the general case of $k$ convex polygons under arbitrary isometries.

All our results hold for both, minimum area and minimum perimeter rectangles.

The different variants of the problem are distinguished by symbols for the different criteria

- $O$ for overlapping and $D$ for disjoint packing.
- $T$ for translation and $G$ for general isometries. Each polygon can be packed by one of these operations independent of the others. By *isometry* we mean a combination of translations, rotations, and reflections. We will only consider translations and rotations in the text. The result follows for additional reflections since for overlapping packing a rectangle contains a polygon $P$ exactly if it contains its reflected copy $P'$. For disjoint packing, we deal with 2 polygons and can solve the problem by just additionally considering the reflected versions of the polygons involved.
- In the case of translations we add a third symbol, namely: $I$ if we allow only *isothetic* (= axis-parallel) rectangles and $A$ for arbitrary rectangles. Observe that under general isometries this distinction does not make sense.

Any combination of these letters stands for a variant of the problem. We add the number "$k-$" of polygons to be packed in front of the combination, so e.g. $2 - DTI$ means finding the smallest size isothetic rectangle containing disjoint translates of two given polygons.

Throughout this paper we will assume that the input polygons are given by a list of their edges sorted in, say, clockwise order.

Due to space limitations some proofs are omitted in this proceedings version of the paper. For a complete version see [1].

## 2   Packing with Overlap

In this section we will assume that convex polygons $A_1, ..., A_k$ with a total of $n$ edges should be packed *allowing overlap* into a minimum *size* (area or perimeter) rectangle $R$.

We will consider axes-prallel and arbitrary rectangles, translations and general isometries for packing the polygons, i.e., the versions $k - OTI$, $k - OTA$, and $k - OG$ of the problem. We will first consider general isometries and then see how the problem can be simplified if we only allow translations.

### 2.1   Preliminary Considerations for General Rigid Motions

Let us identify the rectangle with width $x$ and height $y$ with the point $(x, y)$ in two dimensional space. In this way we can assume that our "space of rectangles" $F$ is the open first quadrant of the Cartesian plane. Rectangles with area $k$ (perimeter $r$) correspond to the points in the positive branch of the hyperbola $xy = k$ (the line segment $2(x + y) = r$).

Consider some polygon $Q$ and some angle $\theta \in [0, \pi]$. Then there is a unique rectangle with the property that two of its edges form the angle $\theta$ with the $x$-axis and the four edges extend to lines supporting $Q$. Let us call this rectangle $\theta$-*optimal* and let $x(\theta)$ and $y(\theta)$ be its width (in direction $\theta$) and its height, respectively. When $\theta$ varies from 0 to $\pi$ we obtain all the directionally optimal rectangles, giving a geometric locus of points $(x(\theta), y(\theta))$ in $F$, parameterized by $\theta$, which we describe precisely in the next lemmas.(In fact it might be more intuitive to keep the orientation of the optimal rectangle isothetic and its lower left corner fixed at the origin, rotate $Q$ instead, and trace the orbit of the rectangle's upper right corner.)

**Lemma 1.** *Consider the family of $\theta$-optimal rectangles of $Q$, for $\theta_1 \leq \theta \leq \theta_2$, which touch $Q$ in the same vertices $a, b, c, d$. The locus of corresponding points in $F$ is an arc of ellipse with center at the origin.*

*Proof.* As we see in Figure 1, we have

$$\begin{cases} x(\theta) = d_1 \cos(\alpha_1 - \theta), \\ y(\theta) = d_2 \cos(\alpha_2 - \theta), \end{cases}$$

for some constants $d_1, d_2, \alpha_1$ and $\alpha_2$. This can be rewritten as

$$\begin{cases} x(\theta) = d_1 \cos(\alpha_1) \cos(\theta) + d_1 \sin(\alpha_1) \sin(\theta), \\ y(\theta) = d_2 \cos(\alpha_2) \cos(\theta) + d_2 \sin(\alpha_2) \sin(\theta). \end{cases}$$

Therefore, the set of points $\{(x(\theta), y(\theta)) | \theta \in [\theta_1, \theta_2]\}$ is the image of the set of points $\{(\cos(\theta), \sin(\theta)) | \theta \in [\theta_1, \theta_2]\}$, i.e., an arc of the unit circle, under a linear mapping. Consequently, it is an arc of an ellipse centered at the origin.   $\square$

Let us consider the optimal rectangle for $\theta = 0$, then allow the variation of $\theta$ from 0 to $\pi$. In a "rotating calipers" fashion [9] the enclosing rectangle rotates

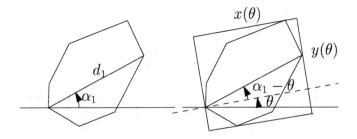

**Figure 1.** Local variation of an enclosing rectangle touching four vertices of the polygon

around $Q$, and all the $\theta$-optimal rectangles are traced. In fact each optimal rectangle is found twice, once for some value $\theta_0$ and again for $\theta_0 + \pi/2$, reversing the roles of $x$ and $y$. As proved in Lemma 1, for each set of contact vertices we obtain an arc of ellipse. Since the set of contact vertices changes each time one of the sides of the rectangle becomes parallel to one of the sides of $Q$, there are $4n$ such sets. Therefore we have proved the following lemma:

**Lemma 2.** *The locus of points corresponding to the directionally optimal rectangles that contain a given polygon $Q$, is a continuous closed curve $C(Q)$, within the first quadrant of $F$, consisting of the concatenation of at most $2n$ arcs of ellipse, where each ellipse is centered at the origin. The curve $C(Q)$ is symmetric with respect to the line $y = x$, and can be computed in $O(n)$ time.*

A polygon $Q$ and its corresponding curve $C(Q)$ are shown in Figure 2.

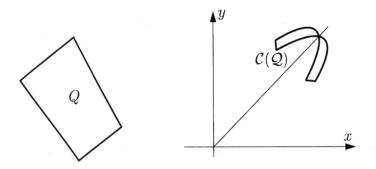

**Figure 2.** $Q$ and $C(Q)$

For each arc of ellipse $\mathcal{E}$ in $C(Q)$ the smallest area (perimeter) rectangle is the point minimizing the product $x(\theta)y(\theta)$ (the sum $x(\theta) + y(\theta)$). But the hyperbolas $xy = k$ and the straight lines $x + y = k$ are convex as seen from the origin, while $\mathcal{E}$ is concave, therefore the minimum value is necessarily attained in an endpoint of $\mathcal{E}$. Since such an endpoint corresponds to an edge of the enclosing rectangle being collinear with an edge of $Q$, this implies that the minimum area (perimeter) rectangle enclosing a convex polygon has a side collinear with an edge of the polygon, a crucial result differently proved in [4].

Our next lemma is concerned with the theory of Davenport-Schinzel-sequences; see [8]. As usual let $\lambda_s(k)$ denote the maximal complexity of the upper or lower envelope of a set of $k$ curves which pairwise may intersect $s$ times.

**Lemma 3.** *Let $f_1,..., f_k$ be curves which are $x$-monotone within some interval $I$. Furthermore, $f_1,...,f_k$ are piecewise composed of arcs which pairwise intersect at most $s$ times for some constant $s$. Let $n$ be the total number of pieces. Then the complexity of the upper or the lower envelope of $f_1,...,f_k$ is $O(n/k\,\lambda_{s+2}(k))$ and the envelopes can be determined in time $O(n/k\,\lambda_{s+1}(k)\log k)$.*

*Proof.* Let $x_0, x_1, ..., x_n$ be the "breakpoints" between the different pieces of $f_1, ..., f_k$ in ascending order. Let $r = \lfloor n/k \rfloor + 1$ and decompose $I$ into the $r$ subintervals $[x_0, x_k], [x_{k+1}, x_{2k}], ..., [x_{rk+1}, x_n]$. Within each of these intervals we have at most $2k$ pieces of the functions $f_1, ..., f_k$. Therefore, as we know from the theory of Davenport-Schinzel-sequences [8] the envelopes within each subinterval have complexity $\lambda_{s+2}(2k) = O(\lambda_{s+2}(k))$ and can be computed in time $O(\lambda_{s+1}(k)\log k)$ with an algorithm due to Hershberger. Since there are $r = O(n/k)$ subintervals, the lemma follows. $\qquad\square$

## 2.2 The Solution for General Rigid Motions

We say that a point (or a vector) $(a_1, b_1)$ *dominates* another one $(a_2, b_2)$ if $a_1 \geq a_2$ and $b_1 \geq b_2$. The dominance relation gives a partial order $\mathcal{D}$ in the plane. A point $(x, y) \in F$ corresponds to a rectangle able to contain polygon $A$ if and only if it dominates some point in $\mathcal{C}(A)$; therefore the locus of such points is a region $R(A)$ as the one shaded in Figure 3, where the zig-zag curve $\mathcal{B}(A)$ bounding it is the lower envelope of $\mathcal{C}(A)$ with respect to the order $\mathcal{D}$. $\mathcal{B}(A)$ can be obtained by first computing the $x$-lower envelope $L_x$ of $\mathcal{C}(A)$, from which we compute next its $y$-lower envelope.

Two of our arcs of ellipse intersect in at most two points. Therefore, the complexity of $L_x$ is $\lambda_4(n)$, which is $\Theta(n \cdot 2^{\alpha(n)})$, and $L_x$ can be computed in time $O(\lambda_3(n)\log n)$ which is $O(n\alpha(n)\log(n))$. Since $L_x$ is an $x$-monotone closed curve its $y$-lower envelope $\mathcal{B}(A)$ can be computed easily by sweeping $L_x$ from left to right in time proportional to its complexity, i.e., $O(\lambda_4(n))$.

For problem $k - OG$ we have $k$ polygons $A_1, ..., A_k$ having $n_1, ..., n_k$ vertices, respectively, with $n_1 + n_2 + ... + n_k = n$. We obtain $k$ curves $\mathcal{C}(A_1), ..., \mathcal{C}(A_k)$ and determine the regions $R(A_1), ..., R(A_k)$ as described. This computation takes time $\sum_{i=1}^k \lambda_3(n_i)\log n_i$ which is $O(\lambda_3(n)\log n)$ which is $O(n\alpha(n)\log n)$.

Each region $R(A_i)$ is bounded by an $x$-monotone chain $\mathcal{B}(A_i)$ of $O(\lambda_4(n_i))$ elliptic or straight segments. In total we have $O(\sum_{i=1}^k \lambda_4(n_i))$, which is $O(\lambda_4(n))$ segments, which pairwise may intersect twice. By Lemma 3 their upper envelope

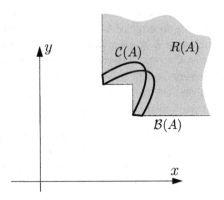

**Figure 3.** Region of points which correspond to rectangles containing a given polygon

$B$, which is the boundary of the intersection $R = R(A_1) \cap \ldots \cap R(A_k)$ can be computed in time $O(\lambda_4(n)/k \, \lambda_3(k) \log k)$, which is $O(n2^{\alpha(n)}\alpha(k) \log k)$.

Finally we have to compute the point $(x, y) \in R$ minimizing the product $xy$ (the sum $x + y$). The hyperbolas $xy = k$ (straight lines $x + y = k$) are convex as seen from the origin, while the region $R$ is bounded by concave pieces; therefore the minimum is attained in a vertex of $B$, which can be obtained by traversing $B$ in time proportional to its complexity. By Lemma 3 this complexity is $O(\lambda_4(n)/k \, \lambda_4(k))$, which is dominated by the runtimes of the previous steps. We have proved the following result:

**Theorem 1.** *The variant $k - OG$ of packing $k$ convex polygons with a total number of $n$ vertices into a rectangle can be solved in*

$$O(n(2^{\alpha(n)} \log k \, \alpha(k) + \log(n) \, \alpha(n)))$$

*time by the algorithm described above.*

There are interesting special cases of Theorem 1: A constant number of convex polygons with a total of $n$ vertices can be packed into a minimum size rectangle in time $O(n \log(n) \, \alpha(n))$. On the other hand, a set of $n$ triangles (or other $m$-gons with constant $m$) can be packed into a minimum size rectangle in time $O(n \log(n) \, 2^{\alpha(n)}\alpha(n))$.

## 2.3   Translations

In the case where only translations are allowed to pack the polygons, the ideas from the previous section can be simplified and slightly faster algorithms can be obtained.

For a convex polygon $A$ let $x_A$ and $y_A$ be the width and height of the minimum isothetic rectangle enclosing $A$. The minimum isothetic rectangle that will contain translated copies of $k$ polygons $A_1, \ldots, A_k$ will obviously have width

$\max\{x_{A_1}, ..., x_{A_k}\}$ and height $\max\{y_{A_1}, ..., y_{A_k}\}$. This simple observation shows that the problem $k - OTI$ can be solved in linear time. Furthermore, it is fundamental for solving the problem $k - OTA$.

For a convex $n$-gon $A$ let $R_A(\theta)$ be the $\theta$-optimal rectangle enclosing $A$ and let $x_A(\theta)$ and $y_A(\theta)$ be its width and height.

We consider now the continuous variation of $\theta$ from 0 to $\pi/2$. The directionally optimal rectangle $R_A(\theta)$ will rotate around $A$ in a "rotating calipers" fashion. In each interval of angles $\theta$ where the same four vertices of $A$ touch the sides of $R(\theta)$ the width $x_A(\theta)$ and the height $y_A(\theta)$ will be functions of the form

$$x_A(\theta) = d_1 \cos(\alpha_1 - \theta),$$
$$y_A(\theta) = d_2 \cos(\alpha_2 - \theta), \tag{1}$$

for some constants $d_1, d_2, \alpha_1$ and $\alpha_2$. The representation (1) of either $x_A$ or $y_A$ changes whenever one of the sides of $R_A(\theta)$ becomes parallel to one of the edges of $A$, so both functions consist of $O(n)$ pieces of the form (1) and can be determined in $O(n)$ time.

Now suppose that $A_1, ..., A_k$ are convex polygons with a total of $n$ vertices for which we would like to solve $k - OTA$. For a given orientation $\theta$ the width and height of the smallest enclosing rectangle are given by $\max(x_{A_1}(\theta), ..., x_{A_k}(\theta))$ and $\max(y_{A_1}(\theta), ..., y_{A_k}(\theta))$, i.e., by the upper envelope of the functions $x_{A_1}, ..., x_{A_k}$ and $y_{A_1}, ..., y_{A_k}$, respectively. Since any two functions of the form (1) intersect at most twice by Lemma 3 these envelopes can be determined in time $O(n/k \; \lambda_3(k) \log k)$ which is $O(n\alpha(k) \log k)$ and their complexity is $O(n/k \; \lambda_4(k))$. Finally, in time proportional to their complexity we can scan and merge both envelopes and determine the orientation $\theta$ where their product (their sum) is minimized in order to obtain the enclosing rectangle of minimum area (perimeter).

Summarizing, we have proved the following theorem:

**Theorem 2.** *The problem $k - OTA$ can be solved in time $O(n \log k \; \alpha(k))$ by the algorithm above.*

If $k$ is a constant, we have a linear time algorithm for packing $k$ polygons into a minimum size arbitrary rectangle.

## 3  Disjoint Packing

In this section we will assume that two convex polygons $A$ and $B$ with a total of $n$ edges should be packed *without overlap* into a minimum size rectangle $R$.

Our algorithms are based on the following property of optimal packings that will significantly simplify the problem. It can be shown for both the area and the perimeter by elementary geometric arguments; see the full version of this paper [1].

**Lemma 4.** *For all variants of the problem there exists an optimal packing where at least one of the sides of the rectangle $R$ touches both bodies $A$ and $B$.*

*Proof.* We will show this property for the simplest version $2 - DTP$, it then easily follows for the other versions.

Clearly, in an optimal placement, each side of the rectangle $R$ touches at least one of the objects.Assume that we have an optimal placement where each side of $R$ is touched by exactly one of the polygons. We will show that this either leads to a contradiction or that another optimal placement satisfying the claim can be constructed easily.

Because of the convexity of $A$ and $B$, there is a straight line $l$ separating them. Then, the following situations or symmetric variants are possible.

a) One body, say $A$, touches all four sides of $R$, which means that $B$ does not touch any side.(see Figure 4 a) ) Then, clearly, $B$ can be moved within $R$ without crossing $l$ so that it touches a side of $R$.

b) $A$ touches three and $B$ one side of $R$ (see Figure 4 b) ) If $l$ is vertical we can

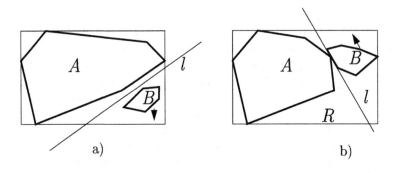

a)                                          b)

**Figure 4.** a) $A$ touches four sides of $R$ b) $A$ touches three and $B$ one side of $R$

translate $B$ in vertical direction until it hits a horizontal side of $R$ without changing $R$. Otherwise it is easy to see that $B$ can be slid along $l$ so that the area of $R$ decreases.

c) $A$ and $B$ both touch two sides of $R$ (see Figure 5). Observe that because of convexity the sides touched by $A$ (or $B$) must be incident. Observe that in situation a) of Figure 5, where the slope of $l$ is nonnegative, R cannot be optimal, since by moving $B$ downward we can decrease the height of $R$ without increasing its width. In situation b) sliding $B$ along $l$ will increase the height $y$ and decrease the width $x$ of $R$, or vice versa. Assume, that the lower left corner of $R$ is the origin, so the upper right corner $C$ has coordinates $(x, y)$, the area is $A(x, y) = x \cdot y$ and the perimeter $p(x, y) = 2(x + y)$. Any translation of $B$ by some (sufficiently short) vector $t$ will translate the upper right corner of $R$ from $C$ to $C + t$, as well.

Let $l'$ be the line through $C$ parallel to $l$. So $l'$ can be represented in the form y=ax+b with $a < 0$.

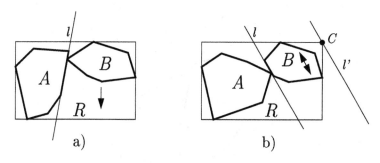

**Figure 5.** $A$ and $B$ touch two sides of $R$

If we slide $B$ along $l$, the area of $R$ as a function of $C's$ $x$-coordinate $x$ will be

$$A(x, ax + b) = ax^2 + bx \qquad (2)$$

Since $a < 0$, this function has no local minimum contradicting the optimality of $R$. The same argument holds for the perimeter, as well, because $p(x, ax + b) = 2(a + 1)x + b$ is a linear function in $x$ and has no local minima.

$\square$

## 3.1   2-DTI

For the problem $2 - DTI$ Lemma 4 reduces the number of candidates of optimal placements of two given convex bodies $A$ and $B$ to 8. In fact, all four sides of $R$ can be candidates for touching both bodies and in each case there are two possible orders for the two tangent points.

These 8 candidates can be determined in linear time. In fact, in order to determine the placements where $A$ and $B$ share the right tangent we consider the orbit of the rightmost point $L$ of $B$ when $B$ is slid along the boundary of $A$ (see Figure 6). This orbit is again the boundary of a convex polygon $P$. The edges of $P$ are obtained in linear time by merging the edges of $A$ and $B$ and translating them accordingly. In fact, $P$ is a translate of the *Minkowski-sum* of $A$ and $-B$.

The placement candidates are determined by intersecting $P$ with the right vertical tangent of $A$ . The other six candidates can be found analogously. For each of the eight candidates we determine the area or perimeter of the enclosing rectangle and select the minimal one. So we have:

**Theorem 3.** *The variant 2-DTI of the packing problem can be solved in $O(n)$ time by the algorithm described above.*

Observe that a bound of $O(n \log n)$ for problem 2-DTI follows from Milenkovic's $O(n^{k-1} \log n)$ bound [5] for packing $k$ objects.

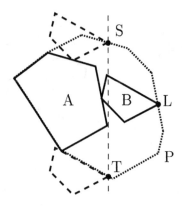

**Figure 6.** The orbit $P$ of $L$

## 3.2   2-DTA

Let us now look for a minimal size rectangle under translation that is not necessarily isothetic.

Our algorithm works as follows: We consider object $A$ as fixed and $B$ may slide around $A$. Fix a certain angle $\vartheta \in [0, 2\pi]$ and consider the oriented line $l$ that forms this angle with the $x$-axis, is tangent to $A$, and has $A$ in its left halfplane; see Figure 7. Then there is a unique placement of $B$ touching both $l$ and $A$ and lying above $A$ with respect to the orientation of $l$.

Let $R_\vartheta$ be the smallest enclosing rectangle of $A$ and $B$ one of whose sides lies on $l$. Then, assuming that we rotate $\vartheta$ from 0 to $2\pi$ we have a rotating callipers situation where the polygon $B$ is pushed forward by one caliper jaw. One of the rectangles $R_\vartheta$ will be a candidate for the optimal solution. The second candidate is obtained in the same manner, only $B$ is placed *below* $A$ in the halfplane left of $l$. In the following, let us number the sides of $R_\vartheta$ counterclockwise from 0 to 3 starting with the side that lies on $l$. Also let $l_i$   $i = 0, 1, 2, 3$ be the line on which side $i$ lies. During the rotation process we will maintain all relevant information to carry out the rotation, in particular for $i = 0, 1, 2, 3$ (see Figure 7):

$a_i$, the vertex of $A$ touching side $i$ of $R_\vartheta$. If side $i$ does not touch $A$, $a_i$ is the vertex of $A$ closest to side $i$.

$b_i$, likewise the points of $B$ closest to the sides of $R_\vartheta$

$e$, the side of $A$ or $B$ on which $B$ is sliding

$v$, the vertex of $B$ or $A$ on which $B$ is sliding

Each angle $\vartheta \in [0, 2\pi)$ where this information changes is an *event point* of the rotation process. We distinguish three types of event points:

**Type a:** Some $a_i$ or $b_i$ changes. This occurs whenever the direction of one of the sides of $R_\vartheta$, i.e., $\vartheta$ or $\vartheta + \pi/2$ becomes parallel to one of the sides of $A$ or $B$.

**Type b:** $e$ and $v$ change. This happens whenever $v$ hits the next vertex of $A$ or $B$.

**Type c:** Some vertex $a_i$ or $b_i$ starts or stops touching $R_\vartheta$.

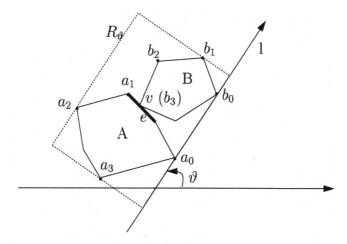

**Figure 7.** Information maintained during the rotation process

The list of type-a events is precomputed and kept in sorted order in a data structure called *event point schedule* (EPS).
At each event point the events of types b and c for $v, e$, and each $a_i, b_i$ are updated and replace the old ones in the EPS. Notice, that many of these events never occur because they are replaced by new values. (For example, $b_2$ starts touching side 2 of $R_\vartheta$ in Figure 7, because $v$ will reach the end of $e$ earlier.)

A detailed consideration (see [1]) shows that computing the new events can be done in constant time and that the number of event points is linear in $n$. The algorithm spends $O(\log n)$ time per event point in order to update the EPS. So we have:

**Theorem 4.** *The variant 2-DTA of packing two convex polygons with a total number of $n$ vertices into a rectangle can be solved in $O(n \log n)$ time by the algorithm described above.*

## 3.3   2-DG

Let us now consider the most general version 2-DG of disjoint packing allowing arbitrary isometries to pack the two polygons. By Lemma 4 we may assume that in the optimal solution the left hand side of the enclosing rectangle $R$ is part of the $y$-axis, the bottom side is part of the $x$-axis, and both, $A$ and $B$ touch the $x$-axis. We will, furthermore assume that $A$ lies left of $B$; we just will have to run our algorithm twice, once for this and once for the opposite situation. Let us number the sides of $R$ from 1 to 4, starting with the bottom side.

Like before, by $a_i$ $(b_i)$ $i = 1, \ldots, 4$ we will denote the vertex of $A$ $(B)$ that is closest to (or touches) side $i$ of $R$ (see Figure 8).

Furthermore, we may assume that in an optimal placement $A$ and $B$ are touching each other; the corresponding feature (edge or vertex) of $A$ and $B$ we will call $a_t$ and $b_t$, respectively. We will call a 10-tuple

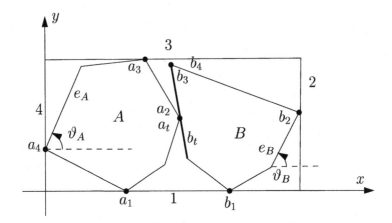

**Figure 8.** A possible placement of $A$ and $B$

$(a_1, a_2, a_3, a_4, b_1, b_2, b_3, b_4, a_t, b_t)$ that is possible under a certain placement of $A$ and $B$, a *combinatorial position* of $A$ and $B$.

Different combinatorial positions can occur through different orientations of $A$ and $B$. We identify an orientation of $A$ with the angle $\vartheta_A$ of some fixed edge $e_A$ of $A$ with the $x$-axis (see Figure 8). Likewise, we define the orientation $\vartheta_B$ of $B$.

Clearly, with a given pair $(\vartheta_A, \vartheta_B)$ of orientations and our assumptions before, the placement of $A$ and $B$ and, thus, its combinatorial position, are uniquely identified. We therefore can identify the space of all possible placements, within which we are looking for the optimal solution, with the two-dimensional box $\Pi = [0, 2\pi] \times [0, 2\pi]$. $\Pi$ is partitioned into finitely many connected cells corresponding to the combinatorial positions of $A$ and $B$. In fact, it holds:

**Lemma 5.** *There are $O(n^3)$ different combinatorial positions of $A$ and $B$.*

*Proof.* Clearly, there are only $O(n)$ 4-tuples $(a_1, a_2, a_3, a_4)$. For increasing $\vartheta_A$ such a 4-tuple changes whenever the orientation $\vartheta_A$ causes one of the edges of $A$ to become vertical or horizontal. These events partition the space $\Pi$ by $O(n)$ vertical lines. Likewise, $\Pi$ is partitioned by $O(n)$ horizontal lines corresponding to the orientations $\vartheta_B$ where the 4-tuple $(b_1, b_2, b_3, b_4)$ changes. Within each of the $O(n^2)$ cells obtained from the horizontal and vertical lines the 8-tuple $(a_1, a_2, a_3, a_4, b_1, b_2, b_3, b_4)$ is invariant. It remains to subdivide $\Pi$ further by the lines separating different pairs $(a_t, b_t)$. First observe that there can be two types of configurations from which the touching pair can change in at least two different ways (see Figure 9).

a) a vertex of $A$ touches a vertex of $B$.

b) an edge of $A$ touches an edge of $B$.

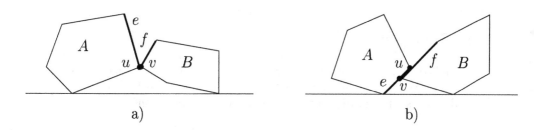

a)                                                                b)

**Figure 9.** Configurations at the boundary between touching pairs $(v, e)$ and $(u, f)$

Clearly, there are $O(n^2)$ type-b configurations which are characterized by equations of the form

$$\vartheta_A = \vartheta_B + \alpha \tag{3}$$

for some constant $\alpha$.

Consequently, $\Pi$ is further partitioned by $O(n^2)$ parallel straight line segments of slope 1.

A detailed analysis shows that type-a configurations are characterized by equations of the form

$$\vartheta_B = \beta - \arcsin(r\sin(\vartheta_A + \alpha)) \tag{4}$$

for some constants $\alpha, \beta$, and $r$.

So by (4) we have $O(n^2)$ additional curve segments which further partition the space $\Pi$.

In order to analyze the complexity of the final partition of $\Pi$, let us count the number of intersection points between the various curve segments. The $O(n)$ vertical and horizontal straight segments intersect the $O(n^2)$ type-a and type-b segments in $O(n^3)$ points. A detailed consideration shows that intersection between two type-a or type-b segments is only possible in very special situations and that, therefore, there are only $O(n^2)$ such intersection points.

So the total arrangement of curves partitioning $\Pi$ has $O(n^3)$ vertices. Since it is a planar graph it has $O(n^3)$ faces as well, by Euler's formula. The faces correspond to the different combinatorial positions.                                                    □

Now our algorithm to find the optimal placement proceeds as follows:

We systematically traverse all cells in the partition of $\Pi$. Within each cell $C$ all $a_i, b_i, a_t$ and $b_t$ are uniquely defined. We compute the optimal placements of the two $k$-gons ($k \le 5$) with vertices $a_1, a_2, a_3, a_4, a_t$ and $b_1, b_2, b_3, b_4, b_t$ in constant time. For each of these placements we check whether it lies in cell $C$ and, if so, keep track of the size (area of perimeter) of the corresponding rectangle. The minimum size rectangle found this way by checking all $O(n^3)$ cells is the optimal solution. So we have:

**Theorem 5.** *The variant 2-DG of packing two convex polygons with a total number of n vertices into a rectangle can be solved in $O(n^3)$ time by the algorithm described above.*

## 4  Summary

The following table summarizes our results. As can be seen all packing problems where overlapping is allowed can be solved in time close to linear. The same holds for disjoint packing of two polygons under translations.

| problem | asymptotic runtime |
|---------|--------------------|
| $k$-OTI | $n$ |
| $k$-OTA | $n \log k\alpha(k)$ |
| $k$-OG  | $n(2^{\alpha(n)} \log k\alpha(k) + \log n\alpha(n))$ |
| 2-DTI   | $n$ |
| 2-DTA   | $n \log n$ |
| 2-DG    | $n^3$ |

## References

1. H .Alt and F. Hurtado. Packing convex polygons into rectangular boxes. Technical Report, Institute for Computer Science, Freie Universität Berlin, 2001.
2. F. Avnaim and J.-D. Boissonnat. Simultaneous containment of several polygons. In *Proc. 3rd Annu. ACM Sympos. Comput. Geom.*, pages 242–250, 1987.
3. K. Daniels and V. J. Milenkovic. Multiple translational containment, Part I: An approximate algorithm. *Algorithmica*, 19(1–2):148–182, September 1997.
4. H. Freeman and R. Shapira. Determining the minimum-area encasing rectangle for an arbitrary closed curve. *Commun. ACM*, 18:409–413, 1975.
5. V. Milenkovic. Translational polygon containment and minimal enclosure using linear programming based restriction. In *Proc. 28th Annu. ACM Sympos. Theory Comput.*, pages 109–118, 1996.
6. J. O'Rourke. Finding minimal enclosing boxes. *Internat. J. Comput. Inform. Sci.*, 14:183–199, June 1985.
7. J. O'Rourke, A. Aggarwal, S. Maddila, and M. Baldwin. An optimal algorithm for finding minimal enclosing triangles. *J. Algorithms*, 7:258–269, 1986.
8. M. Sharir and P. K. Agarwal. *Davenport-Schinzel Sequences and Their Geometric Applications.* Cambridge University Press, New York, 1995.
9. G. T. Toussaint. Solving geometric problems with the rotating calipers. In *Proc. IEEE MELECON '83*, pages A10.02/1–4, 1983.

# On the Number of Views of Polyhedral Scenes

Boris Aronov[1], Hervé Brönnimann[1], Dan Halperin[2], and Robert Schiffenbauer[1]

[1] Polytechnic University, Brooklyn NY 11201, USA
{aronov,hbr,rschiff}@photon.poly.edu
[2] Tel-Aviv University, Tel-Aviv, Israel
halperin@math.tau.ac.il

**Abstract.** It is known that a scene consisting of $k$ convex polyhedra of total complexity $n$ has at most $O(n^4 k^2)$ distinct orthographic views, and that the number of such views is $\Omega((nk^2 + n^2)^2)$ in the worst case. The corresponding bounds for perspective views are $O(n^6 k^3)$ and $\Omega((nk^2 + n^2)^3)$, respectively. In this paper, we close these gaps by improving the lower bounds. We construct an example of a scene with $\Theta(n^4 k^2)$ orthographic views, and another with $\Theta(n^6 k^3)$ perspective views. Our construction can also be used to improve the known lower bounds for the number of silhouette views and for the number of distinct views from a viewpoint moving along a straight line.

## 1 Introduction

*Aspect graphs* have been studied in image analysis as a way to encode all topologically distinct views of a scene [2]. In this paper, we concentrate on simply bounding the *number* of such views, in the case where the scene consists of $k$ disjoint convex polyhedra with total complexity (i.e., number of vertices, edges, and faces) $n$.

We distinguish between two different types of views: *orthographic* views correspond to locating the viewpoint on the plane at infinity and having all *lines of sight* parallel to the fixed viewing direction. *Perspective* views place the viewpoint anywhere in space, outside the objects of the scene. The *viewpoint space* is the space of all allowed placements of the viewpoint. It is (isomorphic to) the unit sphere of directions $\mathbb{S}^2$ in the orthographic model, and all of $\mathbb{R}^3$ (outside the polyhedra) in the perspective model.

In the *dynamic* viewpoint model, we consider a subset of the perspective model in which the viewpoint is restricted to move along a line (*linear motion*) or an algebraic curve (*algebraic motion*). In this model, the viewpoint space is a line in $\mathbb{R}^3$ for the linear motion model, and a curve in $\mathbb{R}^3$ for the algebraic motion model.

Topologically different views in the viewpoint space are separated by *critical events*: these occur at viewpoints where a visible vertex apparently lies on a visible edge (*EV event*) or when three visible edges appear to become concurrent (*EEE event*) [2]. Each event gives rise to a well-behaved curve on $\mathbb{S}^2$ (for the orthographic views) or surface in $\mathbb{R}^3$ (for the perspective views). If the total number of vertices and edges of the scene is $n$, there can be at most $O(n^3)$ such events, and these critical events partition the viewpoint space into $O(n^6)$ regions in the orthographic model, and $O(n^9)$ regions in the perspective model. Matching lower bounds show that these bounds are worst-case tight for a general scene consisting of $n$ triangles whether or not occlusions are considered [5, 2]. For axis-parallel objects in the orthographic model, the bound remains $\Omega(n^6)$ as demonstrated by Snoeyink [6].

The situation changes somewhat when we consider a scene consisting of $k$ disjoint convex polyhedra having a total complexity of $n$, since occlusions no longer remain insignificant in the worst case: the fact that the $n$ edges and vertices form the boundaries of only $k$ convex objects limits the number of non-occluded EV or EEE events that can occur. No more than $O(n^2k)$ of them correspond to unoccluded events, and thus appear in the viewpoint space partition [3]. Hence the upper bounds on the number of distinct orthographic and perspective views become $O(n^4k^2)$ and $O(n^6k^3)$, respectively [3]. The best lower bounds are currently $\Omega((nk^2 + n^2)^2)$ and $\Omega((nk^2 + n^2)^3)$ [3].

An algebraic curve of constant degree may intersect each critical surface in at most a constant number of points, which implies an upper bound of $O(n^2k)$ on the number of distinct views in both the linear and algebraic motion models. The best lower bounds are currently $\Omega(nk + k^3)$ and $\Omega(nk^2 + n^2)$, respectively [4].

Some authors have proposed a restricted notion of topologically different views, one that only takes into account *silhouette* critical events: these occur at viewpoints where a visible silhouette vertex apparently lies on a visible silhouette edge (*silhouette EV event*) or when three visible silhouette edges appear to become concurrent (*silhouette EEE event*) [2]. (Silhouette edges of a polyhedron are those that are incident to both a visible and an occluded facet from a given viewpoint, and a silhouette vertex is incident to a silhouette edge.) The number of distinct silhouette views in the linear motion model has been bounded by $O(nk^2)$, and by $O(n^2k)$ in the algebraic motion model. The best lower bounds are $\Omega(nk + k^3)$ and $\Omega(nk^2 + n^2)$, respectively [4].

*Our results.* The main tool used by de Berg *et al.* [3] to argue their improved bounds is the assertion that no more than $O(n^2k)$ critical events can be met by a straight line, i.e., no more than $O(n^2k)$ distinct views can be observed by a viewpoint moving in a straight line. We argue that this bound is worst-case tight and construct a scene in which a particular straight-line motion of a viewpoint meets $\Theta(n^2k)$ EEE events at distinct points (and thus there are as many different views). This improves the lower bounds to a worst-case tight $\Theta(n^2k)$ for the number of views of $k$ polyhedra having a total complexity of $n$, in both the linear and algebraic motion models.

Using this result, we show how to use two copies of our construction to force the existence of $\Omega(n^4k^2)$ distinct orthographic views. A similar technique using three copies shows a $\Omega(n^6k^3)$ lower bound on the number of distinct perspective views. Both bounds match the upper bounds of [3].

In the proceedings version, we had claimed that similar techniques can also be used to yield worst-case tight lower bounds of $\Omega(n^9)$ on the number of distinct perspective views of $n$ axis-parallel objects. Unfortunately, this cannot be inferred by a direct application of our techniques, and the (conjectured) lower bound of $\Omega(n^9)$ remains open.

Finally, we can modify our construction slightly to prove the following lower bounds on the number of distinct silhouette views: $\Omega(nk^2)$ in the linear motion model, $\Omega(n^2k^4)$ in the orthographic model, and $\Omega(n^3k^6)$ in the perspective model. Unfortunately, we are unable to improve the existing lower bounds in the algebraic motion model.

Our paper is structured as follows. In the next section, we present the main construction. In Section 3, we show how this construction can be used to yield all the lower bounds on the number of views in the various models mentioned above. Finally, in Section 4, we argue by a perturbation argument that although our constructions exhibit degeneracies, our lower bounds are valid as well for non-degenerate constructions.

## 2     The main construction

We present a construction of a scene $\mathcal{C}$ with $k$ polyhedra of total complexity $n$, that has at least $\Omega(n^2k)$ distinct views in the linear motion model (see figure 1). In addition to $n$ and $k$, the scene that we construct depends on a scaling parameter $t$, $0 < t \leq 1$, and on $\varepsilon > 0$ that can be fixed as a function of $n$; $t$ is required to produce scenes with many orthographic or perspective views and can be set to 1 for now.

More precisely, we prove the following.

**Theorem 1.** *As the viewpoint moves along the $y$-axis in the plane $z = +\infty$, it meets $\Omega(n^2k)$ critical surfaces of $\mathcal{C}$ at distinct points, and therefore observes at least as many distinct views of the scene.*

This bound is asymptotically the best possible [3].

Lines and directions not orthogonal to the $z$-direction and oriented downward can be parameterized by $(a, b, -1)$ [6]. The particular line at infinity used in the theorem corresponds to orthographic views in a direction $(0, b, -1)$, for $b \in \mathbb{R}$. Such directions are orthogonal to the $x$-direction. By making the viewpoint move along a line in $\mathbb{R}^3$ sufficiently far from $\mathcal{C}$, we can obtain a result similar to Theorem 1 about perspective views.

### 2.1     Description

Our construction consists of three distinct elements, the *drum*, the *needles* and the *fan*. The drum is a section of a nearly-flat horizontal prism; it has $\Theta(n)$ edges; we consider it completely flat in calculations and then choose its "curvature" small enough so that its slight non-flatness does not affect the construction. The fan is a convex polygon with $\Theta(n)$ edges that can be thickened to a proper convex polyhedron. The needles are a set of $\Theta(k)$ horizontal parallel lines that are eventually replaced by very thin and long tetrahedra.

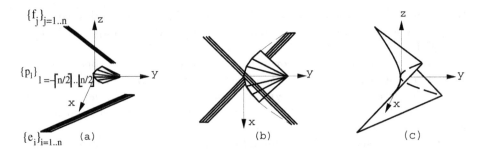

**Fig. 1.** The scene when $t = 1$: (a) a generic orthographic view of the scene, the drum is on the bottom, the fan in the middle, and the needles on the top; (b) another orthographic view from $z = \infty$; (c) the paraboloid and its parabolic section. When $t < 1$ the entire picture is compressed in the $y$-direction.

The intuition behind the construction is to place the elements so that each combination of a drum edge, a needle, and a fan edge yields an unoccluded EEE event. This can be done by placing them in an $\varepsilon$-neighborhood of the hyperbolic paraboloid $\Pi_0 : y = ztx$. We place the drum in the plane $z = -1$. The edges of the drum are parallel to the line $L_{-1} : y = -tx$, and at most $\varepsilon$ away from $L_{-1}$. The needles are in the plane $z = 1$, parallel to and at most $\varepsilon$ away from the line $L_1 : y = tx$. Both lines $L_{-1}$ and $L_1$ belong to $\Pi_0$ (see figure 1c). The plane $x = z$ cuts $\Pi_0$ along the parabola $y = tx^2 = tz^2$; the fan is a polygon in this plane that "grazes" the paraboloid, its vertices are chosen on the parabola.

The exact coordinates of our construction are given in the table below.

|        | Neighborhood | Equation |
|--------|--------------|----------|
| Drum    | $L_{-1}$ | $e_i = \{x \in [-1,1],\ y = -t(x - \frac{i}{n}\varepsilon),\ z = -1\}$, for $i = 1,\dots,n$ |
| Needles | $L_1$    | $f_j = \{x \in [-1,1],\ y = t(x + \frac{j}{k}\varepsilon),\ z = 1,\qquad$ for $j = 1,\dots,n$ |
| Fan     | $\Pi_0 \cap \{x = z\}$ | $[p_l, p_{l+1}]$ with $p_l = \left(\frac{l}{n}, t\left(\frac{l}{n}\right)^2, \frac{l}{n}\right),\qquad$ for $l = -\lceil\frac{n}{2}\rceil,\dots,\lfloor\frac{n}{2}\rfloor$ |

**Table 1.** The elements of the construction. We fix $\varepsilon = \frac{1}{16n^2}$.

## 2.2   Analysis

We now proceed to prove Theorem 1. We show that each triple of edges $e_i$, $f_j$, $[p_l, p_{l+1}]$ gives rise to an unoccluded EEE event, for $i = 1,\dots,n$, $j = 1,\dots,k$, and $l = -\lceil\frac{n}{2}\rceil,\dots,\lfloor\frac{n}{2}\rfloor$. We also show that the corresponding critical lines of sight include lines orthogonal to the $x$-direction.

In order to prove the latter statement, we introduce the surface $\Pi_{i,j}$ ruled by a family of lines orthogonal to the $x$-direction and passing through the drum edge $e_i$ and the needle $f_j$. It is known that $\Pi_{i,j}$ is a hyperbolic paraboloid (e.g., [1,

Cor. 14.4.6]). The equation of $\Pi_{i,j}$, is of the form

$$\Pi_{i,j} \; : \; \pi_{i,j}(x, y, z) = 0,$$

where

$$\pi_{i,j}(x, y, z) = y - zt\left(x + \frac{\varepsilon}{2}\left(\frac{j}{k} - \frac{i}{n}\right)\right) - t\frac{\varepsilon}{2}\left(\frac{j}{k} + \frac{i}{n}\right). \qquad (1)$$

Indeed, this equation is of the form $y = \alpha(x)z + \beta$ (implying that $\Pi_{i,j}$ contains a family of lines orthogonal to the $x$-direction); setting $z = -1$, we find that $\Pi_{i,j} \cap \{z = -1\}$ contains $e_i$, and similarly for $z = 1$, $\Pi_{i,j} \cap \{z = 1\}$ contains $f_j$.

Given a point $q = (x, y, z) \in \mathbb{R}^3$ not on the lines supporting $e_i$ or $f_j$, there is a unique line passing through $q$ and the lines supporting $e_i$ and $f_j$; this line is the intersection of the planes aff($q \cup e_i$) and aff($q \cup f_j$) (see Figure 2). Let the direction of this line be $(a_{i,j}(q), b_{i,j}(q), -1)$. We will use the following property of $\Pi_{i,j}$. Since $\Pi_{ij}$ contains all the lines that simultaneously touch the lines containing $e_i$, $f_i$ and are orthogonal to the x-direction, $a_{i,j}(q) = 0$ if and only if q lies on $\Pi_{ij}$. In fact, the reader will verify easily (for example, by continuity of $a_{i,j}$) that if $\Pi_{i,j}^+$ stands for $\{(x, y, z) \in \mathbb{R}^3 \; : \; \pi_{i,j}(x, y, z) > 0\}$ and $\Pi_{i,j}^-$ for $\{(x, y, z) \in \mathbb{R}^3 \; : \; \pi_{i,j}(x, y, z) < 0\}$, then

$$\begin{cases} a_{i,j}(q) > 0 \text{ iff } q \in \Pi_{i,j}^+, \\ a_{i,j}(q) < 0 \text{ iff } q \in \Pi_{i,j}^-, \\ a_{i,j}(q) = 0 \text{ iff } q \in \Pi_{i,j}. \end{cases} \qquad (2)$$

In view of (2), our goal can now be stated as follows: it suffices to prove that each segment $[p_l, p_{l+1}]$ intersects each hyperbolic paraboloid $\Pi_{i,j}$. Indeed, if $q$ is in the intersection $[p_l, p_{l+1}] \cap \Pi_{i,j}$, considering the line of sight passing through $q$, $e_i$, and $f_j$, proves both that the triple $e_i$, $f_j$, and $[p_l, p_{l+1}]$ induces an EEE event, and (2) implies that the corresponding line of sight is orthogonal to the $x$-direction. Note that there can be at most two such intersections, because $\Pi_{i,j}$ is a hyperbolic paraboloid. Below, we prove the following: if $q_l$ is the midpoint of $p_l$ and $p_{l+1}$, then

$$\begin{cases} p_l, p_{l+1} \in \Pi_{i,j}^-, \\ q_l \in \Pi_{i,j}^+. \end{cases} \qquad (3)$$

If (3) is true, $[p_l, q_l]$ must intersect $\Pi_{i,j}$ exactly once, and so must $[q_l, p_{l+1}]$. It should be clear also, that the corresponding lines of sight are ordered by slope in the $yz$-direction, as $l$ takes on increasing values in $-\lceil\frac{n}{2}\rceil, \ldots, \lfloor\frac{n}{2}\rfloor$, and thus that the corresponding viewpoints on the $y$-axis in the plane $z = +\infty$ are all distinct.

*Proof of (3).* Let us begin by rewriting the equation of $\Pi_{i,j}$ as $\pi_{i,j}(x, y, z) = y - zt(x + \delta_{i,j}) - t\delta'_{i,j} = 0$, with $|\delta_{i,j}| \leq \delta'_{i,j} \leq \varepsilon$, and let us introduce the notation $x_l = \frac{l}{n}$. Note that $p_l = (x_l, tx_l{}^2, x_l)$. We compute

$$\pi_{i,j}(p_l) = tx_l{}^2 - tx_l(x_l + \delta_{i,j}) - t\delta'_{i,j}$$
$$= -t(x_l\delta_{i,j} + \delta'_{i,j}) < 0,$$

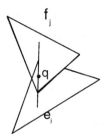

**Fig. 2.** There is a unique line passing through point $q$ and tangent to $e_i$ and $f_j$, and it is obtained as the intersection of the planes $\text{aff}(q, e_i)$ and $\text{aff}(q, f_j)$.

since $|x_l| < 1$ and $0 \le |\delta_{i,j}| \le \delta'_{i,j}$. Thus $p_l \in \Pi^-_{i,j}$, for $l = -\lceil \frac{n}{2} \rceil, \ldots, \lfloor \frac{n}{2} \rfloor$.

For $q_l$, we have $x(q_l) = z(q_l) = \frac{1}{2}(x_{l+1} + x_l)$, and $y(q_l) = \frac{1}{2}(x^2_{l+1} + x^2_l)$, therefore

$$
\begin{aligned}
\pi_{i,j}(q_l) &= \frac{t}{2}\left(x^2_{l+1} + x^2_l\right) - \frac{t}{4}\left(x_{l+1} + x_l\right)^2 - \frac{t}{2}\left(x_{l+1} + x_l\right)\delta_{i,j} - t\delta'_{i,j} \\
&= \frac{t}{4}\left(x_{l+1} - x_l\right)^2 - \frac{t}{2}\left(x_{l+1} + x_l\right)\delta_{i,j} - t\delta'_{i,j} \\
&= \frac{t}{4n^2} - \frac{t}{2}\left(x_{l+1} + x_l\right)\delta_{i,j} - t\delta'_{i,j}
\end{aligned}
$$

since $x_{l+1} - x_l = \frac{1}{n}$. Because $y(q_l) > 0$ and $|\delta_{i,j}| \le \delta'_{i,j} \le \varepsilon$, we have for our choice of $\varepsilon = \frac{1}{16n^2}$

$$
\pi_{i,j}(q_l) > \frac{t}{4n^2} - 2t\varepsilon > 0.
$$

Therefore, we have $q_l \in \Pi^+_{i,j}$. This completes the proof of (3), and by extension, of Theorem 1.

## 3   The lower bounds

We finish by indicating our various lower bound constructions. See Figure 4 for a pictorial explanation of our constructions. First we need a result about the critical curves and surfaces.

### 3.1   The critical curves and surfaces

Each orthographic viewpoint can be represented by a (oriented homogeneous) vector $(a, b, c)$. Note that we are only concerned about the directions such that $c < 0$ here. As in [6], we treat these numbers as homogeneous coordinates, and the set of vectors such that $c < 0$ can be represented by the vector $(a, b, -1)$, or bijectively by a point $(a, b)$ in the $ab$-plane at infinity. In that plane, the $b$-axis has equation $a = 0$.

The directions of the lines passing through $e_i$, $f_j$, and $[p_l, p_{l+1}]$ form a curve $\gamma_{i,j,l}$ in that plane. Each point of this curve corresponds to the point of contact $q_l(s) = (1-s)p_l + sp_{l+1}$ on $[p_l, p_{l+1}]$ (see Figure 2), with the endpoints of $\gamma_{i,j,l}$ corresponding to $p_l$ and $p_{l+1}$. Thus the curve is naturally parameterized by $(a(s), b(s))$ with $s \in [0, 1]$. Property (3) shows in fact that each such curve crosses the $b$-axis twice (see Figure 3a), once for some $s \in [0, \frac{1}{2}]$ and another for some $s \in [\frac{1}{2}, 1]$.[1] Let us consider only the portion $s \in [0, \frac{1}{2}]$ from now on.

In order to argue for complicated patterns of interaction later on (in Sections 3.2 and 3.3), we wish to find a rectangle $A \times B$ such that the critical curves completely traverse $A$ in the $a$-direction, and remain constrained in an interval $tB$ in the $b$-direction. We can compute an interval $A' = A'(n, k)$ such that $a(s)$ spans the entire interval $A'$ as $s \in [0, \frac{1}{2}]$, for every choice of $(i, j, l)$ in $I = \{1 \ldots n\} \times \{1 \ldots k\} \times \{-\lceil \frac{n}{2} \rceil + 1, \ldots, \lfloor \frac{n}{2} \rfloor\}$. Similarly, we can compute an interval $B = B(n, k)$ such that $b(s)$ remains always in $B$. Simply take:

$$A'_{\text{inf}} = \max_{(i,j,l) \in I} a(0) < 0,$$

$$A'_{\text{sup}} = \min_{(i,j,l) \in I} a(\frac{1}{2}) > 0,$$

$$B_{\text{inf}} = \min_{(i,j,l) \in I} \min_{s \in [0,\frac{1}{2}]} b(s),$$

$$B_{\text{sup}} = \max_{(i,j,l) \in I} \max_{s \in [0,\frac{1}{2}]} b(s)$$

and let $A' = [A'_{\text{inf}}, A'_{\text{sup}}]$ and $B = [B_{\text{inf}}, B_{\text{sup}}]$. The sign conditions for $A'$ follow from (2) and (3). Finally, we know that the intersections of the $\gamma_{i,j,l}$ with the $b$-axis are all distinct. We can therefore take a subinterval $A$ of $A'$ containing 0 such that the $\gamma_{i,j,l}$ do not pairwise intersect in the rectangle $A \times B$.

In summary, the critical curves completely traverse $A(n, k)$ in the $a$-direction, remain constrained in an interval $tB(n, k)$ in the $b$-direction, and do not pairwise intersect in the rectangle $A(n, k) \times tB(n, k)$. Thus, taking $t$ small enough, the critical curves $\gamma$ look as in Figure 3b, with the $A$-side much longer than the $B$-side.

Note also that seen from far away in the $z$-direction in the perspective model, the critical surfaces look like Figure 3c. Indeed, we can similarly select intervals $X(n, k)$, $Y(n, k)$ and $Z(n, k)$ of the same length such that the critical surfaces do not pairwise intersect in $X(n, k) \times tY(n, k) \times Z(n, k)$ and do not cross the top and bottom facets of the box. (The construction is straigthforward and omitted; it is possible to express $X$ and $Y$ as functions of $A$ and $B$, and $Z$ a subset of some interval $[z, +\infty)$ for some large enough $z$.) The situation is depicted on Figure 3b for the orthographic model, and 3c for the perspective model.

---

[1] In fact, these curves are the intersection of a hyperbolic paraboloid with a plane at infinity, and it can be shown that they are hyperbolas in that plane [6]. We will not need that characterization here.

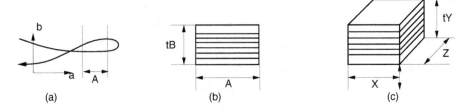

**Fig. 3.** (a) A critical curve and how to choose the interval $A$. (b) The critical curves in the orthographic model. (c) The critical surfaces when the viewpoint is sufficiently far away in the perspective model.

### 3.2    The number of views in the orthographic model

For the orthographic model, place two copies of the scene of Figure 1, one being the image of the other after a rotation of angle of $\pi/2$ about the $z$-axis and a suitable translation to avoid collisions (as in Figure 4a). By the results of the discussion in the previous section, in the viewpoint space, the critical curves overlap in a grid-like fashion. This method, which we call the *skew duplication* method, is already used in [3]. This proves that our scene yields $\Omega((n^2k)^2) = \Omega(n^4k^2)$ critical events.

### 3.3    The number of views in the perspective model

For the perspective model, place the view point at any vertex of a cube with sufficiently large side, and place three copies of the scene of Figure 1 at the three adjacent vertices, as in Figure 4b, with the orientation permuted in such a way that the critical surfaces will overlap in a grid-like fashion, yielding $\Omega((n^2k)^3) = \Omega(n^6k^3)$ critical events. This method, which we call the *skew triplication* method, is also already used in [3].

### 3.4    The number of silhouette views

For silhouette views, we modify the main construction by replacing the $n$ edges of the drum by $k$ needles $e_i'$ in the neighborhood of $L_{-1}$, $i = 1, \ldots, k$. It should be clear that all the EEE events corresponding to $e_i'$, $f_j$, and $[p_l, p_{l+1}]$ are silhouette events, and the same argument as above shows that indeed those $\Omega(nk^2)$ silhouette events are unoccluded, and can be observed as the viewpoint moves along the $y$-axis in the plane $z = +\infty$ (or a line in $\mathbb{R}^3$ sufficiently far from the scene, for perspective views). Thus the bound $\Omega(nk^2)$ for the maximum number of silhouette views in the linear motion model.

The bound extends naturally to $\Omega(n^2k^4)$ for the maximum number of silhouette views in the orthographic model, and to $\Omega(n^3k^6)$ for the maximum number of silhouette views in the perspective model, by using the same skew multiplication methods as in the previous two sections.

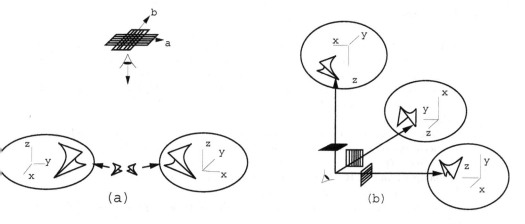

**Fig. 4.** How to arrange the scenes to obtain high complexity viewpoint space partitions in (a) the orthographic model, (b) the perspective model.

## 4   About degeneracies

There are three kinds of degeneracies arising in our construction: the facets of the drum are coplanar, the needles are segments and not polyhedra, and the fan is a polygon and not a polyhedron. We argue now that none of them affects our results.

To compensate for the fact that the facets of the drum are coplanar, we can introduce a perturbation to put these edges in a convex position, depending on a small positive parameter $\eta_1$, by putting edge $e_i$ in the plane $z = -1 - \eta_1 i^2$.

The needles are segments, but we can replace the $x$-smaller endpoint of each needle by a small triangle containing the endpoint, whose sides are smaller than a parameter $\eta_2 n$. A needle is thus a very long and thin tetrahedron. By choosing an appropriate orientation of these tetrahedra, we can make $f_j$ visible on the contour when viewed from all the (near-vertical) directions of interest in this paper; take the vertices of the triangle at the basis of $f_j$ to be $f_j(-1)$, $f_j(-1) + \eta_2 j e_y$, and $f_j(-1) + \eta_2 j e_y + \eta_2^2 j^2 e_z$.

As presented, the fan is a polygon rather than a polyhedron. We can make the fan into a convex polyhedron with non-empty interior by taking the convex hull of the $p_l$'s with an edge of length $\eta_3$, whose endpoints are $(0, 1, -\eta_3)$ and $(0, 1, \eta_3)$.

We argue that the introduction of the $\eta$'s does not affect the veracity of the proof. This can be formally checked by the following approach: check that neither $\eta_1$ nor $\eta_2$ nor $\eta_3$ appear in the asymptotically dominant term of any quantity on which we assert the sign (positive or negative; we do not rely on any quantity to be exactly 0), such as in the grazing property (2), or in the existence of $A(n, k)$, $(n, k)$ and $B(n, k)$ above. Therefore the bounds remain valid for sufficiently small values of $\eta_1$, $\eta_2$, and $\eta_3$.

# References

1. M. Berger. *Geometry (vols. 1-2)*. Springer-Verlag, 1987.
2. K. W. Bowyer and C. R. Dyer. Aspect graphs: An introduction and survey of recent results. *Int. J. of Imaging Systems and Technology*, 2:315–328, 1990.
3. M. de Berg, D. Halperin, M. Overmars, and M. van Kreveld. Sparse arrangements and the number of views of polyhedral scenes. *Internat. J. Comput. Geom. Appl.*, 7:175–195, 1997.
4. A. Efrat, L. J. Guibas, O. A. Hall-Holt, and L. Zhang. On incremental rendering of silhouette maps of a polyhedral scene. In *Proc. 13th ACM-SIAM Sympos. Discrete Algorithms*, 2000.
5. H. Plantinga and C. R. Dyer. Visibility, occlusion, and the aspect graph. *Internat. J. Comput. Vision*, 5(2):137–160, 1990.
6. J. Snoeyink. The number of views of axis-parallel objects. *Algorithms Rev.*, 2:27–32, 1991.

# Problems and Results around the Erdős–Szekeres Convex Polygon Theorem[*]

Imre Bárány[1,2] and Gyula Károlyi[3]

[1] Maths, University College London
[2] Rényi Institute, Budapest
barany@renyi.hu
[3] Eötvös University, Budapest
karolyi@cs.elte.hu

## 1 Introduction

Eszter Klein's theorem claims that among any 5 points in the plane, no three collinear, there is the vertex set of a convex quadrilateral. An application of Ramsey's theorem then yields the classical Erdős–Szekeres theorem [19]: *For every integer $n \geq 3$ there is an $N_0$ such that, among any set of $N \geq N_0$ points in general position in the plane, there is the vertex set of a convex $n$-gon.* Let $f(n)$ denote the smallest such number.

**Theorem 1 ([20,44]).**

$$2^{n-2} + 1 \leq f(n) \leq \binom{2n-5}{n-2} + 2 .$$

A very old conjecture of Erdős and Szekeres is that the lower bound is tight:

**Open Problem 1.** *For every $n \geq 3$, $f(n) = 2^{n-2} + 1$.*

Similarly, let $f_d(n)$ denote the smallest number such that, in any set of at least $f_d(n)$ points in general position in Euclidean $d$-space, there is the vertex set of a convex polytope with $n$ vertices, that is, $n$ points in convex position. A simple projective argument [47] shows that $f_d(n) \leq f(n)$. It is conjectured by Füredi [22] that $f_d(n)$ is essentially smaller if $d > 2$, namely that $\log f_d(n) = O(n^{1/(d-1)})$. A lower bound that matches this conjectured upper bound was given recently in [33]. On the other hand, Morris and Soltan [34] contemplate about an exponential lower bound on $f_d(n)$.

In this paper we survey recent results and state some open questions that are related to Theorem 1. In particular, we consider "homogeneous", "partitional", and "modular" versions of the Erdős–Szekeres theorem. We will discuss the question whether empty convex polygons (and then how many of them) can be found among $N$ points in the plane. We will also describe how the convex

[*] The authors gratefully acknowledge that they have been partially supported by Hungarian Research Grants OTKA T032452 and F030822, respectively.

J. Akiyama, M. Kano, and M. Urabe (Eds.): JCDCG 2000, LNCS 2098, pp. 91–105, 2001.

position condition can be strengthened or relaxed in order to arrive at well-posed questions, and present the results obtained so far.

For further aspects of the Erdős–Szekeres theorem we refer to the very recent and comprehensive survey article [34].

## 2   Homogeneous Versions

From now on we assume that $X \subset \mathbb{R}^2$ is a finite set of points in general position. We assume further that $X$ has $N$ elements. By the Erdős–Szekeres theorem, any subset of $X$ of size $f(n)$ contains the vertices of a convex $n$-gon. As a fixed $n$-set is contained in $\binom{N}{f(n)-n}$ subsets of size $f(n)$, a positive fraction of all the $n$-tuples from $X$ are in convex position. This is a well-known principle in combinatorics. Maybe one can say more in the given geometric situation, for instance, the many convex position $n$-tuples come with some structure. The following theorem, due to Bárány and Valtr [7], shows that these $n$-tuples can be chosen homogeneously:

**Theorem 2 ([7]).** *Given $n \geq 4$, there is a constant $C(n)$ such that for every $X \subset \mathbb{R}^2$ of $N$ points in general position the following holds. There are subsets $Y_1, \ldots, Y_n$ of $X$, each of size at least $C(n)N$ such that for every transversal $y_1 \in Y_1, \ldots, y_n \in Y_n$, the points $y_1, \ldots, y_n$ are in convex position.*

We call this result the "homogeneous" Erdős–Szekeres theorem. The proof in [7] is based on another homogeneous statement, the so called same type lemma. We state it in dimension $d$, but first a definition: Two $n$-tuples $x_1, \ldots, x_n$ and $y_1, \ldots, y_n$ are said to be of the *same type* if the orientations of the simplices $x_{i_1}, \ldots, x_{i_{d+1}}$ and $y_{i_1}, \ldots, y_{i_{d+1}}$ are the same for every $1 \leq i_1 < i_2 < \ldots < i_{d+1} \leq n$.

**Theorem 3 ([7]).** *Given $d \geq 2$ and $k \geq d+1$, there is a constant $C(k, d)$ such that for all finite sets $X_1, \ldots, X_k \subset \mathbb{R}^d$ of points such that $\cup_1^k X_i$ is in general position the following holds. For every $i = 1, \ldots k$, the set $X_i$ contains a subset $Y_i$ of size at least $C(k, d)|X_i|$ such that all transversals $y_1 \in Y_1, \ldots, y_k \in Y_k$ are of the same type.*

The proof is based on the center-point theorem of Rado [7], or on Borsuk's theorem [37]. It uses a reformulation of the definition of same type: all transversals of $Y_1, \ldots, Y_k$ are of the same type if no hyperplane meets the convex hulls of any $d+1$ of these sets. The same type lemma implies the homogeneous version of the Erdős–Szekeres theorem in the following way. Choose $k = f(n)$ and partition $X \subset \mathbb{R}^2$ by vertical lines, say, into sets $X_1, \ldots, X_k$ of almost equal size. Apply the same type lemma to them. All transversals of the resulting subsets $Y_1, \ldots, Y_k$ are of the same type. Fix a transversal $y_1, \ldots, y_k$. As $k = f(n)$, the Erdős–Szekeres theorem implies that some $n$ points of this transversal, $y_{j_1}, \ldots, y_{j_n}$, say, are in convex position. Then by the same type lemma, all transversals of $Y_{j_1}, \ldots, Y_{j_n}$ are in convex position.

This proof gives a doubly exponential lower bound for $C(n)$. An alternative proof, with a better bound for $C(n)$ was found by Solymosi [39]. A sketch of Solymosi's neat argument goes as follows. As we have seen above, a positive fraction of the $2n$ element subsets of $X$ are in convex position. Write such a $2n$ element subset as $x_1, y_1, x_2, y_2, \ldots, x_n, y_n$ with the points coming in this order on the boundary of their convex hull. Choose $a_1 = x_1, a_2 = x_2, \ldots, a_n = x_n$ so that the number of possible convex extensions $a_1, y_1, a_2, y_2, \ldots, a_n, y_n$ is maximal. Averaging shows that the number of such extensions is at least $\mathrm{const}\binom{N}{n}$ where $|X| = N$. A simple geometric argument explains that the possible $y_i$s all lie in the triangle $T_i$ formed by the lines through the pairs of points $(a_{i-1}, a_i)$, $(a_i, a_{i+1})$ and $(a_{i+1}, a_{i+2})$. It is not hard to check then that $Y_i = X \cap T_i$ $(i = 1, \ldots, n)$ satisfies the requirements.

This proof gives $C(n) \approx 2^{-16n^2}$, while the lower bound for $f(n)$ shows that $C(n)$ is at least $2^{n-2}$. Better bounds are available for $n = 4, 5$ [7]: $C(4) \geq 1/22$ and $C(5) \geq 1/352$. The reader is invited to prove or improve these bounds.

Pach [37] uses the same type lemma to prove a homogeneous version of Caratheodory's theorem that was conjectured in [6]: Given $X_i \subset \mathbb{R}^d$ $(i = 1, \ldots, d + 1)$, there is a point $z \in \mathbb{R}^d$ and there are subsets $Z_i$, each of size $c_d |X_i|$ at least $(i = 1, \ldots, d + 1)$, such that the convex hull of each transversal $z_1 \in Z_1, \ldots, z_{d+1} \in Z_{d+1}$ contains the point $z$. (Here $c_d$ is a constant depending only on $d$.) Pach's nice argument uses, besides the same type lemma, a quantitative version of Szemerédi's regularity theorem.

We expect that the same type lemma will have many more applications. Also, several theorems from combinatorial convexity extend to positive fraction or homogeneous versions. For instance, a positive fraction Tverberg theorem is proved in [7]. One question of this type concerns Kirchberger's theorem [14]. The latter says that finite sets $A, B \subset \mathbb{R}^d$ can be separated by a hyperplane if and only if for every $S \subset A \cup B$ of size $d + 2$ there is a hyperplane separating $A \cap S$ and $B \cap S$. This suggests the following question:

**Open Problem 2.** *Let $A, B \subset \mathbb{R}^d$ be finite sets, each of size $N$, with $A \cup B$ in general position. Assume that for $(1 - \varepsilon)$ fraction of the $\binom{2N}{d+2}$ $(d+2)$-tuples $S \subset A \cup B$ there is a hyperplane separating $A \cap S$ from $B \cap S$. Does it follow then that there are subsets $A' \subset A$ and $B' \subset B$ that are separated by a hyperplane and $|A'|, |B'| \geq (1 - g(\varepsilon))N$ with $g(\varepsilon)$ tending to zero as $\varepsilon \to 0$?*

Partial results in this direction are due to Attila Pór [40]. One word of caution is in place here: the condition $g(\varepsilon) \to 0$ is important since by the ham-sandwich theorem (or Borsuk's theorem, if you like) any two finite sets $A, B$ in $\mathbb{R}^d$ can be simultaneously halved by a hyperplane $H$. Then half of $A$ is on one side of $H$ while half of $B$ is on the other side.

## 3  Partitional Variants

Let $P$ be any set of points in general position in the plane. Let $C_1, C_2$ be subsets of $P$, each in convex position. We say that the *convex polygons* $C_1$ and $C_2$ are *vertex disjoint* if $C_1 \cap C_2 = \emptyset$. If, moreover, their convex hulls are also disjoint, we simply say that the two polygons are *disjoint*. A polygon is called *empty* if its convex hull does not contain any point of $P$ in its interior.

Eszter Klein's theorem implies that $P$ can be partitioned into vertex disjoint convex quadrilaterals plus a remainder set of size at most 4. The following result answers a question posed by Mitchell.

**Theorem 4 ([30]).** *Let $P$ be any set of $4N$ points in general position in the plane, $N$ sufficiently large. Then there is a partition of $P$ into $N$ vertex disjoint convex quadrilaterals if and only if there is no subset $A$ of $P$ such that the size of $A$ is odd but the the size of $A \cap C$ is even for every convex quadrilateral $C$.*

There is also an $N \log N$-time algorithm [30] which decides if such a partition exists. The following problem seems to be more difficult.

**Open Problem 3.** *Is there a fast algorithm which decides if a given set of $4N$ points in general position in the plane admits a partition into disjoint convex quadrilaterals?*

For $k \geq 3$ the *Ramsey-remainder $rr(k)$* was defined by Erdős et al. [21] as the smallest integer such that any sufficiently large set of points in general position in the plane can be partitioned into vertex disjoint polygons, each of size $\geq k$, and a remaining set of size $\leq rr(k)$. Thus, $rr(k) < f(k)$ for every $k$. In particular, $rr(3) = 0$ and $rr(4) = 1$. Partial results on $rr(k)$ in general were proved in [21]. It is known, for example, that $rr(k) \geq 2^{k-2} - k + 1$. The solution of the following problem could make an essential step to settle Problem 1, see [21].

**Open Problem 4.** *Is it true that $rr(k) = 2^{k-2} - k + 1$?*

There is no Ramsey-remainder in higher dimensions. The following result is due to Károlyi [30].

**Theorem 5.** *Let $k > d \geq 3$. If $N$ is sufficiently large, then every set of $N$ points in general position in $\mathbb{R}^d$ can be partitioned into subsets of size at least $k$ each of which is in convex position.*

The main observation here is that, for large enough $N$, every point of $P$ belongs to some $k$-element subset which is in convex position.

A problem in close relation to Problem 3 is the following. Given natural numbers $k$ and $n$, let $F_k(n)$ denote the maximum number of pairwise disjoint empty convex $k$-gons that can be found in every $n$-element point set in general position in the plane. The study of this function was initiated in [27]. Horton's result mentioned in Section 5 implies $F_7(n) = 0$ for every $n$. Thus, the interesting functions are $F_4, F_5$ and $F_6$. Nothing is known about $F_6$, in fact Problem 6 is equivalent to asking whether $F_6(n) > 0$ for some $n$. Since every 5-point set determines an empty convex quadrilateral, obviously $F_4(n) \geq \lfloor n/5 \rfloor$. Similarly, it follows from a result of Harborth [23] that $F_5(n) \geq \lfloor n/10 \rfloor$ for every $n$.

The non-trivial lower bound $F_4(n) \geq \lfloor 5n/22 \rfloor$ is presented in [27], based on the following observation. Suppose $P$ is any set of $2m + 4$ points in general position in the plane. Then there is a partition of the plane into 3 convex regions such that one region contains 4 points of $P$ in convex position, and the other regions contain $m$ points of $P$ each. There is no counterpart of this lemma for pentagons, and in fact no lower bound is known about $F_5$ beyond what is said above. As for $F_4$, an even stronger lower bound $F_4(n) \geq (3n - 1)/13$ has been proved for an infinite sequence of integers $n$.

Concerning upper bounds, a construction in [27] shows that $F_5(n) \leq 1$ if $n \leq 15$. It is not too difficult to prove that $F_5(n) < n/6$, but no nontrivial upper bound is known for $F_4(n)$ in general.

For any positive integer $n$ let $F(n)$ denote the smallest integer such that every set of $n$ points in general position in the plane can be partitioned into $F(n)$ empty convex polygons, with the convention that point sets consisting of at most two points are always considered as empty convex polygons. Urabe [45] proved $\lceil (n - 1)/4 \rceil \leq F(n) \leq \lceil 2n/7 \rceil$. The upper bound follows from the fact that every 7-point set can be partitioned into an empty triangle and an empty convex quadrilateral.

An improved upper bound $F(n) \leq \lceil 5n/18 \rceil$ is presented in [27] along with an infinite sequence of integers $n$ for which also $F(n) \leq (3n + 1)/11$.

An other function $H(n)$ was also introduced in [45] as the smallest number of vertex disjoint convex polygons into which any $n$-element point set can be partitioned in the plane. An application of Theorem 1 gives that the order of magnitude of this function is $n/\log n$.

Finally we mention that the functions $F$ and $H$ can be naturally defined in any dimension; denote the corresponding functions in $d$-space by $F_d$ and $H_d$. Urabe [46] proves that $\Omega(n/\log n) \leq F_3(n) \leq \lceil 2n/9 \rceil$ and that $H_3(n) = o(n)$. The proof technics of [45] coupled with the bounds given in Section 1 on $f_d$ in fact yield $\Omega(n/(\log^{d-1} n) \leq F_d(n) \leq O(n/\log n)$.

## 4   Matrix Partitions

Assume $X_1, \ldots, X_n$ in $\mathbb{R}^d$, are pairwise disjoint sets, each of size $N$, with $\cup X_i$ in general position. A matrix partition, or $\mu$-partition for short, of the $X_i$s with $m$ columns is the partition $X_i = \cup_{k=1}^m M_{ik}$ for $i = 1, \ldots, n$ if $|M_{ik}| = |M_{jk}|$ for every $i, j = 1, \ldots, n$ and every $k = 1, \ldots, m$. In other words, a $\mu$-partition of $X_1, \ldots, X_n$ with $m$ columns is an $n \times m$ matrix $M$ whose $(i, k)$ entry is a subset $M_{ik}$ of $X_i$ such that row $i$ forms a partition of $X_i$ and the sets in column $k$ are of the same size. Gil Kalai asked [28] whether the homogeneous Erdős–Szekeres theorem admits a partitioned extension:

**Open Problem 5.** *Show that for every $n \geq 4$ there is an integer $m = g(n)$ such that for every finite set $X \subset \mathbb{R}^2$ of $N$ points in general position there is a subset $X_0 \subset X$, of size less than $f(n)$, (this is the Erdős–Szekeres function from*

*Section 1), and there exists a partition of $X \setminus X_0$ into sets $X_1, \dots, X_n$ of equal size such that the following holds. The sets $X_1, \dots, X_n$ admit a $\mu$-partition $M$ with $m$ columns so that every transversal $x_1 \in M_{1k}, x_2 \in M_{2k}, \dots, x_n \in M_{nk}$ is in convex position, for all $k = 1, \dots, m$.*

By the homogeneous version one can choose the sets for the first column of a $\mu$-partition, each of size $C(n)N/n$, then for the second, third, etc columns from the remaining part of $X$, but this would result in a suitable $\mu$-partition with too many, namely $\log N$, columns. The remainder set $X_0$ is needed for two simple reasons: when $N$ is smaller than $f(n)$ there may not be a convex $n$-gon at all, and when $N$ is not divisible by $n$.

Partial solution to Problem 2 is due to Attila Pór [41]. He first proved a partitioned extension of the same type lemma. To state this result we define the sets $Y_1, \dots, Y_n \subset \mathbb{R}^d$ with $n \geq d + 1$ *separated* if every hyperplane intersects at most $d$ sets of the convex hulls of $Y_1, \dots, Y_n$. As we mentioned in Section 2, the sets $Y_1, \dots, Y_n$ are separated if and only if every transversal $y_1 \in Y_1, \dots, y_n \in Y_n$ is of the same type.

**Theorem 6 ([41]).** *For all natural numbers $n, d$ with $n \geq d + 1$ there is a natural number $m = m(n, d)$ such that if finite sets $X_1, \dots, X_n \subset \mathbb{R}^d$ have the same size and $\cup_1^n X_i$ is in general position, then there exists a $\mu$-partition with $m$ columns such that the sets $M_{1k}, \dots, M_{nk}$ in every column are separated.*

This is exactly the partitioned version of the same type lemma. The proof is based on a clever induction argument and a third characterization for sets $Y_1, \dots, Y_n$ being separated. The result is used by A. Pór [41] to solve the first interesting case, $n = 4$ of Problem 2.

**Theorem 7 ([41]).** *Assume $X \subset \mathbb{R}^2$ is a finite set of $N$ points in general position. Then there is an $X_0 \subset X$ of size at most 4, and a partition of $X \setminus X_0$ into sets $X_1, X_2, X_3, X_4$ of equal size such that they admit a $\mu$-partition $M$ with 30 columns so that every transversal $x_1 \in M_{1k}, \dots, x_4 \in M_{4k}$ is in convex position.*

The proof starts by cutting up $X$ into four sets of almost equal size by vertical lines, say. Then the same type lemma (matrix partition version) is applied to these four sets giving a matrix partition with few columns. The columns are of two types: either every transversal is a convex quadrangle and there is nothing to do, or every transversal is a triangle with the fourth point inside it. In the latter case one has to partition the column further. This can be done with a topological argument: the interested reader should consult the paper [41]. The method does not seem to work for $n \geq 5$, apparently new ideas are needed.

## 5   Empty Convex Polygons

For a long time it had been conjectured that every sufficiently large point set, in general position in the plane contains the vertex set of an *empty* convex $n$-gon, that is, $n$ points which form the vertex set of a convex polygon with no other point of the set in its interior. Harborth [23] showed that every 10-element point set determines an empty convex *pentagon*, and that here 10 cannot be replaced by any smaller number. Finally, in 1983 a simple recursive construction of arbitrarily large finite point sets determining no empty convex *heptagons* was found by Horton [24]. The corresponding problem for *hexagons* is still open:

**Open Problem 6.** *Is it true that every sufficiently large set of points in general position in the plane contains the vertex set of an empty convex hexagon?*

We strongly believe that the answer is yes, but there is no proof in sight.

Several algorithms had been designed [4,15,36] to determine if a given set of points contains an empty 6-gon, and to construct large point sets without any empty hexagon. The current world record, a set of 26 points that does not contain an empty convex 6-gon was discovered by Overmars et al. [36] in 1989.

A surprising number of questions can be related to this seemingly particular problem. The first one, due to Solymosi [43], relates it to a Ramsey type problem for geometric graphs. A *geometric graph* is a graph drawn in the plane such that the vertices are represented by points in general position while the edges are straight line segments that connect the corresponding vertices.

**Open Problem 7.** *Let G be a complete geometric graph on n vertices whose edges are colored with two different colors. Assume that n is sufficiently large. Does it follow then that G contains an empty monochromatic triangle?*

Were the answer to this question negative, it would imply that there are arbitrarily large point sets without an empty convex 6-gon. For assume, on the contrary, that every sufficiently large point set contains such an empty polygon. Color the edges of the corresponding complete geometric graph with two colors, it induces a coloring of the edges that connect the vertices of the empty 6-gon. It follows from Ramsey's theorem that this two-colored graph on 6 vertices contains a monochromatic triangle (which is also empty), a contradiction.

An other related problem has been studied recently by Hosono et al. [26]. Let $P$ denote a simple closed polygon together with its interior. A *convex subdivision* of $P$ is a 2-dimensional cell complex in the plane whose vertex set coincides with the vertex set of $P$, whose body is $P$, and whose faces are all convex polygons. Denote by $F'(n)$ the smallest integer for which any set of $n$ points in general position in the plane can be connected with a closed simple polygon that admits a convex subdivision with at most $F'(n)$ faces. Since each face in a convex subdivision is an empty convex polygon, it follows from Horton's construction that $F'(n) \geq n/4$ for an infinite sequence of $n$. It is proved for every $n$ in [26] where an upper bound $F'(n) \leq \lceil 3n/5 \rceil$ is also presented.

**Open Problem 8.** *Is it true that $F'(n) \geq (n-2)/3$?*

A negative answer would give an affirmative solution to the empty hexagon problem.

Essential combinatorial properties of Horton's construction were studied and extended into higher dimensions by Valtr [47], resulting in constructions that yield the following general result. Denote by $h(d)$ the largest integer $h$ with the following property: every sufficiently large point set in general position in $\mathbb{R}^d$ contains an $h$-*hole*, that is, $h$ points which are vertices of an empty convex $d$-polytope. Thus, $5 \leq h(2) \leq 6$.

**Theorem 8 ([47]).** *The integer $h(d)$ exist for any $d \geq 2$ and satisfies*

$$2d + 1 \leq h(d) \leq 2^{d-1}(P_{d-1} + 1) \ ,$$

*where $P_i$ denotes the product of the first $i$ positive prime numbers.*

It is also known that $h(3) \leq 22$.

We close this section by turning back to the plane: there are certain nontrivial classes of point sets where large empty convex polygons can be found. For example, if every triple in the point set determines a triangle with *at most one* point in its interior, then it is said to be *almost convex*.

**Theorem 9 ([32]).** *For any $n \geq 3$, there exists an integer $K(n)$ such that every almost convex set of at least $K(n)$ points in general position in the plane determines an empty convex $n$-gon. Moreover, we have $K(n) = \Omega(2^{n/2})$.*

This result has been extended recently by Valtr [50] to point sets where every triple determines a triangle with at most a fixed number of points in its interior. It also must be noted that Bisztriczky and Fejes Tóth [10] proved the following related result.

**Theorem 10.** *Let $l, n$ denote natural numbers such that $n \geq 3$. Any set of at least $(n-3)(l+1)+3$ points in general position in the plane, with the property that every triple determines a triangle with at most $l$ of the points in its interior, contains $n$ points in convex position. Namely, its convex hull has at least $n$ vertices, and in this respect this bound cannot be improved upon.*

## 6    The Number of Empty Polygons

Let $X \subset \mathbb{R}^2$ be a set of $N$ points in general position, and write $g_n(X)$ for the number of empty convex $n$-gons with vertices from $X$. Of course, $n \geq 3$. Define $g_n(N)$ as the minimum of $g_n(X)$ over all planar sets $X$ with $N$ points in general position. Horton's example shows that $g_n(N) = 0$ when $n \geq 7$. Problem 6 is, in fact, to decide whether $g_6(N) = 0$ or not.

The first result on $g_n(N)$ is due to Katchalski and Meir [29] who showed $g_3(N) \leq 200N^2$. In Bárány and Füredi [5] lower and upper bounds for $g_n(N)$ are given. The lower bounds are:

**Theorem 11 ([5]).**

$$g_3(N) \geq N^2 - O(N \log N) \,,$$

$$g_4(N) \geq \frac{1}{2}N^2 - O(N \log N) \,,$$

$$g_5(N) \geq \lfloor \frac{N}{10} \rfloor \,.$$

The last estimate can be easily improved to $g_5(N) \geq \lfloor \frac{N-4}{6} \rfloor$.

Of these inequalities, the most interesting is the one about $g_3$. Its proof gives actually more than just $g_3(N) \geq N^2(1 + o(1))$. Namely, take any line $\ell$ and project the points of $X$ onto $\ell$. Let $z_1, \ldots, z_N$ be the projected points on $\ell$ in this order, and assume $z_i$ is the projection of $x_i \in X$. We say that pair $z_i, z_j$ supports the empty triangle $x_i, x_k, x_j$ if this triangle is empty and $i < k < j$. Now the proof of the lower bound on $g_3(N)$ follows from the observation that all but at most $O(N \log N)$ pairs $z_i, z_j$ support at least two empty triangles. (This fact implies, further, the lower bound on $g_4$ as well.) It is very likely that a small but positive fraction of the pairs supports three or more empty triangles but there is no proof in sight. If true, this would solve the next open problem in the affirmative:

**Open Problem 9.** *Assume $X$ is a finite set of $N$ points in general position in $\mathbb{R}^2$. Show that $g_3(N) \geq (1 + \varepsilon)N^2$ for some positive constant $\varepsilon$.*

The upper bounds from [5] have been improved upon several times, [48], [16], and [8]. The constructions use Horton sets with small random shifts. We only give the best upper bounds known to date [8].

**Theorem 12.**

$$g_3(N) \leq (1 + o(1))1.6195...N^2 \,,$$

$$g_4(N) \leq (1 + o(1))1.9396...N^2 \,,$$

$$g_5(N) \leq (1 + o(1))1.0205...N^2 \,,$$

$$g_6(N) \leq (1 + o(1))0.2005...N^2 \,.$$

It is worth mentioning here that the function $g_n(X)$ satisfies two linear equations. This is a recent discovery of Ahrens et al. [1] and Edelman-Reiner [17]. Since the example giving the upper bounds in the last theorem is the same point set $X$ and $g_7(X) = 0$, only two of the numbers $g_n(X)$ ($n = 3, 4, 5, 6$) have to be determined.

There is a further open problem due to the first author, that appeared in a paper by Erdős [18]. Call the degree of a pair $e = \{x, y\}$ (both $x$ and $y$ coming from $X$) the number of triples $x, y, z$ with $z \in X$ that are the vertices of an empty triangle, and denote it by $\deg(e)$.

**Open Problem 10.** *Show that the maximal degree of the pairs from $X$ goes to infinity as the size of $X$, $N \to \infty$.*

The lower bound on $g_3$ implies that the average degree is at least $6 + o(1)$ in the following way. Write $T$ for the set of triples from $X$ that are the vertices of an empty triangle. We count the number, $M$, of pairs $(e, t)$ where $t \in T$, $e \subset T$ and $e$ consists of two elements of $X$ in two ways. First $M = \sum \deg(e)$ the sum taken over all two-element subsets of $X$. Secondly, as every triangle has three sides, $M = 3|T| = 3g_3(X) \geq (3 + o(1))N^2$ from the lower bound on $g_3(N)$, showing indeed that the average degree is at least $6 + o(1)$.

We show next that the maximal degree is at least 10 when $N$ is large enough, a small improvement that is still very far from the target. Choose first a vertical line $\ell_1$ having half of the points of $X$ on its left, the other half on its right. (Throw away the leftmost or rightmost point if $N$ is odd.) Then choose a line $\ell_2$, by the ham-sandwich theorem, halving the points on the left and right of $\ell_1$ simultaneously (throwing away, again, one or two points if necessary). We have now four sectors, $S_1, S_2, S_3, S_4$ each containing $m$ points from $X$ with $m = \lfloor N/4 \rfloor$. ($S_1, S_4$ are on the left of $\ell_1$ and $S_1, S_2$ are below $\ell_2$, say.) Let $e = \{x, y\}$ with $x, y \in X$ and define $\deg(e; S_i)$ as the number of points $z \in X \cap S_i$ such that $\{x, y, z\} \in T$. The observation following the lower bounds for $g_n$ gives that, when $e = \{x, y\}$ with $x \in X \cap S_1$ and $y \in X \cap S_2$, then for all but at most $O(m \log m)$ of the possible pairs $\deg(e; S_1 \cup S_2) \geq 2$, so

$$\sum_{x \in S_1} \sum_{y \in S_2} \deg(\{x, y\}; S_1 \cup S_2) \geq (2 + o(1))m^2.$$

On the other hand,

$$\sum_{x \in S_1} \sum_{y \in S_2} \deg(\{x, y\}; S_1 \cup S_2) = \sum_{x, z \in S_1} \deg(\{x, z\}; S_2) + \sum_{y, z \in S_2} \deg(\{y, z\}; S_1).$$

The analogous identities and inequalities for pairs in $S_2 \times S_3$, $S_3 \times S_4$, and $S_4 \times S_1$ together yield that

$$\sum_{i=1}^{4} \sum_{x, y \in S_i} \deg(\{x, y\}; S_{i-1} \cup S_{i+1}) \geq (8 + o(1))m^2,$$

where $i + 1$ and $i - 1$ are to be taken modulo 4. This means that, in at least one of the sectors, the average degree of a pair is at least $4 + o(1)$ in the neighboring two sectors. As we have seen, the average degree of a pair is at least $6 + o(1)$ within each sector. This proves the claim.

## 7    The Modular Version

Bialostocki, Dierker, and Voxman [9] proposed the following elegant "modular" version of the original problem.

**Open Problem 11.** *For any $n \geq 3$ and $p \geq 2$, there exists an integer $B(n, p)$ such that every set of $B(n, p)$ points in general position in the plane determines a convex $n$-gon such that the number of points in its interior is 0 mod $p$.*

Bialostocki et al. proved this conjecture for every $n \geq p + 2$. Their proof goes as follows. Assume, for technical simplicity, that $n = p + 2$. Choose an integer $m$ that is very large compared to $n$. Consider a set $P$ of $f(m)$ points in general position in the plane, by Theorem 1 it contains an $m$-element set $S$ in convex position. Associate with every triple $\{a, b, c\} \subseteq S$ one of the $p$ colors $0, 1, 2, \ldots, p - 1$; namely color $i$ if triangle $abc$ contains $i$ points of $P$ in its interior modulo $p$. As a consequence of Ramsey's theorem we can select an $n$-element subset $S'$ of $S$ all of whose triples are of the same color, given that $m$ is sufficiently large. Consider any triangulation of the convex hull of $S'$, it consists of $p$ triangles. Consequently, the number of points inside this convex $n$-gon is divisible by $p$.

This proof implies a triple exponential upper bound on $B(n, p)$, a bound which was later improved essentially by Caro [12], but his proof also relied heavily on the assumption $n \geq p+2$. Recently the conjecture was proved in [32] for every $n \geq 5p/6 + O(1)$. A key factor in this improvement is Theorem 9.

The situation changes remarkably in higher dimensions. For example, a 3-polytope with 5 vertices admits two essentially different triangulations: one into two simplices and an other into three simplices. Based on this observation Valtr [49] proved the following result.

**Theorem 13.** *For any $n \geq 4$ and $p \geq 2$, there exists an integer $C(n, p)$ such that every set of $C(n, p)$ points in general position in 3-space determines a convex polytope with $n$ vertices such that the number of points in its interior is $0 \bmod p$.*

Indeed, let $P$ be any sufficiently large set of points in general position in 3-space. As in the planar case, we can use the Erdős–Szekeres theorem and then Ramsey's theorem to find at least $n$ and not less than 5 points in convex position such that every tetrahedron determined by these points contains the same number of points, say $i$, in its interior modulo $p$. Consider any 5 of these points and triangulate their convex hull in two different ways: first into two tetrahedra, then into three tetrahedra. It follows that $2i \equiv 3i$, and thus $i \equiv 0 \pmod{p}$.

The same argument can be used to extend Theorem 9, and also its generalization by Valtr, to 3-space:

**Theorem 14.** *Given any natural numbers $k$ and $n \geq 3$, there exists an integer $K_3(k, n)$ such that the following holds. Every set of at least $K_3(k, n)$ points in general position in 3-space, with the property that any tetrahedron determined by these points contains at most $k$ points in its interior, contains an $n$-hole.*

Similar results are proved also in every odd dimension. First we recall the following strengthening of the Erdős–Szekeres theorem, which seems to be folklore. See [13] or [11, Proposition 9.4.7] for a proof.

**Theorem 15.** *Let $d \geq 2$. For every $n \geq d + 1$ there is an integer $N_d(n)$ such that, among any set of $N \geq N_d(n)$ points in general position in $\mathbb{R}^d$ there is the vertex set of a cyclic $d$-polytope with $n$ vertices.*

Note that in the above theorem we cannot replace the cyclic polytopes with any class of polytopes of different combinatorial kind: one may select any number of points on the moment curve yet every $n$-element subset will determine a cyclic polytope.

Next, suppose that $d$ is odd. In general, any cyclic polytope with $d + 2$ vertices admits a triangulation into $(d + 1)/2$ simplices, and also a different one into $(d + 3)/2$ simplices. Thus, Theorems 13 and 14 have counterparts in every odd dimension [50].

These arguments however cannot be extended to even dimensions: it is known [42] that every triangulation of a cyclic $d$-polytope, $d$ even, consists of the same number of simplices.

## 8    Further Problems

Let $h(n, k)$ denote the smallest number such that among at least $h(n, k)$ points in general position in the plane there is always the vertex set of a convex $n$-gon such that the number of points in its interior is at most $k$. Horton's result says that $h(n, 0)$ does not exist for $n \geq 7$. In general, Nyklová [35], based on Horton's construction, established that $h(n, k)$ does not exist for $k \leq c \cdot 2^{n/4}$. She also determined that $h(6, 5) = 19$, yet another step towards the solution of Problem 6.

The following problem was motivated in [30]. For integers $n \geq k \geq 3$, let $g(k, n)$ be the smallest number with the property that among any $g(k, n)$ points in general position in the plane, there exist $n$ points whose convex hull has at least $k$ vertices. Clearly $g(k, n)$ exists and satisfies $f(k) \leq g(k, n) \leq f(n)$. Based on the results of Section 2 one can easily conclude that $g(k, n) < c_1 n + c_2$, where the constants $c_1, c_2$ (dependent only on $k$) are exponentially large in $k$. The true order of magnitude of $g(k, n)$ was found by Károlyi and Tóth [31]. It is not difficult to see that $g(4, n) = \lceil 3n/2 \rceil - 1$. In general the following bounds are known.

**Theorem 16 ([31]).** *For arbitrary integers $n \geq k \geq 3$,*

$$\frac{(k - 1)(n - 1)}{2} + 2^{k/2-4} \leq g(k, n) \leq 2kn + 2^{8k} .$$

To obtain the upper bound, peel off convex layers from a set $P$ of at least $2kn + 2^{8k}$ points as follows. Let $P_1 = P$ and $Q_1$ the vertex set of its convex hull. Having $P_i, Q_i$ already defined, set $P_{i+1} = P_i \setminus Q_i$ and let $Q_{i+1}$ be the set of vertices of the convex hull of $P_{i+1}$. If there is an integer $i \leq 2n$ such that $|Q_i| \geq k$, then we are ready. Otherwise we have $2n$ convex layers $Q_1, Q_2, \ldots, Q_{2n}$, and at least $4^{4k}$ further points inside $Q_{2n}$. Thus, by Theorem 1, $P_{2n+1}$ contains the vertex set of a convex $4k$-gon $C$, and the desired configuration of $n$ points whose convex hull has at least $k$ vertices can be selected from the nested arrangement of the convex sets $Q_1, Q_2, \ldots, Q_{2n}, C$.

**Open Problem 12.** *Is it true that $g(5, n) = 2n - 1$?*

**Open Problem 13.** *Is it true for any fixed value of $k$ that*

$$\lim_{n \to \infty} \frac{g(k, n)}{n} = \frac{k - 1}{2} ?$$

An *interior* point of a finite point set is any point of the set that is not on the boundary of the convex hull of the set. For any integer $k \geq 1$, let $g(k)$ be the smallest number such that every set of points $P$ in general position in the plane, which contains at least $g(k)$ interior points has a subset whose convex hull contains exactly $k$ points of $P$ in its interior. Avis, Hosono, and Urabe [2] determined that $g(1) = 1$, $g(2) = 4$ and $g(3) \geq 8$. It is not known if $g(k)$ exists for $k \geq 3$. It was pointed out by Pach (see [2]) that if $P$ contains at least $k$ interior points, then it has a subset such that the number of interior points of $P$ inside its convex hull is between $k$ and $\lfloor 3k/2 \rfloor$. A similar problem was studied also in [3].

**Open Problem 14.** *Prove or disprove that every point set in general position in the plane with sufficiently many interior points contains a subset in convex position with exactly 3 interior points.*

A first step towards the solution may be the following result of Hosono, Károlyi, and Urabe [25]. Let $g_\triangle(k)$ be the smallest number such that every set of points $P$ in general position in the plane whose convex hull is a triangle which contains at least $g(k)$ interior points also has a subset whose convex hull contains exactly $k$ points of $P$ in its interior.

**Theorem 17.** *If $g_\triangle(k)$ is finite then so is $g(k)$.*

The proof is based on a result of Valtr [50] which extends Theorem 9.

**Note Added in Proof.** The answer to Open Problem 2 is yes and the proof is quite simple. Open Problem 5 was solved very recently by Pór and Valtr: the answer is again yes, but the proof is not that simple.

# References

1. C. Ahrens, G. Gordon, and E.W. McMahon, Convexity and the beta invariant, *Discr. Comp. Geom.* **22** (1999), 411–424.
2. D. Avis, K. Hosono, and M. Urabe, On the existence of a point subset with a specified number of interior points, to appear in *Discr. Math.*
3. D. Avis, K. Hosono, and M. Urabe, On the existence of a point subset with 4 or 5 interior points, in: Discrete and Computational Geometry (J. Akiyama, M. Kano, and M. Urabe, eds.), *Lecture Notes in Comp. Sci.* **1763**, Springer (2000), pp. 56–64.

4. D. Avis and D. Rappaport, Computing the largest empty convex subset of a set of points, *Proc. 1st ACM Symp. Comp. Geom.*, Baltimore (1985), pp. 161–167.

5. I. Bárány and Z. Füredi, Empty simplices in Euclidean spaces, *Canad. Math. Bull.* **30** (1987), 436–445.

6. I. Bárány, Z. Füredi, and L. Lovász, On the number of halving planes, *Combinatorica* **10** (1990), 175–183.

7. I. Bárány and P. Valtr, A positive fraction Erdős–Szekeres theorem, *Discr. Comp. Geom.* **19** (1998), 335–342.

8. I. Bárány and P. Valtr, Planar point sets with a small number of empty convex polygons, *manuscript* (1999).

9. A. Bialostocki, P. Dierker, and B. Voxman, Some notes on the Erdős–Szekeres theorem, *Discr. Math.* **91** (1991), 231–238.

10. T. Bisztriczky and G. Fejes Tóth, On general Erdős–Szekeres numbers, *manuscript* (1994).

11. A. Björner, M. Las Vergnas, B. Sturmfels, N. White, and G.M. Ziegler, Oriented Matroids, *Encyclopedia Math. Appl.* **46**, Cambridge University Press (1993).

12. Y. Caro, On the generalized Erdős–Szekeres conjecture – A new upper bound, *Discr. Math.* **160** (1996), 229–233.

13. R. Cordovil and P. Duchet, Cyclic polytpes and oriented matroids, *European J. Comb.* **21** (2000) 49–64.

14. L. Danzer, B. Grünbaum, and V. Klee, Helly's theorem and its relatives, *Proc. Symp. Pure Math.* Vol 7, AMS Providence RI (1963), pp. 101–138.

15. D.P. Dobkin, H. Edelsbrunner, and M. Overmars, Searching for empty convex polygons, *Algorithmica* **5** (1990), 561–571.

16. D. Dumitrescu, Planar point sets with few empty polygons, *Studia Sci. Math. Hungar.* **36** (2000), 93–107.

17. P. Edelman and V. Reiner, Counting the interior points of a point configuration, *Discr. Comp. Geom.* **23** (2000), 1–14.

18. P. Erdős, On some unsolved problems in elementary geometry (in Hungarian), *Mat. Lapok* **2** (1992), 1–10.

19. P. Erdős and G. Szekeres, A combinatorial problem in geometry, *Comp. Math.* **2** (1935), 463–470.

20. P. Erdős and G. Szekeres, On some extremum problems in elementary geometry, *Ann. Univ. Sci. Budapest. Eötvös, Sect. Math.* **3/4** (1960–61), 53–62.

21. P. Erdős, Zs. Tuza, and P. Valtr, Ramsey-remainder, *European J. Comb.* **17** (1996) 519–532.

22. Z. Füredi, Private communication (1989).

23. H. Harborth, Konvexe Fünfecke in ebenen Punktmengen, *Elem. Math.* **33** (1978), 116–118.

24. J.D. Horton, Sets with no empty 7-gons, *Canad. Math. Bull.* **26** (1983), 482–484.

25. K. Hosono, Gy. Károlyi, and M. Urabe, On the existence of a convex polygon with a specified number of interior points, submitted for publication in *Comput. Geom. Theory Appl.*

26. K. Hosono, D. Rappaport, and M. Urabe, On convex decomposition of points, submitted to the *Japanese Conf. Discr. Comput. Geom.* (2000).

27. K. Hosono and M. Urabe, On the number of disjoint convex quadrilaterals for a planar point set, submitted for publication in *Comp. Geom. Theory Appl.*

28. G. Kalai, Private communication (1997).

29. M. Katchalski and A. Meir, On empty triangles determined by points in the plane, *Acta Math. Hung.* **51** (1988), 323–328.

30. Gy. Károlyi, Ramsey-remainder for convex sets and the Erdős–Szekeres theorem, to appear in *Discr. Appl. Math.*

31. Gy. Károlyi and G. Tóth, An Erdős–Szekeres type problem in the plane, *Period. Math. Hung.* **39** (1999), 153–159.

32. Gy. Károlyi, J. Pach, and G. Tóth, A modular version of the Erdős–Szekeres theorem, to appear in *Studia Sci. Math. Hungar.*

33. Gy. Károlyi and P. Valtr, Point configurations in $d$-space without large subsets in convex position, submitted for publication in *Discr. Comp. Geom.*

34. W. Morris and V. Soltan, The Erdős–Szekeres problem on points in convex position – a survey, *Bull. Amer. Math. Soc.* **37** (2000), 437–458.

35. H. Nyklová, Almost empty polygons, preprint, *KAM-DIMATIA Series* 2000–498.

36. M. Overmars, B. Scholten, and I. Vincent, Sets without empty convex 6-gons, *Bull. European Assoc. Theor. Comp. Sci.* **37** (1989), 160–168.

37. J. Pach, A Tverberg type result on multicolored simplices, *Comp. Geom. Theory Appl.* **10** (1998), 71–76.

38. J. Pach, The Happy End problem – The beginnings of combinatorial geometry (in Hungarian), *manuscript* (2000).

39. J. Pach and J. Solymosi, Canonical theorems for convex sets, *Discr. Comp. Geom.*, **19** (1998), 427–436.

40. A. Pór, Combinatorial properties of finite point sets (in Hungarian) *Diploma Thesis*, Budapest (1996).

41. A. Pór, A partitioned version of the Erdős–Szekeres theorem, submitted for publication in *Discr. Comp. Geom.*

42. J. Rambau, Triangulations of cyclic polytopes and higher Bruhat orders, *Mathematika* **44** (1997), 162–194.

43. J. Solymosi, Combinatorial problems in finite Ramsey theory (in Hungarian), *Diploma Thesis*, Budapest (1988).

44. G. Tóth and P. Valtr, Note on the Erdős–Szekeres theorem, *Discr. Comp. Geom.* **19** (1998), 457–459.

45. M. Urabe, On a partition into convex polygons, *Discr. Appl. Math.* **64** (1996), 179–191.

46. M. Urabe, Partitioning point sets in space into disjoint convex polytopes, *Comp. Geom. Theory Appl.* **13** (1999), 173–178.

47. P. Valtr, Sets in $\mathbb{R}^d$ with no large empty convex subsets, *Discr. Math.* **108** (1992), 115–124.

48. P. Valtr, On the minimum number of empty polygons in planar point sets, *Studia Sci. Math. Hungar.* **30** (1995), 155–163.

49. P. Valtr, Private communication (2000).

50. P. Valtr, A sufficient condition for the existence of large empty convex polygons, *manuscript* (2000).

# On Finding Maximum-Cardinality Symmetric Subsets

Peter Brass

Free University Berlin, Institute of Computer Science,
Takustrasse 9, D-14195 Berlin, Germany,
brass@inf.fu-berlin.de

**Abstract.** In this paper I study the complexity of the problem of finding a symmetric subset of maximum cardinality among $n$ point in the plane, or in three-dimensional space, and some related problems like the largest repeated or $k$-fold repeated subsets. For the maximum-cardinality symmetric subset problem in the plane I show a connection to the maximum number of isosceles triangles among $n$ points in the plane; if this number is denoted by $I(n)$ this gives an algorithm of complexity

$$O\Big((n^2 + I(n))\log n\Big) = O(n^{2.136+\varepsilon}\log n).$$

## 1 Results

Finding the symmetries of a set of $n$ points, or more general testing two sets for congruence and finding all congruence mappings between them, is an old and well-studied problem [3, 2, 4, 7–9, 11], which is solved satisfactorily in dimensions two and three ($O(n\log n)$) and remains an interesting problem in higher dimensions. There are at least two ways to make the problem more realistic: allowing for errors in the points (Hausdorff-approximate symmetry) and for errors in the sets (large symmetric subsets). The Hausdorff-approximate symmetry recognition, however, is NP-complete [12], whereas the identification of large symmetric parts in the exact model leads to interesting problems, which are related to combinatorial geometry in a way already apparent in several other exact point pattern matching problems [1, 6].

There are several ways to formalize the notion of 'large symmetric parts' contained in a point set. The most obvious is to determine the largest-cardinality subset with a nontrivial symmetry (Figure 1 shows a set, the largest-cardinality symmetric subset, and another symmetric subset). For this problem Eades [8] gave an $O(n^4 \log n)$-algorithm.

Let $I(n)$ denote the maximum number of isosceles triangles that can occur among $n$ points in the plane. Then holds

**Theorem 1.** *The largest-cardinality symmetric subset of a set of $n$ points in the plane can be determined in $O\Big((n^2 + I(n))\log n\Big)$ time.*

A classical and very simple bound is $I(n) = O(n^{2+\frac{1}{3}})$, obtained by counting incidences of points and mid-perpendiculars [15], with a lower bound $I(n) =$

J. Akiyama, M. Kano, and M. Urabe (Eds.): JCDCG 2000, LNCS 2098, pp. 106–112, 2001.

$\Omega(n^2 \log n)$ given by the integer lattice. The upper bound was recently improved to $I(n) = O(n^{2.136+\varepsilon})$ for every positive $\varepsilon$ [16]. This implies

**Corollary 1.** *The largest-cardinality symmetric subset of a set of $n$ points in the plane can be determined in $O(n^{2.136+\varepsilon})$ time, for every positive $\varepsilon$.*

Figure 1

Our algorithm lists as an intermediate result all regular polygons contained in that set. It is remarkable that this can indeed be done in that time, since for each fixed $k$ there are sets of $n$ points containing $c_k n^2$ regular $k$-gons [10, 14].

A different formalization is to ask for the largest subset $Y$ of the given set $X$ that is repeated: there is a nontrivial motion $\varphi$ with $Y \subset X$ and $\varphi(Y) \subset X$; or that is $r$-fold repeated: $Y \subset X$, $\varphi(Y) \subset X, \ldots, \varphi^{r-1}(Y) \subset X$. (Figure 2 shows a set, a 8-fold repeated subset, and a once repeated subset.) This notion captures parts of some bigger symmetric structure, e.g. some finite part of an infinite frieze group symmetry. The special case of equidistant collinear rows of points ($\varphi$ a translation, $Y$ only one point) was also studied previously [5, 13, 17].

**Theorem 2.** *The largest $r$-fold repeated subset of a set of $n$ points in the plane can be determined in $O(n^{3+\frac{1}{6}} \log n)$ for $r = 1$ and $O(n^{2+\frac{1}{3}} \log n)$ for $r \geq 2$.*

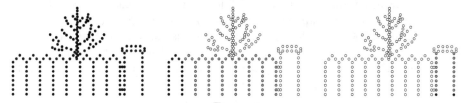

Figure 2

Essentially the same algorithm works for both problems also in three-dimensional space (but not in higher dimensions) where we get a time bound $O(n^3 \log n)$.

## 2   The basic algorithm

In all the above cases symmetries by reflections are simple, and can be enumerated trivially in $O(n^2 \log n)$ time, since we have to look only at the $\binom{n}{2}$ possible pairs of points that can be exchanged by a reflection, and see which reflection line occurs most frequently. So in the following we will only look for rotation symmetries. Also, the algorithms for the different problems are almost the same (with an important difference only in the case of finding one time repeated sets), so we will give only the first, and state the necessary modifications later.

Given a set $X$ of $n$ points in the plane, the algorithm maintains two search tree data structures, one ($\mathcal{T}$) for isosceles triangles in $X$ (point triples $(a, b, c)$ with $d(a, b) = d(b, c)$) and the other ($\mathcal{S}$) for possible symmetry operations (pairs $(p, k)$ of a centerpoint and a rotation order, with the current number of points $\#(p, k)$ in that symmetric subset and their list appended. The algorithm does the following:

1. Determine for each pair $x_1, x_2 \in X$ the distance $d(x_1, x_2)$, and collect the point pairs having the same distance $d$ to get a partition of $\binom{X}{2}$ into distance graphs $G_d$ ($d$ a distance occuring in $X$).
2. For each distance $d$ and each point $x \in X$, take each pair of neighbours $y_1, y_2$ of $x$ in $G_d$, and insert the triple $(y_1, x, y_2)$ in a search tree $\mathcal{T}$.
3. As long as $\mathcal{T}$ is not empty, repeat
   **3.1** Choose any triple $(a, b, c)$ from $\mathcal{T}$, delete it from $\mathcal{T}$.
   **3.2** Determine the rotation $\psi$ that maps $a \mapsto b$, $b \mapsto c$.
   This rotation determines a polygonal arc $p_0 p_1 p_2 \ldots$ with $p_0 = a$, $p_1 = \psi(a) = b$, $p_2 = \psi^2(a) = c$, and generally $p_i = \psi^i(a)$. This may be a regular polygon, an orbit under $\psi$.
   **3.3** Construct the sequence of isosceles triangles $(p_i, p_{i+1}, p_{i+2})$, checking for each of these triangles whether it is contained in $\mathcal{T}$, and deleting it from $\mathcal{T}$, until we either find a triangle that is not contained in $\mathcal{T}$, or arrive at the starting triangle $(p_k, p_{k+1}, p_{k+2}) = (a, b, c)$.
      **3.3.1** In the first case (the polygonal arc remained incomplete), discard $(a, b, c)$ and the polygonal arc and return to step 3.
      **3.3.2** In the second case (the polygonal arc closed to a regular $k$-gon in $X$), determine the center $p$ of this regular polygon.
         **3.3.2.1** Insert $(p, k)$ in $\mathcal{S}$, if it does not already exist, increase $\#(p, k)$ by $k$, and append the regular polygon to the list.
         **3.3.2.2** For each $j \in \{2, \ldots, k\}$ that is coprime to $k$ and each $i = 1, \ldots, k$ delete $(p_i, p_{i+j}, p_{i+2j})$ (all indices mod $k$) from $\mathcal{T}$.
4. Traverse $\mathcal{S}$ and determine the element $(p, i)$ for which $\#(p, i)$ is maximal. Output the list of all regular $i$-gons with center $p$.

## 3    Correctness

To determine the maximum cardinality of a subset that has a nontrivial rotation symmetry, we use that each set with a $k$-fold rotation symmetry is the union of concentric regular $k$-gons, the orbits of the points under the symmetry. Thus we have only to find all regular polygons, collect those polygons which have the same center and the same order, and determine the point which occurs as common center of the largest group.

To find all regular polygons $r_1 \ldots r_k$ contained in the set, we use that any three vertices $r_i, r_{i+a}, r_{i+2a}$ (e.g. three consecutive vertices $r_i, r_{i+1}, r_{i+2}$) form an isosceles triangle, and for each isosceles triangle in our set there is at most one minimal regular polygon containing them in this way. And this polygon can be found by just following the polygonal arc defined by the rotation around the intersection point of the midperpendiculars of that triangle that maps the first leg of the isosceles triangle on the second. In each path-following step we remove the isosceles triangle we just used from the set of all isosceles triangles, so in the end we either find a regular polygon, or we have removed some partial polygonal arc which does not extend to any regular polygon in our set, and which therefore can be removed.

It remains to avoid that we find the same polygon several times, since each isosceles triangle completes to at most one minimal regular polygon containing that triangle, but the same polygon will be obtained with different numberings of vertices from different isosceles triangles. E.g. a regular pentagon $p_1 \ldots p_5$ will be found by following $p_1 p_2 p_3$ and completing that arc, but will again be found by following $p_1 p_3 p_5$ and completing that arc $(p_1 p_3 p_5 p_2 p_4)$. So after we found the regular polygon $p_1 \ldots p_r$ we have to remove all other isosceles triangles $p_i p_{i+a} p_{i+2a}$ which generate the same polygon, which is done in step 3.3.2.2. The same polygon is found exactly for those $a$ which are coprime to the vertex number $k$ of the regular polygon; if $k$ has a nontrivial divisor $\kappa$, then the regular $k$-gon can also be interpreted as union of $\frac{k}{\kappa}$ regular $\kappa$-gons, and will be found and stored in $S$ as set with $\kappa$-fold rotational symmetry again. Thus in the structure $S$ the same set is stored for each symmetry order exactly once, and a simple traversal of $S$ gives the largest subset with a nontrivial rotational symmetry.

## 4    Analysis

The construction of the distance graphs in step 1. can be trivially done in $O(n^2 \log n)$ time. If the distance graphs are given, the construction of all isosceles triangles (step 2) can be done in time $O(n^2 + I \log n)$ where $I$ denotes the number of isosceles triangles that are constructed (trivially $I < n^3$). In each of the following steps 3.* one of the isosceles triangles is removed from $\mathcal{T}$, which takes $O(\log n)$, and some further operation of complexity at most $O(\log n)$ is done. Thus the total complexity of step 3 is $O(I \log n)$. Step 4 finally also takes at most $O(I \log n)$ time, since we touch each regular polygon at most once, and there are at most $O(I)$ regular polygons. Thus the total complexity is $O((n^2 + I) \log n) \leq O\left((n^2 + I(n)) \log n\right)$, as claimed in Theorem 1.

A further speedup (perhaps to $O(n^{2+\varepsilon})$) would be possible if one could avoid inserting all isosceles triangles into $\mathcal{T}$. Only those triangles are really needed that can occur in a regular polygon; so one really needs only those isosceles triangles with an angle of form $(1 - \frac{2}{k})\pi$ at the apex, all others can never lead to a regular polygon. For a fixed vertex $x$, having a given list of $y_1, \ldots, y_m$ neighbours at a fixed distance, it seems probable that there are much less than $\binom{m}{2}$ pairs $y_i, y_j$ which determine an angle of that form ($|\angle y_i x y_j| \in \{\frac{1}{3}\pi, \frac{1}{2}\pi, \frac{3}{5}\pi, \ldots, \frac{k-2}{k}\pi, \ldots\}$). I have a construction of points $y_1, \ldots, y_m$ with $cm \log m$ such pairs, and believe this to be near the correct order. If those pairs could be determined in $O(m^{1+\varepsilon})$, it would allow a speedup of the whole algorithm to $O(n^{2+\varepsilon})$.

The same algorithm works also in three-dimensional space, since the possible symmetries there are also reflections (determined by one point pair, so can be checked in $O(n^2 \log n)$ time) and rotations around a line; so the nontrivial orbits are regular polygons in space. Unfortunately, the bound for the number of isosceles triangles in three-dimensional space is only $O(n^3)$, and that order can be reached (take half of the points on a circle and the other half on the mid-perpendicular of that circle, than any triangle of two points on the circle and one point on the mid-perpendicular is isosceles).

## 5    Variants for repeated sets

If we are looking for repeated sets it does make a big difference whether we are also interested in sets $Y$ which are once repeated ($Y \subset X$ and $\varphi(Y) \subset X$ for a nontrivial motion $\varphi$), or accept only those $Y$ that occur at least $r$ further times, $r \geq 2$. If $Y \subset X$, $\varphi(Y) \subset X$, $\varphi^2(Y) \subset X, \ldots \varphi^r(Y) \subset X$, then for each $y \in Y$ the triangle $y, \varphi(y), \varphi^2(y)$ is isosceles, and we can again just follow the paths determined by the motion $\varphi$, where we determine the motion from the isosceles triangle. There are, however, two important differences:

– *Two motions are possible*
  If we are looking for complete orbits of isometries, then the only possibe isometries are rotations (and reflections, which have only two-point orbits). If we are also interested in pieces of infinite orbits, then additionally translations and glide-reflections become possible. By this the continuation of an isosceles triangle as an orbit is not anymore unique, but can be a regular polygon (by a rotation) or a zig-zag path (by a glide-reflection).
  To overcome this, we have to insert two copies of each isosceles triangle in $\mathcal{T}$, marked as 'rotation' and 'glide-reflection', and remove the right copy when extending a path.
– *The paths do not close*
  As long as we were looking only for complete regular polygons, we found the whole polygon by just going around. If we also take polygonal arcs, we have to make sure that we remove *maximal* polygonal arcs from $\mathcal{T}$. So we have to follow the path generated by $\varphi$ from the starting triangle in both directions, forward ($\varphi$) as well as backward ($\varphi^{-1}$).

In this way we obtain all subsets which are partial orbits of at least three points of some isometry. Each regular polygon found this way should be inserted in $\mathcal{S}$ under the appropriate isometry, and with its full number of points, and all copies of that regular polygon should be deleted as in step 3.3.2.2. If the partial orbit is only a path of length $k$, and we are looking for an $r$-fold repeated subset, then it should be discarded if $k \leq r$, otherwise the first $k - r$ points of the path should be inserted in $\mathcal{S}$ under the appropriate isometry. Then in the end again a simple traversal of $\mathcal{S}$ is sufficient to find the maximum-cardinality $r$-fold repeated subset.

If we are looking for sets $Y \subset X$ which are only once repeated ($Y \subset X$, $\varphi(Y) \subset X$), then the partial orbit of a point consists only of two points, and does not anymore determine the motion. Instead we have to look at the possible images of pairs of points, and count how often which motion $\varphi$ is determined by them, obtaining essentially the same algorithm as Akutsu, Tamaki and Tokuyama [1] for the 'maximum congruent subsets' problem.

For a pair $y_1, y_2 \in X$ the possible image pairs $z_1 = \varphi(y_1), z_2 = \varphi(y_2)$ are the pairs that have the same distance, so we construct the distance graphs of $X$, take each pair of edges $(y_1, y_2), (z_1, z_2)$ of the same length, determine the motion that maps the first pair on the second, and increase the count of this motion and its reflected counterpart in a search structure $S$ for isometries. Then the nontrivial isometry with the largest count gives the maximum-cardinality subset $Y \subset X$ that is once repeated.

The analysis of that algorithm is the same as in [1], giving an $O(n^{3.2} \log n)$ complexity.

# References

1. Akutsu, T., Tamaki, H., Tokuyama, T.: Distribution of distances and triangles in a point set and algorithms for computing the largest common point set. Discrete Comput. Geom. **20** (1998) 307–331
2. Alt, H., Mehlhorn, K., Wagener, H., Welzl, E.: Congruence, similarity, and symmetries of geometric objects. Discrete Comput. Geom. **3** (1988) 237–256
3. Alt, H., Guibas, L.: Resemblance of geometric objects. In J.-R. Sack, J. Urrutia (Eds.): Handbook of Computational Geometry Elsevier 1999, 121–153
4. Atallah, M.J.: On symmetry detection. IEEE Trans. Comput. **34** (1985) 663–666
5. Boxer, L., Miller, R.: Parallel algorithms for all maximal equally-spaced collinear sets and all maximal regular coplanar lattices. Pattern Recog. Letters **14** (1993) 14–20
6. Braß, P.: Exact point pattern matching and the number of congruent triangles. Algorithms — ESA 2000 (Mike Paterson, Ed.) LNCS 1879, 112–119
7. Braß, P., Knauer, C.: Testing the congruence of $d$-dimensional point sets. ACM Symposium on Comput. Geom. 2000, 310–314
8. Eades, P.: Symmetry finding algorithms. In: Computational Morphology. A computational geometric approach to the analysis of form. (Toussaint, Godfried T., Ed.) Machine Intelligence and Pattern Recognition Series 6 (1988) North-Holland, 41–51
9. Eades, P., Ng, H.C.: An algorithm for detecting symmetries in drawings. Ars Comb. **23A** (1987) 95–104
10. Elekes, G., Erdős, P.: Similar configurations and pseudo grids. In: Intuitive geometry 1991 (K. Böröczky et al, Eds.) North-Holland, Colloq. Math. Soc. Janos Bolyai. **63** (1994) 85–104
11. Highnam, P.T.: Optimal algorithms for finding the symmetries of a planar point set. Inf. Proc. Letters **22** (1986) 219–222
12. Iwanowski, S.: Testing approximate symmetry in the plane is NP-hard. Theor. Comput. Sci. **80** (1991) 227–262
13. Kahng, A.B., Robins, G.: Optimal algorithms for extracting spatial regularity in images. Inf. Proc. Letters **12** (1991) 757–764

14. Laczkovich, M., Ruzsa, I.Z.: The number of homothetic subsets. In: The mathematics of Paul Erdős. Vol. II. (R.L. Graham, et al, Eds.) Springer, Algorithms Comb. 14 (1997) 294–302
15. Pach, J., Agarwal, P.K.: Combinatorial Geometry. Wiley, New York 1995
16. Pach, J., Tardos, G.: personal communication
17. Robins, G., Robinson, B.L., Sethi, B.S.: On detecting spatial regularity in noisy images. Inf. Proc. Letters **69** (1999) 189–195

# Folding and Unfolding
# Linkages, Paper, and Polyhedra

Erik D. Demaine

Department of Computer Science, University of Waterloo,
Waterloo, Ontario N2L 3G1, Canada, eddemaine@uwaterloo.ca

## 1 Introduction

Folding and unfolding problems have been implicit since Albrecht Dürer in the early 1500's [Dür77], but have not been studied extensively until recently. Over the past few years, there has been a surge of interest in these problems in discrete and computational geometry. This paper gives a brief survey of some of the recent work in this area, subdivided into three sections based on the type of object being folded: linkages, paper, or polyhedra. See also [O'R98] for a related survey from this conference two years ago.

In general, we are interested in how objects (such as linkages, paper, and polyhedra) can be moved or reconfigured (folded) subject to certain constraints depending on the type of object and the problem of interest. Typically the process of *unfolding* approaches a more basic shape, whereas *folding* complicates the shape. We can also generally define the *configuration space* as the set of all configurations or states of the object, with paths in the space corresponding to motions (foldings) of the object.

## 2 Linkages

A *linkage* or *framework* consists of a collection of rigid line segments (*bars*) joined at their endpoints (*vertices*) to form a particular graph. A linkage can be *folded* by moving the vertices around in $\mathbb{R}^d$ in any way that preserves the length of each bar. Such linkages have been studied extensively in the case that bars are allowed to cross; see, for example, [KM95,LW95,Sal73,Whi92]. Recently there has been much work on the case that the linkage must remain *simple*, never crossing any two bars.[1] This additional constraint is the type of linkage folding considered in this section. Such linkage folding has applications in hydraulic tube bending [O'R98] and motion planning of robot arms. There are also connections to protein folding in molecular biology. See [CDR00,O'R98,Tou99a] for other surveys on this area.

Perhaps the most fundamental question we can ask about folding linkages is whether it is possible to fold between any two simple configurations of the

---

[1] Typically, bars are allowed to touch, provided they do not properly intersect. However, requiring bars to touch only at common endpoints does not change the results.

same linkage (with matching graphs, combinatorial embeddings, and bar lengths) while preserving the bar lengths and not crossing any bars during the folding. Because folding motions can be reversed and concatenated, this fundamental question is equivalent to whether every simple configuration can be folded into some *canonical configuration*.

In this context, three general types of linkages are commonly studied, characterized by the structure of their associated graphs: a *polygonal arc* or *open polygonal chain* (a single path); a *polygonal cycle, polygon,* or *closed polygonal chain* (a single cycle); and a *polygonal tree* (a single tree). The canonical configuration of an arc is the *straight configuration*, all vertex angles equal to 180°. A canonical configuration of a cycle is a *convex configuration*, planar and having all interior vertex angles less than or equal to 180°. It is relatively easy to show that convex configurations are indeed "canonical" in the sense that any one can be folded into any other; this result was first proved in the literature in [ADE+01]. Finally, a canonical configuration of a tree is a *flat configuration*: all vertices lie on a horizontal line, and all bars point "rightward" from a common root. Again it is easy to fold any flat configuration into any other.

The fundamental questions thus become whether every arc can be straightened, every cycle can be convexified, and every tree can be flattened. The answers to these questions depend on the dimension of the space. Over the past few years, this collection of questions has been completely resolved. A summary is shown in Table 1. In the remainder of this section, we describe the historical progress of these results, and describe other results on linkage folding not captured by this categorization.

| Dimension | Can all arcs be straightened? | | Can all cycles be convexified? | | Can all trees be flattened? |
|---|---|---|---|---|---|
| 2 | Yes | [CDR00] | Yes | [CDR00] | No [BDD+01] |
| 3 | No [CJ98,BDD+99] | | No [CJ98,BDD+99] | | No |
| 4 & above | Yes | [CO99] | Yes | [CO99] | Yes |

**Table 1.** Answers to main linkage-folding problems.

The questions of whether every polygonal arc can be straightened in the plane and whether every polygon can be convexified in the plane have arisen in many contexts over the last quarter-century. In particular, they were posed independently by Stephen Schanuel and George Bergman in the early 1970's, Ulf Grenander in 1987, William Lenhart and Sue Whitesides in 1991, and Joseph Mitchell in 1992. In the discrete and computational geometry community, the arc-straightening problem has become known as the *carpenter's rule problem* because a carpenter's rule folds like a polygonal arc.

Many people devoted time to these two problems over the past 10 years. It was widely conjectured, particularly by those unfamiliar with the problem, that the answers were yes. On the other hand, several people proposed examples of

**Fig. 1.** Two views of convexifying a "doubled tree." The top snapshots are all scaled the same, and the bottom snapshots are scaled differently to improve visibility. See `http://daisy.uwaterloo.ca/~eddemain/linkage/` for more animations.

polygonal arcs and cycles that might be "locked" (unstraightenable and unconvexifiable), but eventually every proposed example was unlocked by hand. It was not until early in the year 2000 that the problems were solved in the positive by Connelly, Demaine, and Rote [CDR00]. See [CDR00] for a more detailed history.

More generally, the result in [CDR00] shows that a collection of nonintersecting polygonal arcs and cycles in the plane may be simultaneously folded so that the outermost arcs are straightened and the outermost cycles are convexified. The "outermost" proviso is necessary because arcs and cycles cannot always be straightened and convexified when they are contained in other cycles. The key idea for the solution, introduced by Günter Rote, is to look for *expansive* motions in which no vertex-to-vertex distance decreases. Expansive motions automatically preserve simplicity, so the difficult noncrossing aspect of the problem can be ignored by guaranteeing expansiveness. This idea allowed applying theorems in rigidity theory and tensegrity theory to show that, *infinitesimally*, arcs and cycles can be folded expansively. These infinitesimal motions are combined by flowing along a vector field implicitly defined by an optimization problem. As a result, the motion is piecewise-differentiable, and the configuration space of arcs and cycles is contractible. In addition, any symmetries present in the initial configuration of the linkage are preserved throughout the motion. Similar techniques show that the area of each cycle increases by this motion and furthermore by any expansive motion [CDR00].

Ileana Streinu [Str00] has demonstrated another motion for straightening arcs and convexifying polygons that is piecewise-algebraic, made up of $O(n^2)$ mechanisms each with one degree of freedom. As a result, the motion is possible to compute in principle. On the other hand, an approximation to the motion in [CDR00] is easy to implement, and has resulted in animations such as the one in Fig. 1.

The pursuit of the arc-and-cycle problems in 2D inspired research on several related problems. For example, it was shown that starshaped polygons [ELR+98] and monotone polygons [BDL+99] can be convexified by particularly simple motions. Biedl et al. [BDD+01] showed that a positive answer to the arc-and-cycle problem could not be generalized to flattening trees. Recently, Connelly, Demaine, and Rote have shown that even a tree with one degree-3 vertex and the remaining vertices degree-2 can be locked (manuscript in preparation, October 2000), so the result in [CDR00] is tight.

Linkage folding in 3D was initiated earlier, by Paul Erdős in 1935 [Erd35]. He asked whether a particular "flipping" algorithm for folding a planar polygon through three dimensions (preserving edge lengths and simplicity) would convexify the polygon in a finite number of steps. With a slight modification, this question was answered positively by Nagy [Nag39]. This problem and result have been rediscovered several times; see [Tou99b,Grü95] for the history. Unfortunately, Erdős's algorithm (or more precisely, Nagy's modification) can require arbitrarily many moves, even for a quadrangle [Grü95,Tou99b,BDD+99]. Recently, algorithms that convexify planar polygons through 3D in a linear number of "simple moves" have been developed [BDD+99,AGP99]. More generally, if a polygonal arc or cycle in 3D has a simple orthogonal projection, then it can be straightened or convexified [CKM+01]; interestingly, this result is based on the 2D result [CDR00]. But if we start with a general polygonal arc or an unknotted polygon in 3D, it is not always possible to straighten or convexify it [CJ98,Tou01,BDD+99]; see Fig. 2 for an example of a locked arc in 3D. Other problems related to Erdős flips include flipturns, described elsewhere in this proceedings [ACD+00,ABC+00,Grü95], and deflations [FHM+01,Grü95].

**Fig. 2.** A locked polygonal arc in 3D with 5 bars [CJ98,BDD+99].

Finally, analogous to the nonexistence of knots in dimensions higher than 3, polygonal arcs can be straightened and polygonal cycles can be convexified in 4D and higher dimensions [CO99]. Intuitively, this result holds because the number of degrees of freedom of any vertex is much higher than the dimensionality of the obstacles imposed by any bar. It would be interesting to explore scaling the dimension of the object to be folded together with the dimension of the space in which it is folded. For example, how can solid polygons connected at their edges be folded in dimensions higher than 2?

# 3 Paper

Paper folding (origami) has lead to several interesting mathematical and computational questions over the past fifteen years or so. A piece of paper, normally a (solid) polygon such as a square or rectangle, can be folded by any continuous motion that preserves the distances on the surface and does not properly self-intersect. Informally, paper cannot tear, stretch, or cross itself, but may otherwise bend freely. Formally, a folding is a continuum of isometric embeddings of the piece of paper in $\mathbb{R}^3$. However, the use of the term "embedding" is weak: paper is permitted to touch itself provided it does not properly cross itself. In particular, a *flat folding* folds the piece of paper back into the plane, and so the paper must necessarily touch itself. We frequently identify the continuous motion of a folding with the final folded state of the paper; in the case of a flat folding, the flat folded state is called a *flat origami*.

Some of the pioneering work in origami mathematics [Hul94,Jus94,Kaw89] studies the *crease pattern* that results from unfolding a flat origami, that is, the graph of edges on the paper that fold to edges of a flat origami. Stated in reverse, what crease patterns have flat foldings? Necessary and sufficient conditions are known [Hul94,Jus94,Kaw89], but there is little hope for a polynomial characterization: Bern and Hayes [BH96] have shown that this decision problem is NP-hard. The key difficulty is the non-self-intersection property, more precisely, in finding an overlap order of faces that avoids self-intersection in the folded state. If such nonlocal interactions are ignored, the existence of a flat origami can be tested in linear time [BH96].

A more recent trend, as in [BH96], is to explore *computational origami*, the algorithmic aspects of paper folding. This aspect was pioneered by Lang [Lan96], who has shown how to design a wide class of origami "bases" from which real origami models are folded. In the past two years, computational geometry techniques have been applied to computational origami; we briefly survey these results in the remainder of this section. See also [DD01].

One result involves the *fold-and-cut problem*: given a sheet of paper, fold it flat, make one complete straight cut, and unfold the pieces. What shapes can be achieved? Surprisingly, we can arrange the folds and the cut in order to make any desired plane graph (planar graph embedded with straight edges). See Fig. 3 for some examples. Two solutions to this problem have been developed. Demaine, Demaine, and Lubiw [DDL98] presented a solution based on the straight skeleton at this conference two years ago. Bern, Demaine, Eppstein, and Hayes [BDEH98] developed a different solution based on disk packing.

Another surprising "universality" result in paper folding is about folding silhouettes and wrapping polyhedra. Given a polygon in the plane, possibly with holes, can we fold a sufficiently large piece of paper into that silhouette? This question is implicit throughout origami design, and was first formally stated in [BH96]. If the paper has a different color on each side, and the polygon is partitioned into differently colored regions (as in Fig. 4), can we fold the paper into that shape with the appropriate colors showing at the appropriate regions? More generally, if the desired shape is not a flat silhouette but a general con-

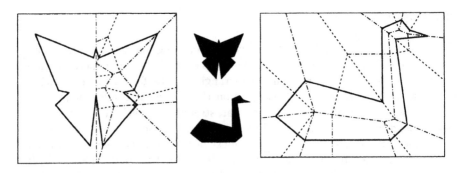

**Fig. 3.** Crease patterns for folding a rectangle of paper flat so that one complete straight cut makes a butterfly (left) or a swan (right). Valley creases are drawn with dotted lines, and mountain creases are drawn dash-dotted.

nected union of polygons in 3-space (a "polyhedron"), can such a package always be tightly wrapped by a sufficiently large piece of paper, possibly matching a 2-color pattern? Demaine, Demaine, and Mitchell [DDM00] have shown that the answers to all of these questions are yes, and describe three algorithms for solving these problems. Several problems concerning the efficiency of the foldings remain open.

**Fig. 4.** A flat folding of a square of paper, black on one side and white on the other side, designed by John Montroll [Mon91, pp. 94–103].

Returning to the problem of recognizing flat-foldable crease patterns, an interesting special case is *map folding*. More precisely, a *map* is a rectangle with horizontal and vertical creases, each marked either mountain or valley. While map folding is normally only studied from the combinatorial perspective [Gar83, Lun71], Jack Edmonds (personal communication, August 1997) posed two attractive decision questions: (1) does a given map have a flat folded state, and (2) can a given map be folded flat by a sequence of simple folds (each folding along one line)? The complexity of the first problem remains open; an NP-hardness result would be an interesting strengthening of [BH96]. Recently, Arkin, Bender, Demaine, Demaine, Mitchell, Sethia, and Skiena [ABD+00] resolved the complexity of the second problem. The exact results depend on the model of simple folds: if the paper can be folded one layer at a time, then foldability can be decided in linear time; if all layers must be folded at once (a more restrictive model), then foldability can be decided in near-linear time, e.g., $O(n \log n)$. Surprisingly, however, map folding is on the border of computational intractability: the same question with folds allowed at 45 degrees, or with a nonrectangular piece of paper, is (weakly) NP-complete [ABD+00].

## 4 Polyhedra

Unlike the other problems, there are several different models of folding that arise in the context of polyhedra.

A classic open problem is whether (the surface of) every convex polyhedron can be cut along some of its edges and unfolded into the plane without overlap [She75,O'R98]. Such unfoldings go back to Dürer [Dür77], and have important practical applications in manufacturing, such as sheet-metal bending. It is widely conjectured that the answer to this question is yes, but all attempts at a solution have so far failed. Experiments by Schevon [Sch89,O'R98] suggest that a random unfolding of a random polytope overlaps with probability 1, but this does not preclude the existence of at least one nonoverlapping unfolding for all polyhedra.

Instead of answering this difficult question directly, we can examine to what extent it can be generalized. In particular, define a polyhedron to be *topologically convex* if its 1-skeleton (graph) is the 1-skeleton of a convex polyhedron. Does every topologically convex polyhedron have such an edge unfolding? Bern, Demaine, Eppstein, Kuo, Mantler, and Snoeyink [BDE+01] have shown that the answer is no: there is a polyhedron homeomorphic to a sphere and with every face a triangle that has no (one-piece, nonoverlapping) edge unfolding. It is shown in Fig. 5. The complexity of deciding whether a given topologically convex polyhedron can be edge-unfolded remains open.

Another intriguing open problem in this area is whether every polyhedron homeomorphic to a sphere has *some* one-piece unfolding, not necessarily using cuts along edges. It is known that every convex polyhedron has an unfolding in this model, allowing cuts across the faces of the polytope [AO92,MMP87]. But many nonconvex polyhedra also have such unfoldings. For example, Fig. 5 illustrates one for the polyhedron described above. Biedl, Demaine, Demaine, Lubiw, Overmars, O'Rourke, Robbins, and Whitesides [BDD+98] have shown how to unfold many orthogonal polyhedra, even with holes and knotted topology, although it remains open whether all orthogonal polyhedra can be unfolded. The only known scenario that prevents unfolding altogether [BDE+01] is a polyhedron with a single vertex of negative curvature (more than 360° of material), but this requires the polyhedron to have boundary, edges incident to only one face.

In addition to unfolding polyhedra into simple planar polygons, we can consider the reverse problem of folding polygons into polyhedra. Lubiw and O'Rourke [LO96] have shown how to test in polynomial time whether a polygon

**Fig. 5.** (Left) Simplicial polyhedron with no edge unfolding. (Right) An unfolding when cuts are allowed across faces.

has an edge-to-edge gluing that can be folded into a convex polyhedron, and how to list all such edge-to-edge gluings in exponential time. The exponential time is necessary because some examples have that many gluings, as described elsewhere in this proceedings [DDLO00a,DDLO00b]. This work shows several other enumerative and structural results about foldings and unfoldings. We are also working on efficient algorithms for detecting the existence of and enumerating non-edge-to-edge gluings, generalizing [LO96]. An intriguing open problem remains relatively unexplored: a theorem of Aleksandrov implies that any gluing found can be folded into a unique convex polyhedron, but how efficiently can this polyhedron be constructed?

A different kind of polyhedron folding comes from extending the fold-and-cut problem from the previous section to one higher dimension. Given any polyhedral complex, can $\mathbb{R}^3$ be folded (through $\mathbb{R}^4$) "flat" into $\mathbb{R}^3$ so that the surface of the polyhedral complex maps to a common plane, and nothing else maps to that plane? While the applicability of four dimensions is difficult to imagine, the problem's restriction to the surface of the complex is quite practical, e.g. in packing: *flatten* the surface of a polyhedron into a flat folded state, without cutting or stretching the paper. Demaine, Demaine, and Lubiw [DDL00] have shown that convex polyhedra and orthogonal polyhedra can be flattened, among other classes. An example is shown in Fig. 6. We conjecture further that every polyhedral complex can be flattened.

## 5  Conclusion

The area of folding and unfolding offers many beautiful mathematical and computational problems. Much progress has been made recently in the many problems outlined above, and many more important problems remain open. For ex-

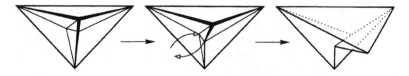

**Fig. 6.** Flattening a tetrahedron, from left to right. Note that the faces are not flat in the middle picture.

ample, most aspects of unfolding polyhedra remain unsolved, including the original problem in the area, edge-unfolding convex polyhedra. A variety of results suggest that paper folding possesses a vast power, but what is known is certainly not the whole story of what is possible. And while the described class of linkage problems has been resolved, there are several other aspects that remain unstudied. For example, protein folding is a domain of great practical importance in biology that should be the source of many interesting geometric problems, with connections to linkages. But even more exciting are the avenues of folding and unfolding that have not yet been explored or even conceived.

## Acknowledgments

I thank my co-authors with whom I have enjoyed collaborating in the area of folding and unfolding.

## References

[ABC+00]  H.-K. Ahn, P. Bose, J. Czyzowicz, N. Hanusse, E. Kranakis, and P. Morin. Flipping your lid. *Geombinatorics*, 10(2):57–63, 2000.

[ABD+00]  E. M. Arkin, M. A. Bender, E. D. Demaine, M. L. Demaine, J. S. B. Mitchell, S. Sethia, and S. S. Skiena. When can you fold a map? Computing Research Repository cs.CG/0011026, Nov. 2000. http://www.arXiv.org/abs/cs.CG/0011026.

[ACD+00]  O. Aichholzer, C. Cortés, E. D. Demaine, V. Dujmović, J. Erickson, H. Meijer, M. Overmars, B. Palop, S. Ramaswami, and G. T. Toussaint. Flip-turning polygons. In *Proc. Japan Conf. Discrete Comput. Geom.*, Lecture Notes in Comput. Sci., Tokyo, Japan, Nov. 2000. To appear in *Discrete and Computational Geometry*.

[ADE+01]  O. Aichholzer, E. D. Demaine, J. Erickson, F. Hurtado, M. Overmars, M. A. Soss, and G. T. Toussaint. Reconfiguring convex polygons. *Comput. Geom. Theory Appl.*, 2001. To appear.

[AGP99]  B. Aronov, J. E. Goodman, and R. Pollack. Convexification of planar polygons in $\mathbb{R}^3$. Manuscript, Oct. 1999. http://www.math.nyu.edu/faculty/pollack/convexifyingapolygon10-27-99.ps.

[AO92]  B. Aronov and J. O'Rourke. Nonoverlap of the star unfolding. *Discrete Comput. Geom.*, 8(3):219–250, 1992.

[BDD+98] T. Biedl, E. Demaine, M. Demaine, A. Lubiw, M. Overmars, J. O'Rourke, S. Robbins, and S. Whitesides. Unfolding some classes of orthogonal polyhedra. In *Proc. 10th Canadian Conf. Comput. Geom.*, Montréal, Canada, Aug. 1998. http://cgm.cs.mcgill.ca/cccg98/proceedings/cccg98-biedl-unfolding.ps.gz.

[BDD+99] T. Biedl, E. Demaine, M. Demaine, S. Lazard, A. Lubiw, J. O'Rourke, M. Overmars, S. Robbins, I. Streinu, G. Toussaint, and S. Whitesides. Locked and unlocked polygonal chains in 3D. Technical Report 060, Smith College, 1999. A preliminary version appeared in the *Proc. 10th ACM-SIAM Sympos. Discrete Algorithms*, Baltimore, Maryland, Jan. 1999, pages 866–867.

[BDD+01] T. Biedl, E. Demaine, M. Demaine, S. Lazard, A. Lubiw, J. O'Rourke, S. Robbins, I. Streinu, G. Toussaint, and S. Whitesides. A note on reconfiguring tree linkages: Trees can lock. *Discrete Appl. Math.*, 2001. To appear.

[BDE+01] M. Bern, E. D. Demaine, D. Eppstein, E. Kuo, A. Mantler, and J. Snoeyink. Ununfoldable polyhedra with convex faces. *Comput. Geom. Theory Appl.*, 2001. To appear.

[BDEH98] M. Bern, E. Demaine, D. Eppstein, and B. Hayes. A disk-packing algorithm for an origami magic trick. In *Proc. Internat. Conf. Fun with Algorithms*, Isola d'Elba, Italy, June 1998.

[BDL+99] T. C. Biedl, E. D. Demaine, S. Lazard, S. M. Robbins, and M. A. Soss. Convexifying monotone polygons. In *Proc. Internat. Symp. Algorithms and Computation*, volume 1741 of *Lecture Notes in Comput. Sci.*, pages 415–424, Chennai, India, Dec. 1999.

[BH96] M. Bern and B. Hayes. The complexity of flat origami. In *Proc. 7th ACM-SIAM Sympos. Discrete Algorithms*, pages 175–183, Atlanta, Jan. 1996.

[CDR00] R. Connelly, E. D. Demaine, and G. Rote. Straightening polygonal arcs and convexifying polygonal cycles. In *Proc. 41st IEEE Sympos. Found. Comp. Sci.*, pages 432–442, Redondo Beach, California, Nov. 2000.

[CJ98] J. Cantarella and H. Johnston. Nontrivial embeddings of polygonal intervals and unknots in 3-space. *J. Knot Theory Ramifications*, 7(8):1027–1039, 1998.

[CKM+01] J. A. Calvo, D. Krizanc, P. Morin, M. Soss, and G. Toussaint. Convexifying polygons with simple projections. *Infor. Process. Lett.*, 2001. To appear.

[CO99] R. Cocan and J. O'Rourke. Polygonal chains cannot lock in 4D. In *Proc. 11th Canadian Conf. Comput. Geom.*, Vancouver, Canada, Aug. 1999. http://www.cs.ubc.ca/conferences/CCCG/elec_proc/c17.ps.gz.

[DD01] E. D. Demaine and M. L. Demaine. Recent results in computational origami. In *Proc. 3rd Internat. Meeting of Origami Science, Math, and Education*, Monterey, California, March 2001. To appear.

[DDL98] E. D. Demaine, M. L. Demaine, and A. Lubiw. Folding and cutting paper. In J. Akiyama, M. Kano, and M. Urabe, editors, *Revised Papers from the Japan Conf. Discrete Comput. Geom.*, volume 1763 of *Lecture Notes in Comput. Sci.*, pages 104–117, Tokyo, Japan, Dec. 1998.

[DDL00] E. D. Demaine, M. L. Demaine, and A. Lubiw. Flattening polyhedra. Manuscript, 2000.

[DDLO00a] E. Demaine, M. Demaine, A. Lubiw, and J. O'Rourke. Examples, counterexamples, and enumeration results for foldings and unfoldings between polygons and polytopes. Technical Report 069, Smith College, Northampton, MA, July 2000.

[DDLO00b] E. D. Demaine, M. L. Demaine, A. Lubiw, and J. O'Rourke. Enumerating foldings and unfoldings between polygons and polytopes. In *Proc. Japan Conf. Discrete Comput. Geom.*, Lecture Notes in Comput. Sci., Tokyo, Japan, Nov. 2000.

[DDM00] E. D. Demaine, M. L. Demaine, and J. S. B. Mitchell. Folding flat silhouettes and wrapping polyhedral packages: New results in computational origami. *Comput. Geom. Theory Appl.*, 16(1):3–21, 2000.

[Dür77] A. Dürer. *The Painter's Manual: A Manual of Measurement of Lines, Areas, and Solids by Means of Compass and Ruler Assembled by Albrecht Dürer for the Use of All Lovers of Art with Appropriate Illustrations Arranged to be Printed in the Year MDXXV*. Abaris Books, Inc., New York, 1977. English translation of *Unterweysung der Messung mit dem Zirkel un Richtscheyt in Linien Ebnen uhnd Gantzen Corporen*, 1525.

[ELR⁺98] H. Everett, S. Lazard, S. Robbins, H. Schröder, and S. Whitesides. Convexifying star-shaped polygons. In *Proc. 10th Canadian Conf. Comput. Geom.*, Montréal, Canada, Aug. 1998. `http://cgm.cs.mcgill.ca/cccg98/proceedings/cccg98-everett-convexifying.ps.gz`.

[Erd35] P. Erdős. Problem 3763. *Amer. Math. Monthly*, 42:627, 1935.

[FHM⁺01] T. Fevens, A. Hernandez, A. Mesa, M. Soss, and G. Toussaint. Simple polygons that cannot be deflated. *Beiträge Algebra Geom.*, 2001. To appear.

[Gar83] M. Gardner. The combinatorics of paper folding. In *Wheels, Life and Other Mathematical Amusements*, chapter 7, pages 60–73. W. H. Freeman and Company, 1983.

[Grü95] B. Grünbaum. How to convexify a polygon. *Geombinatorics*, 5:24–30, July 1995.

[Hul94] T. Hull. On the mathematics of flat origamis. *Congr. Numer.*, 100:215–224, 1994.

[Jus94] J. Justin. Towards a mathematical theory of origami. In K. Miura, editor, *Proc. 2nd Internat. Meeting of Origami Science and Scientific Origami*, pages 15–29, Otsu, Japan, November–December 1994.

[Kaw89] T. Kawasaki. On the relation between mountain-creases and valley-creases of a flat origami. In H. Huzita, editor, *Proc. 1st Internat. Meeting of Origami Science and Technology*, pages 229–237, Ferrara, Italy, Dec. 1989. An unabridged Japanese version appeared in *Sasebo College of Technology Report*, 27:153–157, 1990.

[KM95] M. Kapovich and J. Millson. On the moduli space of polygons in the Euclidean plane. *J. Differential Geom.*, 42(1):133–164, 1995.

[Lan96] R. J. Lang. A computational algorithm for origami design. In *Proc. 12th Sympos. Comput. Geom.*, pages 98–105, Philadelphia, PA, May 1996.

[LO96] A. Lubiw and J. O'Rourke. When can a polygon fold to a polytope? Technical Report 048, Smith College, June 1996.

[Lun71] W. F. Lunnon. Multi-dimensional map-folding. *The Computer Journal*, 14(1):75–80, Feb. 1971.

[LW95] W. J. Lenhart and S. H. Whitesides. Reconfiguring closed polygonal chains in Euclidean $d$-space. *Discrete Comput. Geom.*, 13:123–140, 1995.

[MMP87] J. S. B. Mitchell, D. M. Mount, and C. H. Papadimitriou. The discrete geodesic problem. *SIAM J. Comput.*, 16(4):647–668, Aug. 1987.

[Mon91] J. Montroll. *African Animals in Origami*. Dover Publications, 1991.

[Nag39] B. Nagy. Solution to problem 3763. *Amer. Math. Monthly*, 46:176–177, Mar. 1939.

[O'R98]    J. O'Rourke. Folding and unfolding in computational geometry. In *Revised Papers from the Japan Conf. Discrete Comput. Geom.*, volume 1763 of *Lecture Notes in Comput. Sci.*, pages 258–266, Tokyo, Japan, Dec. 1998.

[Sal73]    G. T. Sallee. Stretching chords of space curves. *Geom. Dedicata*, 2:311–315, 1973.

[Sch89]    C. Schevon. *Algorithms for Geodesics on Polytopes*. PhD thesis, Johns Hopkins University, 1989.

[She75]    G. C. Shephard. Convex polytopes with convex nets. *Math. Proc. Cambridge Philos. Soc.*, 78:389–403, 1975.

[Str00]    I. Streinu. A combinatorial approach to planar non-colliding robot arm motion planning. In *Proc. 41st IEEE Sympos. Found. Comp. Sci.*, pages 443–453, Redondo Beach, California, Nov. 2000.

[Tou99a]   G. Toussaint. Computational polygonal entanglement theory. In *Proceedings of the VIII Encuentros de Geometria Computacional*, Castellon, Spain, July 1999.

[Tou99b]   G. Toussaint. The Erdős-Nagy theorem and its ramifications. In *Proc. 11th Canadian Conf. Comput. Geom.*, Vancouver, Canada, Aug. 1999. `http://www.cs.ubc.ca/conferences/CCCG/elec_proc/fp19.ps.gz`.

[Tou01]    G. Toussaint. A new class of stuck unknots in $pol_6$. *Beiträge Algebra Geom.*, 2001. To appear.

[Whi92]    S. Whitesides. Algorithmic issues in the geometry of planar linkage movement. *Australian Computer Journal*, 24(2):42–50, May 1992.

# On the Skeleton of the Metric Polytope

Antoine Deza[1,5], Komei Fukuda[2], Dmitrii Pasechnik[3], and Masanori Sato[4]

[1] Institute of Statistical Mathematics, Prediction and Control, Tokyo, Japan
deza@ism.ac.jp
[2] ETH Zürich, Institute for Operations Research, Zürich, Switzerland
fukuda@ifor.math.ethz.ch
[3] Delft University of Technology, Operations Research, TWI, Netherlands
d.pasechnik@twi.tudelft.nl
[4] Tokyo Institute of Technology, Math. and Comput. Sci., Tokyo, Japan
[5] EHESS, Centre d'Analyse et de Mathématique Sociales, Paris, France.
deza@ehess.fr

**Abstract.** We consider convex polyhedra with applications to well-known combinatorial optimization problems: the metric polytope $m_n$ and its relatives. For $n \leq 6$ the description of the metric polytope is easy as $m_n$ has at most 544 vertices partitioned into 3 orbits; $m_7$ - the largest previously known instance - has 275 840 vertices but only 13 orbits. Using its large symmetry group, we enumerate orbitwise 1 550 825 600 vertices of the 28-dimensional metric polytope $m_8$. The description consists of 533 orbits and is conjectured to be complete. The orbitwise incidence and adjacency relations are also given. The skeleton of $m_8$ could be large enough to reveal some general features of the metric polytope on $n$ nodes. While the extreme connectivity of the cuts appears to be one of the main features of the skeleton of $m_n$, we conjecture that the cut vertices do not form a cut-set. The combinatorial and computational applications of this conjecture are studied. In particular, a heuristic skipping the highest degeneracy is presented.

## 1 Introduction

Combinatorial polytopes, i.e. polytopes arising from combinatorial optimization problems, are often trivial for the very first cases and then suddenly the so-called combinatorial explosion occurs even for small instances. While these polytopes turn out to be quickly intractable for enumeration algorithm designed for general polytopes, tailor-made algorithms using their rich combinatorial features can exhibit surprisingly strong performances. For example, CHRISTOF AND REINELT [2] computed large instances of the *traveling salesman polytope*, the *linear ordering polytope* and the *cut polytope* exploiting their symmetry groups. In a similar vein, in addition to its symmetry group, we used its combinatorial structure to orbitwise enumerate the vertices of another combinatorial polytope: *the metric polytope*. Let first recall basic definitions and present some applications to well-known combinatorial optimization problems.

J. Akiyama, M. Kano, and M. Urabe (Eds.): JCDCG 2000, LNCS 2098, pp. 125-136, 2001.
© Springer-Verlag Berlin Heidelberg 2001

The $\binom{n}{2}$-dimensional cut polytope $c_n$ is usually introduced as the convex hull of the incidence vectors of all the cuts of $K_n$. More precisely, given a subset $S$ of $V_n = \{1, 2, \ldots, n\}$, the *cut* determined by $S$ consists of the pairs $(i, j)$ of elements of $V_n$ such that exactly one of $i$, $j$ is in $S$. By $\delta(S)$ we denote both the cut and its incidence vector in $\mathbb{R}^{\binom{n}{2}}$; that is, $\delta(S)_{ij} = 1$ if exactly one of $i$, $j$ is in $S$ and 0 otherwise for $1 \leq i < j \leq n$. By abuse of notation, we use the term cut for both the cut itself and its incidence vector, so $\delta(S)_{ij}$ are considered as coordinates of a point in $\mathbb{R}^{\binom{n}{2}}$. The cut polytope $c_n$ is the convex hull of all $2^{n-1}$ cuts, and the *cut cone* $C_n$ is the conic hull of all $2^{n-1} - 1$ nonzero cuts. The cut polytope and one of its relaxation - the metric polytope - can also be defined in terms of a finite metric space in the following way. For all 3-sets $\{i, j, k\} \subset \{1, \ldots, n\}$, we consider the following inequalities:

$$x_{ij} - x_{ik} - x_{jk} \leq 0, \tag{1}$$
$$x_{ij} + x_{ik} + x_{jk} \leq 2. \tag{2}$$

(1) induce the $3\binom{n}{3}$ facets which define the *metric cone* $M_n$. Then, bounding the latter by (2) we obtain the metric polytope $m_n$. The $3\binom{n}{3}$ (resp. $\binom{n}{3}$) facets defined by (1) (resp. by (2)) can be seen as triangle (resp. perimeter) inequalities for distance $x_{ij}$ on $\{1, 2, \ldots, n\}$. While the cut cone is the conic hull of all, up to a constant multiple, $\{0, 1\}$-valued extreme rays of the metric cone, the cut polytope $c_n$ is the convex hull of all $\{0, 1\}$-valued vertices of the metric polytope. The link with finite metric spaces is the following: there is a natural $1 - 1$ correspondence between the elements of the metric cone and all the semi-metrics on $n$ points, and the elements of the cut cone correspond precisely to the semi-metrics on $n$ points that are isometrically embeddable into some $l_1^m$. It is easy to check that such minimal $m$ is smaller or equal to $\binom{n}{2}$.

One of the motivations for the study of these polyhedra comes from their applications in combinatorial optimization, the most important being the max-cut and multicommodity flow problems. Given a graph $G = (V_n, E)$ and nonnegative weights $w_e$, $e \in E$, assigned to its edges, the *max-cut* problem consists in finding a cut $\delta(S)$ whose weight $\sum_{e \in \delta(S)} w_e$ is as large as possible. It is a well-known $NP$-complete problem. By setting $w_e = 0$ if $e$ is not an edge of G, we can consider without loss of generality $K_n$ the complete graph on $V_n$. Then the max-cut problem can be stated as a linear programming problem over the cut polytope $c_n$, as follows: max $w^T x$ subject to $x \in c_n$. Since the metric polytope is a relaxation of the cut polytope, optimizing $w^T x$ over $c_n$ instead of $m_n$ provides an upper bound for the max-cut problem. Consider now the complete graph $K_n$; an instance of the *multicommodity flow* problem is given by two nonnegative vectors indexed by $E$: a capacity $c(e)$ and a requirement $r(e)$ for each $e \in E$. Let $U = \{e \in E : r(e) > 0\}$. If $T$ denotes the subset of $V_n$ spanned by the edges in $U$, then we say that the graph $G = (T, U)$ denotes the *support* of $r$. For each edge $e = (s, t)$ in the support of $r$, we seek a flow of $r(e)$ units between $s$ and $t$ in the complete graph. The sum of all flows along any edge $e' \in E$ must not exceed $c(e')$. If such a set of flows exists, we call $c, r$ *feasible*. A necessary and sufficient condition for feasibility is: a pair $c, r$ is feasible if and only if $(c - r)^T x \geq 0$ is valid over the metric cone, see [7]. For example, the triangle facet induced by (1) can be seen as an elementary solvable flow problem with $c(ij) = r(ik) = r(jk) = 1$ and $c(e) = r(e) = 0$ otherwise, so (1) corresponds to $(c - r)^T x \geq 0$ for $x$ in the metric cone. Therefore, the metric cone is the dual cone to the cone of feasible multicommodity flow problems. For a detailed study of those polytopes and their applications in combinatorial optimization we refer to DEZA AND LAURENT [4] and POLJAK AND TUZA [9].

# 2   Vertices of the Metric Polytope

## 2.1   Combinatorial and Geometric Properties

The polytope $c_n$ is a $\binom{n}{2}$ dimensional $0-1$ polyhedron with $2^{n-1}$ vertices and $m_n$ is a polytope of the same dimension with $4\binom{n}{3}$ facets inscribed in the cube $[0, 1]^{\binom{n}{2}}$. We have $c_n \subseteq m_n$ with equality only for $n \leq 4$. It is easy to see that the point $\omega_n = (\frac{1}{2}, \frac{1}{2}, \ldots, \frac{1}{2})$ is the center of gravity of both $c_n$ and $m_n$ and is also the center of the sphere of radius $r = \frac{1}{2}\sqrt{n(n-1)}$ where all the cuts lie. Any facet of the metric polytope contains a face of the cut polytope and the vertices of the cut polytope are vertices of the metric polytope. In fact, the cuts are precisely the integral vertices of the metric polytope. The metric polytope $m_n$ wraps the cut polytope $c_n$ very tightly. Indeed, in addition to the vertices, all edges and 2-faces of $c_n$ are also faces of $m_n$, for 3-faces it is false for $n \geq 4$. Any two cuts are adjacent both on $c_n$ and on $m_n$; in other words $m_n$ is *quasi-integral*; that is, the skeleton of the convex hull of its integral vertices, i.e. the skeleton of $c_n$, is an induced subgraph of the skeleton of the metric polytope itself. We recall that the skeleton of a polytope is the graph formed by its vertices and edges. While the diameters of the cut polytope and the dual metric polytope satisfy $\delta(c_n) = 1$ and $\delta(m_n^*) = 2$, the diameters of their dual are conjectured to be $\delta(c_n^*) = 4$ and $\delta(m_n) = 3$.

One important feature of the metric and cut polytopes is their very large symmetry group. We recall that the *symmetry group* $Is(P)$ of a polytope $P$ is the group of isometries preserving $P$. More precisely, for $n \geq 5$, $Is(m_n) = Is(c_n)$ and both are induced by permutations on $V_n = \{1, \ldots, n\}$ and *switching reflections by a cut* and, for $n \geq 5$, we have $|Is(m_n)| = 2^{n-1}n!$. Given a cut $\delta(S)$, the switching reflection $r_{\delta(S)}$ is defined by $y = r_{\delta(S)}(x)$ where $y_{ij} = 1 - x_{ij}$ if $(i, j) \in \delta(S)$ and $y_{ij} = x_{ij}$ otherwise. As these symmetries preserve the adjacency relations and the linear independency, all faces of $m_n$ are partitioned into orbits of faces equivalent under permutations and switchings.

## 2.2   Vertices of the Metric Polytope

We recall some results on the vertices of the metric polytope and the LAURENT-POLJAK *dominant clique conjecture*. The cuts are the only integral vertices of $m_n$. All other vertices with are not fully fractional are so-called *trivial extensions* of a vertex of $m_{n-1}$. Consider the following two mappings

$$\mathbb{R}^{\binom{n-1}{2}} \longrightarrow \mathbb{R}^{\binom{n}{2}} \qquad \mathbb{R}^{\binom{n-1}{2}} \longrightarrow \mathbb{R}^{\binom{n}{2}}$$
$$v \longrightarrow \phi_0(v) \qquad\qquad v \longrightarrow \phi_1(v)$$

$$\begin{array}{lll} \phi_0(v)_{ij} = v_{ij} & \phi_1(v)_{ij} = v_{ij} & \text{for } 1 \leq i < j \leq n-1 \\ \phi_0(v)_{i,n} = v_{1,i} & \phi_1(v)_{i,n} = 1 - v_{1,i} & \text{for } 2 \leq i \leq n-1 \\ \phi_0(v)_{1,n} = 0 & \phi_0(v)_{1,n} = 1. & \end{array}$$

The vertices $\phi_0(v)$ and $\phi_1(v)$ are called trivial extensions of $v$. Note that $\phi_1(v) = r_{\delta(\{n\})}(\phi_0(v))$. In other words, the new vertices are the fully fractional ones. The $(\frac{1}{3}, \frac{2}{3})$-valued fully fractional vertices are well studied and include the anticut orbit formed by the $2^{n-1}$ anticuts $\bar{\delta}(S) = \frac{2}{3}(1,\ldots,1)-\frac{1}{3}\delta(S)$. If $G = (V_n, E)$ is a connected graph, we denote by $d_G$ its path metric, where $d_G(i,j)$ is the length of a shortest path from $i$ to $j$ in $G$ for $i \neq j \in V_n$. Then $\tau(d_G) = max(d_G(i,j) + d_G(i,k) + d_G(j,k) : i,j,k \in G)$ is called the *triameter of $G$* and we set $x_G = \frac{2}{\tau(d_G)}d_G$. Any vertex of $m_n$ of the form $x_G$ for some graph is called a *graphic* vertex, see Fig. 1 for the graphs of 2 graphic $(\frac{1}{3}, \frac{2}{3})$-valued vertices of $m_8$. Note that for any connected graph $G = (V_n, E)$, we have $\tau(d_G) \leq 2(n-1)$ and that any $(\frac{1}{3}, \frac{2}{3})$-valued vertex of $m_n$ is (up to switching) graphic. Let the incidence $Icd_v$ denotes the number of facets containing the vertex $v$ and the adjacency $Icd_v$ denotes the number of vertices adjacent to $v$ (i.e. forming an edge with $v$). The following is straigtforward to prove.

**Proposition 1.** *The vertices of the metric polytope $m_n$ are partitioned into orbits of its symmetry group. Let $v$ be a vertex of $m_n$, $Icd_v$ its incidence, $Adj_v$ its adjacency, $O_v$ the orbit generated by the action of $Is(m_n)$ on $v$, and $\tilde{v}$ the canonical representative of $O_v$. Then $Icd_v = Icd_{\tilde{v}}$, $Adj_v = Adj_{\tilde{v}}$ and $O_v = O_{\tilde{v}}$.*

Since $m_3 = c_3$ and $m_4 = c_4$, the vertices of $m_3$ and $m_4$ are made of 4 and 8 cuts forming 1 orbit. The 32 vertices of $m_5$ are 16 cuts and 16 anticuts, i.e., form 2 orbits. The metric polytope $m_6$ has 544 vertices, see [8], partitioned into 3 orbits: cuts, anticuts and 1 orbit of trivial extensions; and $m_7$ has 275 840 vertices, see [3], partitioned into 13 orbits: cuts, anticuts, 3 orbits of trivial extensions, 3 $(\frac{1}{3}, \frac{2}{3})$-valued orbits and 5 other fully fractional orbits. See Table 1, where the 13 canonical representative vertices of the metric polytope on 7 nodes are given with their incidence and adjacency.

*Property 1.* Let $v$ be a vertex of $m_n$ and $\delta(S)$ any cut. Then one has: $Icd_v \leq Icd_{\delta(S)} = 3\binom{n}{3}$ with equality only for $v = \delta(S)$. Moreover, if $v$ is a trivial extension, $Icd_v \leq 2\binom{n}{3}$ and, if $v$ is fully fractional, $Icd_v \leq Icd_{\bar{\delta}(S)} = \binom{n}{3}$ with equality only for $v = \bar{\delta}(S)$.

Property 1 is illustrated in Table 1 where the orbits $O_{\tilde{v}_i}$ are ordered by decreasing values of the incidence $Icd_{\tilde{v}_i}$. The first orbit $O_{\tilde{v}_1}$ is the cut orbit and all fully fractional orbits are after the anticut orbit $O_{\tilde{v}_5}$. The trivial extension orbits are $O_{\tilde{v}_i}$ for $i = 2, 3$ and $4$.

**Table 1.** The 13 orbits of vertices of $m_7$

| Orbit $O_{\tilde{v}_i}$ | Canonical representative vertex $\tilde{v}_i$ | $Icd_{\tilde{v}_i}$ | $Adj_{\tilde{v}_i}$ | $|O_{\tilde{v}_i}|$ |
|---|---|---|---|---|
| $O_{\tilde{v}_1}$ | $(0,0,0,0,0,0,0,0,0,0,0,0,0,0,0,0,0,0,0,0,0)$ | 105 | 55 226 | 64 |
| $O_{\tilde{v}_2}$ | $\frac{2}{3}(1,1,1,1,0,0,1,1,1,1,1,1,1,1,1,1,1,1,1,1,0)$ | 49 | 496 | 2 240 |
| $O_{\tilde{v}_3}$ | $\frac{2}{3}(1,1,1,1,0,1,1,1,1,1,0,1,1,1,1,1,1,1,1,1,1)$ | 45 | 594 | 6 720 |
| $O_{\tilde{v}_4}$ | $\frac{2}{3}(1,1,1,1,1,0,1,1,1,1,1,1,1,1,1,1,1,1,1,1,1)$ | 40 | 763 | 1 344 |
| $O_{\tilde{v}_5}$ | $\frac{2}{3}(1,1,1,1,1,1,1,1,1,1,1,1,1,1,1,1,1,1,1,1,1)$ | 35 | 896 | 64 |
| $O_{\tilde{v}_6}$ | $\frac{1}{4}(1,2,3,1,2,1,1,2,2,1,2,1,1,2,3,2,3,2,1,2,1)$ | 30 | 96 | 20 160 |
| $O_{\tilde{v}_7}$ | $\frac{2}{5}(2,1,1,1,1,2,2,1,1,1,1,2,1,1,1,2,1,1,2,1,2)$ | 28 | 57 | 23 040 |
| $O_{\tilde{v}_8}$ | $\frac{1}{3}(1,1,1,1,1,1,2,2,1,1,1,2,1,1,1,1,1,1,2,2,2)$ | 26 | 76 | 4 480 |
| $O_{\tilde{v}_9}$ | $\frac{2}{7}(1,2,3,2,1,2,1,2,1,2,1,1,2,1,1,2,2,1,1,1)$ | 25 | 30 | 40 320 |
| $O_{\tilde{v}_{10}}$ | $\frac{1}{5}(3,2,3,3,1,1,1,2,2,2,2,3,3,3,3,4,4,2,2,4,2)$ | 25 | 27 | 16 128 |
| $O_{\tilde{v}_{11}}$ | $\frac{1}{3}(1,1,1,1,1,1,2,2,1,1,1,2,1,1,1,2,1,1,2,2,2)$ | 23 | 39 | 40 320 |
| $O_{\tilde{v}_{12}}$ | $\frac{1}{6}(1,2,4,2,2,2,1,3,3,3,3,2,2,2,4,2,2,2,4,4,4)$ | 23 | 24 | 80 640 |
| $O_{\tilde{v}_{13}}$ | $\frac{1}{3}(2,2,1,1,1,2,2,1,1,1,1,2,1,1,1,2,1,1,2,1,2)$ | 22 | 46 | 40 320 |
| Total | | | | 275 840 |

*Conjecture 1.* [8] Any vertex of the metric polytope is adjacent to a cut.

Conjecture 1 underlines the extreme connectivity of the cuts. Recall that the cuts form a clique in both the cut and metric polytopes. Therefore, if Conjecture 1 holds, the cuts would be a dominant clique in the skeleton of $m_n$ implying that its diameter would satisfy $\delta(m_n) \leq 3$.

## 3    Orbitwise Enumeration Algorithm

As stated in Proposition 1, the *neighborhood*, that is, the set of vertices adjacent to a given vertex, is equivalent up to permutations and switchings for all vertices belonging to the same orbit. This property leads to the following *orbitwise enumeration algorithm*. The main two subroutines are the computation of the canonical representative $\tilde{v}$ of the orbit generated by a vertex $v$ and the enumeration of the neighborhood $N_{\tilde{v}}$ of the vertex $\tilde{v}$. Starting from an initial vertex $v_{start}$ the algorithm computes the canonical representative $\tilde{v}_{start}$, enumerates its neighborhood $N_{\tilde{v}_{start}}$, identifies new orbits contained in $N_{\tilde{v}_{start}}$, updates the list $L$ of canonical representatives and then picks up the next canonical representative in $L$ whose neighborhood is not yet computed. The algorithm terminates when there is no more such canonical representative in $L$ and outputs $L$. Since the skeleton of a polytope is connected, this algorithm finds all orbits.

*Orbitwise Enumeration Algorithm*
**begin**
    find an initial vertex $v_{start}$;
    compute the canonical representative $\tilde{v}_{start}$ of the orbit $O_{\tilde{v}_{start}}$;
    mark $\tilde{v}_{start}$ with 0;   /* neighborhood not yet computed */
    initialized the list of canonical representatives $L := \{\tilde{v}_{start}\}$;
    **while** $L$ contains a 0-marked vertex $\tilde{v}_i$ **do**
        **begin**
            compute the neighborhood $N_{\tilde{v}_i}$ of $\tilde{v}_i$;
            **for** each vertex $v$ adjacent to $\tilde{v}_i$
                compute the canonical representative $\tilde{v}$ of the orbit $O_{\tilde{v}}$;
                **if** $\tilde{v} \notin L$ then mark $\tilde{v}$ with 0 and $L := L \cup \{\tilde{v}\}$; **endif**;
            **endfor**;
            mark $\tilde{v}_i$ with 1;   /* neighborhood computed */ ;
    **endwhile**;
    sort $L$ by decreasing values of $Icd_{\tilde{v}_i}$, decreasing $Adj_{\tilde{v}_i}$ and increasing $|O_{\tilde{v}_i}|$;
    output $L$;
**end.**

**Lemma 1.** *Let $I$ be the number of orbits, $\mathrm{Icd}_{\tilde{v}_i}$ and $\mathrm{Adj}_{\tilde{v}_i}$ the incidence and the adjacency of the orbit $O_{\tilde{v}_i}$ for $i = 1, \ldots, I$. The neighborhood enumeration subroutine is called exactly $I$ times and each neighborhood is generated by $\mathrm{Icd}_{\tilde{v}_i}$ facets. The canonical representative computation subroutine is called exactly $\sum_{\tilde{v}_i} \mathrm{Adj}_{\tilde{v}_i}$ times.*

*Remark 1.*

1. The orbitwise enumeration algorithm performs $I$ classic vertex enumerations for smaller sub-polytopes (one for each orbit of neighborhoods) instead of performing one large classic vertex enumeration (the whole polytope).
2. The computation is independent of the choice of the initial vertex $v_{start}$. Among the known vertices of $m_n$; an easy choice for $v_{start}$ is the anticut $\bar{\delta}(\emptyset) = \frac{2}{3}(1, \ldots, 1)$.
3. In case of very high degeneracy, the subroutine computing the canonical representative has to be called a large number of times and some of the neighborhoods might represent a large fraction of the whole polytope. It is the case for the neighborhood of a cut $N_{\delta(S)}$ as $Icd_{\delta(S)} = 3\binom{n}{3}$.

For $i = 1, \ldots, I$, the algorithm gets the orbitwise incidence $Icd_{\tilde{v}_i}$ (by simply checking which inequality is satisfied with equality) as input for the neighborhood enumeration subroutine and produces the adjacency $Adj_{\tilde{v}_i}$ as output of this subroutine. By counting the number of times a vertex equivalent to the canonical representative $\tilde{v}_j$ is found in $N_{\tilde{v}_i}$, we get the *orbitwise adjacency table*, that is, the $I \times I$ matrix $Adj$ with $Adj_{i,j} = Adj_{\tilde{v}_i, \tilde{v}_j}$ the number of vertices of the orbit $O_{\tilde{v}_j}$ adjacent to $\tilde{v}_i$. The orbits $O_{\tilde{v}_i}$ are ordered first by decreasing values of the incidence $Icd_{\tilde{v}_i}$, then by decreasing adjacency $Adj_{\tilde{v}_i}$ and then by increasing orbitsize $|O_{\tilde{v}_i}|$. Let us assume we know the size of one orbit; for example, we have

$|O_{\bar{\delta}(S)}| = 2^{n-1}$. Then, from the matrix $Adj$, we can usually get the size of the other orbits using the following easy relation: $Adj_{\tilde{v}_i,\tilde{v}_j} \times |O_{\tilde{v}_i}| = Adj_{\tilde{v}_j,\tilde{v}_i} \times |O_{\tilde{v}_j}|$. See, for example, Table 2 where the orbitwise adjacency table is given for the metric polytope on 7 nodes. The first row of Table 2 lists orbitwise $N_{\tilde{v}_1}$; that is, the 55 226 neighboors of a vertex belonging to $O_{\tilde{v}_1}$, that is a cut. For example, $Adj_{\tilde{v}_1,\tilde{v}_4} = 945$ in the fourth column means that a cut is adjacent to 945 vertices belonging to the orbit $O_{\tilde{v}_4}$. Since all the facets incident to the origin $\delta(\emptyset)$ are precisely the $3\binom{n}{3}$ triangle facets, an extreme ray of the metric cone $M_n$ corresponds to each vertex adjacent to $\tilde{v}_1 = \delta(\emptyset)$. In other words, the adjacency $Adj_{\delta(S)} = Adj_{\tilde{v}_1}$ of a cut equals the number of extreme rays of the metric cone $M_n$. We recall that the 41 orbits under permutations of the extreme rays of $M_7$ were found by GRISHUKHIN [6].

**Table 2.** Orbitwise adjacency table of the skeleton of $m_7$

|  | $O_{\tilde{v}_1}$ | $O_{\tilde{v}_2}$ | $O_{\tilde{v}_3}$ | $O_{\tilde{v}_4}$ | $O_{\tilde{v}_5}$ | $O_{\tilde{v}_6}$ | $O_{\tilde{v}_7}$ | $O_{\tilde{v}_8}$ | $O_{\tilde{v}_9}$ | $O_{\tilde{v}_{10}}$ | $O_{\tilde{v}_{11}}$ | $O_{\tilde{v}_{12}}$ | $O_{\tilde{v}_{13}}$ | $Adj_{\tilde{v}_i}$ |
|---|---|---|---|---|---|---|---|---|---|---|---|---|---|---|
| $O_{\tilde{v}_1}$ | 63 | 980 | 3 570 | 945 | 56 | 7 560 | 5 400 | 1 120 | 6 930 | 2 772 | 6 930 | 10 080 | 8 820 | 55 226 |
| $O_{\tilde{v}_2}$ | 28 | 24 | 132 | 45 | 3 | 18 | 0 | 12 | 0 | 36 | 72 | 108 | 18 | 496 |
| $O_{\tilde{v}_3}$ | 34 | 44 | 126 | 42 | 4 | 60 | 48 | 8 | 48 | 12 | 48 | 48 | 72 | 594 |
| $O_{\tilde{v}_4}$ | 45 | 75 | 210 | 20 | 3 | 60 | 0 | 20 | 0 | 60 | 60 | 180 | 30 | 763 |
| $O_{\tilde{v}_5}$ | 56 | 105 | 420 | 63 | 0 | 0 | 0 | 0 | 0 | 252 | 0 | 0 | 0 | 896 |
| $O_{\tilde{v}_6}$ | 24 | 2 | 20 | 4 | 0 | 8 | 8 | 2 | 4 | 0 | 8 | 8 | 8 | 96 |
| $O_{\tilde{v}_7}$ | 15 | 0 | 14 | 0 | 0 | 7 | 7 | 0 | 7 | 0 | 0 | 0 | 7 | 57 |
| $O_{\tilde{v}_8}$ | 16 | 6 | 12 | 6 | 0 | 9 | 0 | 0 | 0 | 0 | 9 | 18 | 0 | 76 |
| $O_{\tilde{v}_9}$ | 11 | 0 | 8 | 0 | 0 | 2 | 4 | 0 | 0 | 0 | 1 | 0 | 4 | 30 |
| $O_{\tilde{v}_{10}}$ | 11 | 5 | 5 | 5 | 1 | 0 | 0 | 0 | 0 | 0 | 0 | 0 | 0 | 27 |
| $O_{\tilde{v}_{11}}$ | 11 | 4 | 8 | 2 | 0 | 4 | 0 | 1 | 1 | 0 | 0 | 4 | 4 | 39 |
| $O_{\tilde{v}_{12}}$ | 8 | 3 | 4 | 3 | 0 | 2 | 0 | 1 | 0 | 0 | 2 | 0 | 1 | 24 |
| $O_{\tilde{v}_{13}}$ | 14 | 1 | 12 | 1 | 0 | 4 | 4 | 0 | 4 | 0 | 4 | 2 | 0 | 46 |

*Remark 2.* The output, that is, the list $L$ of canonical representatives $\tilde{v}_i$ for $i = 1, \ldots, I$, is extremely compact. Apart from vertex enumeration, the algorithm computes the orbits invariants $Adj_{\tilde{v}_i}$, $Icd_{\tilde{v}_i}$ and $|O_{\tilde{v}_i}|$. The orbitwise adjacency table $Adj$ reveals the skeleton. The total number of vertices is simply $\sum_{\tilde{v}_i} |O_{\tilde{v}_i}|$ and the full list of vertices can be generated by the action of the symmetry group on each representative $\tilde{v}_i$.

## 4    Generating Vertices of the Metric Polytope

The heuristics presented in this section are valid for other combinatorial polytopes, but for convenience we restrict ourselves to the metric polytope. Insertion algorithms usually handle high degeneracy better than pivoting algorithms, see [1] for a detailed presentation of the main vertex enumeration methods. The metric polytope $m_n$ is quite degenerate (the cut incidence $Icd_{\delta(S)} = 3\binom{n}{3}$ is much larger than the dimension $d = \binom{n}{2}$). Thus we choose an insertion algorithm for the neighborhood enumeration subroutine: the *cddlib* implementation of the double description method [5]. In the remainder, we always assume that the neighborhood enumeration subroutine is performed by an insertion algorithm. Item 3 of Remark 1 indicates that even the neighborhoods of highly degenerate polytopes might lie beyond the range of problems currently solvable by insertion algorithms. In Sect. 4.2, we present heuristics addressing this issue.

## 4.1    A Conjecture on the Skeleton of the Metric Polytope

If true, the LAURENT-POLJAK Conjecture 1 would give the following computational implication: the enumeration of the extreme rays of $M_n$ gives all the orbits of the $m_n$. Since the number of extreme rays of the metric cone $|M_n| = Adj_{\delta(S)}$ might be a large fraction of the number of vertices of the metric polytope, the computational gain would be limited. Therefore, we propose a *no cut-set conjecture* which can be seen as complementary to the LAURENT-POLJAK conjecture both graphically and computationally.

*Conjecture 2.* For $n \geq 6$, the restriction of the skeleton of the metric polytope $m_n$ to the non-cut vertices is connected.

For any pair of vertices, while Conjecture 1 implies that there is a path made of cuts joining them, Conjecture 2 means that there is a path made of non-cuts vertices joining them. In other words, the cut vertices would form a *dominating set* but not a *cut-set* in the skeleton of $m_n$. On the other hand, while Conjecture 1 means that the enumeration of the metric cone $M_n$ is enough to obtain the metric polytope $m_n$; Conjecture 2 means that we can obtain $m_n$ without enumerating $M_n = N_{\delta(\emptyset)}$, see Sect. 4.2. Note that for arbitrary graphs these are clearly independent. Both are strongly believed to be true and hold for $n \leq 7$.

## 4.2    Heuristic: Skipping High Degeneracy

If Conjecture 2 holds, all orbits can be found by the following metric cone skipping heuristic: disregard $\tilde{v}$ if $\tilde{v} = \delta(S)$. In other words, disregard the neighborhood of the cuts, that is, essentially the metric cone $M_n$. This neighborhood is believed to be by a large margin the largest, as we expect that $Adj_{\delta(S)} \gg Adj_{v \neq \delta(S)}$, see Property 1 and Item 1 of Remark 3. In other words, the heuristic removes the hardest neighborhood enumeration. Note that cuts are easy to recognize as $\delta(S)$ is uniquely characterized by its incidence: $Icd_{\delta(S)} = 3\binom{n}{3}$. Therefore, disregarding the metric cone $M_n$, consists simply in choosing a non-cut as initial vertex $v_{start}$ and modifying the main loop of the orbitwise enumeration algorithm in the following way:

*Metric Cone Skipping Heuristic*
> **if** $\tilde{v} \notin L$ then $L := L \cup \{\tilde{v}\}$;
>> **if** $Icd_v = 3\binom{n}{3}$ then mark $\tilde{v}$ with 1
>> else mark $\tilde{v}$ with 0; **endif**;
> **endif**;

For the metric on 8 nodes, while $Adj_{\delta(S)} \geq 119\,269\,588$ - i.e. 7.7% of the total number of vertices of $m_8$ - the enumeration of the other 532 neighborhoods generates (with multiplicity) $\sum_{\tilde{v}_i \neq \delta(S)} Adj_{\tilde{v}_i} = 780\,711$ vertices - i.e. less than 0.05% of the total number of vertices. One can easily get the neighborhood of the cut from the orbitwise adjacency table $Adj$. Taking the column and the row corresponding to the cuts as we have: $Adj_{\delta(S),\tilde{v}_i} \times 2^{n-1} = Adj_{\tilde{v}_i,\delta(S)} \times |O_{\tilde{v}_i}|$ where $Adj_{\tilde{v}_i,\delta(S)}$ is the number of cuts adjacent to $\tilde{v}_i$.

**Proposition 2.** *If true, Conjecture 2 would be a certificate that the "Metric Cone Skipping Heuristic" gives a complete description of the metric polytope by generating only a very small fraction of the vertices.*

One can further decrease the computation time by skipping not only the orbit with the highest incidence (the cuts) but all orbits with arbitrarily set in advance upper bound $Icd_{max}$ on the incidence. Skipping high degeneracy consists simply in the following modification of the main loop of the orbitwise enumeration algorithm:

*Skipping High Degeneracy Heuristic*
> **if** $\tilde{v} \notin L$ then $L := L \cup \{\tilde{v}\}$;
>> **if** $Icd_v > Icd_{max}$ then mark $\tilde{v}$ with 1
>> else mark $\tilde{v}$ with 0; **endif**;
> **endif**;

In this case, a certificate for a complete description is that the restriction of the skeleton of $m_n$ to $O_{\tilde{v}_{start}}$ and the low incidence orbits $O_{\tilde{v}_i} : Icd_{\tilde{v}_i} \leq Icd_{max}$ is connected. This heuristic is particularly suitable for partial enumeration purpose and the choice of the initial vertex $v_{start}$ could become a critical factor, see Item 2 of Proposition 5. For the metric polytope on 8 nodes, we can take $v_{start} = \tilde{v}_4$; that is, the trivial extension with the fourth highest incidence $Icd_{\tilde{v}_4} = 74$. We expected this type of vertex to be connected to many orbits and, indeed, the neighborhood $N_{\tilde{v}_4}$ contains representatives of 450 different orbits out of 533. Another choice is $v_{start} = \tilde{v}_{41}$ with $Icd_{\tilde{v}_{41}} = 42$ and $Adj_{\tilde{v}_{41}} = 533$. The neighborhood computation subroutine was restricted to vertices satisfying $Icd_{\tilde{v}_i} \leq 40 < \frac{1}{2}\binom{n+1}{3}$ - i.e. halfway from the dimension $\binom{n}{2}$ to the anticut incidence $Icd_{\delta(S)} = \binom{n}{3}$. Besides $N_{\tilde{v}_{41}}$, the algorithm computed 485 neighborhoods with low incidences generating $\sum_{\tilde{v}_i : Icd_{\tilde{v}_i} \leq 40} Adj_{\tilde{v}_i} = 63\,095$ vertices - i.e. less than 0.005% of the total number of vertices. Still, this heuristic approach proved to be enough as this tiny number of vertices contains representatives of all 533 orbits. Another remarkable feature is that since all the 485 neighborhoods are generated by few facets - 389 have even less than 34 facets - the neighborhood enumeration subroutine is performing extremely well. Similarly to the previous *metric cone skipping heuristic* case, missing entries of the table *Adj* (i.e. the rows corresponding to the orbits with high incidence $Icd_{\tilde{v}_i} > Icd_{max}$) can be computed using the relations: $Adj_{\tilde{v}_i,\tilde{v}_j} \times |O_{\tilde{v}_i}| = Adj_{\tilde{v}_j,\tilde{v}_i} \times |O_{\tilde{v}_j}|$. In particular, we can first compute $N_{\bar{\delta}(S)}$; that is, all nonzero values of $Adj_{\bar{\delta}(S),\tilde{v}_j}$ and, using the fact that $|O_{\bar{\delta}(S)}| = 2^{n-1}$, get all corresponding $|O_{\tilde{v}_j}|$ and then use them iteratively to obtain some of the remaining unknown $|O_{\tilde{v}_i}|$.

## 5    Vertices of the Metric Polytope on 8 Nodes

Using the metric cone skipping heuristic presented in Sect. 4.2, we enumerate 533 orbits of $m_8$. The list of canonical representative with their adjacency and incidence and, especially, the adjacency table *Adj* being too large to be included in this paper, we refer to http://www.is.titech.ac.jp/~deza/deza.html where a detailed presentation is available. For example, the anticut row of *Adj* has only 15 nonzero entries $Adj_{\bar{\delta}(S),\tilde{v}_j}$. A summary description is given in Proposition 4.

**Proposition 3.**

1. *The metric polytope $m_8$ has at least 1 550 825 600 vertices and the metric cone $M_8$ has at least 119 269 588 extreme rays; we conjecture that both descriptions are complete.*
2. *For $i = 1,\ldots,533$ each orbit representative $\tilde{v}_i$ is adjacent to at least 2 cuts implying that the LAURENT-POLJAK dominant clique conjecture holds for these 533 orbits of $m_8$.*

**Proposition 4.** *The 1 550 825 600 vertices of the metric polytope on 8 nodes are partitioned into 533 orbits:*

(i)  *1 cut orbit $O_{\delta(S)}$ with $Icd_{\delta(S)} = 168$, $Adj_{\delta(S)} \geq 119\,269\,588$ and $|O_{\delta(S)}| = 128$*

(ii)  *28 trivial extensions orbits $O_{\tilde{v}_i}$ with $Icd_{\tilde{v}_i} = 88, 79, 74, \ldots, 42$, $Adj_{\tilde{v}_i} = 137\,758, \ldots, 127$ and $|O_{\tilde{v}_i}| = 1\,290\,240, \ldots, 3\,584$*

(iii)  *504 fully fractional orbits $O_{\tilde{v}_i}$:*

*1 anticut orbit $O_{\bar{\delta}(S)}$ with $Icd_{\bar{\delta}(S)} = 56$, $Adj_{\bar{\delta}(S)} = 52\,367$ and $|O_{\bar{\delta}(S)}| = 128$*

*37 $(\frac{1}{3}, \frac{2}{3})$-valued orbits $O_{\tilde{v}_i}$ with $Icd_{\tilde{v}_i} = 44, 40, \ldots, 28$, $Adj_{\tilde{v}_i} = 6\,285, 5\,247, \ldots, 28$ and $|O_{\tilde{v}_i}| = 5\,160\,960, \ldots, 35\,840$*

*466 non $(\frac{1}{3}, \frac{2}{3})$-valued fully fractional orbits $O_{\tilde{v}_i}$ with $Icd_{\tilde{v}_i} = 48, 45, \ldots, 29$, $Adj_{\tilde{v}_i} = 22\,300, 4\,906, \ldots, 29$ and $|O_{\tilde{v}_i}| = 5\,160\,960, \ldots, 40\,320$.*

**Proposition 5.**

1. *Exactly two of the 533 orbits of $m_8$ described in Proposition 4 are orbits of simple vertices; that is, satisfying $Icd_{\tilde{v}_i} = Adj_{\tilde{v}_i} = \binom{n}{2}$. Both representative vertices $\tilde{v}_{532}$ and $\tilde{v}_{533}$ are graphic $(\frac{1}{3}, \frac{2}{3})$-valued vertices, see Fig. 1.*

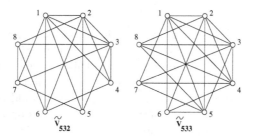

**Fig. 1.** Two simple graphic vertices of $m_8$

2. The $O_{\tilde{v}_{532}}$ row of the adjacency table has only 3 nonzero entries: $\text{Adj}_{\tilde{v}_{532},\delta(S)} = 14$, $\text{Adj}_{\tilde{v}_{532},\tilde{v}_4} = 7$ and $\text{Adj}_{\tilde{v}_{532},\tilde{v}_{41}} = 7$ with $\text{Icd}_{\tilde{v}_{41}} = 42$. It implies that, among the 532 non-cut orbits, the vertices of $O_{\tilde{v}_4}$ and $O_{\tilde{v}_{41}}$ are the only initial vertices such that the restriction of the skeleton of $m_n$ to $O_{\tilde{v}_{start}}$ and $O_{\tilde{v}_i} : \text{Icd}_{\tilde{v}_i} \leq 41$ could be connected. One can easily check that it holds for both $v_{start} = \tilde{v}_4$ and $v_{start} = \tilde{v}_{41}$.

3. One can easily check that the set $N_{\tilde{v}_{532}}$ of the neighbors of $\tilde{v}_{532}$ and the set $N_{r_{\delta(\{1\})}(\tilde{v}_{533})}$ of the neighbors of the switching of $\tilde{v}_{533}$ by the cut $\delta(\{1\})$ are disjoint. It implies that the diameter $\delta(m_8^{533})$ of the restriction of the skeleton of $m_8$ to the 533 orbits described in Proposition 4 satisfies $\delta(m_8^{533}) \geq 3$. Since the LAURENT-POLJAK conjecture holds for these 533 orbits, see Item 2 of Proposition 3, we have $\delta(m_8^{533}) = 3$.

*Remark 3.*

1. For $n \leq 7$, we have $Adj_{\tilde{v}} < Adj_{\delta(S)}$ for $\tilde{v} \neq \delta(S)$ and it is conjectured in [3] to be true for any $n$. For $n = 8$, it holds for the 533 orbits described in Proposition 4 as we have $865 \times Adj_{\tilde{v}_i} < Adj_{\delta(S)}$ for $i = 2 \ldots 533$.

2. For any $n$, we have $|O_{\tilde{v}}| \leq |Is(m_n)| = 2^{n-1} n!$. While for $n \leq 7$, this inequality is strict for all orbits, it is satisfied with equality for the largest (fully fractional) orbits of $m_8$.

3. It is conjectured, see [3], that for $n$ large enough, at least one vertex of $m_n$ is simple. While it is false $n = 6$ and 7, Item 1 of Proposition 5 implies that it holds for $n = 8$.

## 6    Conclusions

We presented an orbitwise enumeration algorithm for combinatorial polytopes with large symmetry group. In particular, we computed 1 550 825 600 vertices of a highly degenerate 28-dimensional polytope defined by its 224 facets: the metric polytope on 8 nodes. The description consists of only 533 canonical representatives and we conjecture it is complete. The orbitwise incidence, adjacency and skeleton are also given. While the extreme connectivity of the cuts (LAURENT-POLJAK conjectured they form a dominating set) appears to be one of the main features of the skeleton of $m_n$, we conjecture that the cut vertices do not form a cut-set in the skeleton of $m_n$. The combinatorial and computational applications of this conjecture are studied. In particular, a heuristic skipping the metric cone is presented. The algorithm can be parallelized very easily and, combined with the heuristic, higher-dimensional instances of the metric polytope and other combinatorial polyhedra vertex enumeration problems could be solvable. While the largest previously computed metric polytope $m_7$ has only 13 orbits of vertices, $m_8$ has at least 533 orbits and therefore could be large enough to reveal some general features of the metric polytope on $n$ nodes. In particular, the skeleton of $m_8$ suggests the orbitwise adjacency relations between the cuts, anticuts, the trivial extensions and the fully fractional orbits: The row $Adj_{\delta(S),j}$ and the column $Adj_{i,\delta(S)}$ should have only nonzero entries (LAURENT-POLJAK dominating set conjecture). The anticuts are mainly orbitwise adjacent to (few) trivial extensions and the fully fractional orbits are badly orbitwise connected among themselves. The trivial extensions are well connected among themselves and not so well to the fully fractional orbits but still the restriction to the non-cut orbits is connected (no cut-set conjecture).

*Acknowledgments.* We would like to thank DAVID AVIS for many helpful suggestions.

# References

1. Avis D., Bremner D., Seidel R.: How good are convex hull algorithms? Computational Geometry: Theory and Applications **7** (1997) 265–301
2. Christof T., Reinelt G.: Decomposition and Parallelization Techniques for Enumerating the Facets of 0/1-Polytopes. Preprint, Heidelberg University (1998)
3. Deza A, Deza M., Fukuda F.: On skeletons, diameters and volumes of metric polyhedra. Lecture Notes in Computer Science, Vol. 1120. Springer-Verlag, Berlin Heidelberg New York (1996) 112–128
4. Deza M., Laurent M.: Geometry of cuts and metrics. Algorithms and Combinatorics Vol. 15 Springer-Verlag, Berlin Heidelberg New York (1997)
5. Fukuda K.: cddlib reference manual, version 0.86, IFOR, ETHZ, Zürich, Switzerland (1999)
6. Grishukhin V. P.: Computing extreme rays of the metric cone for seven points. European Journal of Combinatorics **13** (1992) 153–165
7. Iri M.: On an extension of maximum-flow minimum-cut theorem to multicommodity flows. Journal of the Operational Society of Japan **13** (1970-1971) 129-135
8. Laurent M, Poljak S.: The metric polytope. In: Balas E., Cornuejils G., Kannan R. (eds.): Integer Programming and Combinatorial Optimization (1992) 247–286
9. Poljak S., Tuza Z.: Maximum Cuts and Large Bipartite Subgraphs. In: Cook W., Lovasz L., Seymour P. D. (eds.): DIMACS **20** (1995) 181–244

# Geometric Dissections that Swing and Twist

Greg N. Frederickson*

Department of Computer Science
Purdue University, West Lafayette, IN 47907, USA
gnf@cs.purdue.edu, http://www.cs.purdue.edu/people/gnf

**Abstract.** Two ways of hinging geometric dissections of 2-dimensional figures are explored. *Swing hinges* allow rotation in the plane. *Twist hinges* allow rotation by 180° through the third dimension. Techniques are presented and analyzed for designing hingeable dissections that use either only swing hinges or only twist hinges. For swing hinges these include the superposition of tessellations, the crossposition of T-strips, and the exploitation of the structure of regular polygons and stars. For twist hinges these include the conversion of swing hinges, the P-twist for parallelograms, and completing the pseudo-tessellation. Open problems relating to the possible universality of such hingings are posed.

## 1 Introduction

A *geometric dissection* is a cutting of a geometric figure into pieces that can be rearranged to form another figure [11, 20]. Dissections date back to Arabian mathematicians a millennium ago and Greek mathematicians more than two millennia ago [11]. As mathematical puzzles they enjoyed great popularity a century ago, in newspaper and magazine columns written by the American Sam Loyd [21] and the Englishman Henry Ernest Dudeney [6]. Loyd and Dudeney set as a goal the minimization of the number of pieces. After presenting the remarkable 4-piece dissection of an equilateral triangle to a square, Dudeney wrote [7]:

I add an illustration showing the puzzle in a rather curious practical form, as it was made in polished mahogany with brass hinges for use by certain audiences. It will be seen that the four pieces form a sort of chain, and that when they are closed up in one direction they form a triangle, and when closed in the other direction they form a square.

This hinged model (Figure 1) has been described subsequently in [4, 5, 9, 11, 12, 20, 24–26, 29, 30]. It uses the simplest of the lower pairs of linkages, namely the revolute joint (or pin joint), which permits only relative rotation [17]. Hinge-ability of dissections addresses basic issues related to the role of movement in the transformation of rigid objects.

---

* Supported in part by the National Science Foundation under grant CCR-9731758.

J. Akiyama, M. Kano, and M. Urabe (Eds.): JCDCG 2000, LNCS 2098, pp. 137-148, 2001.

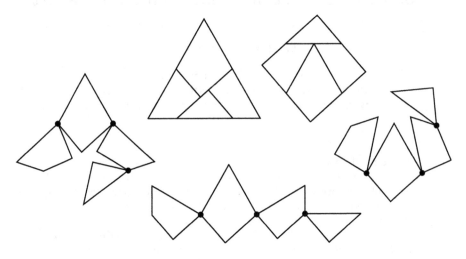

**Fig. 1.** Hinged dissection of a triangle to a square

We focus here on dissections of 2-dimensional (planar) figures and consider cuts that are along straight line segments. We investigate two ways to hinge dissections: The first uses *swing hinges*, illustrated in Figure 1, which allow rotation in the plane. A fundamental question in regard to dissections and movement is:

**Open Problem I**: For any two figures of equal area and bounded by straight line segments, is a swing-hingeable dissection possible?

Wallace [28], Bolyai [2], and Gerwien [13] proved the analogous property with respect to normal (unhinged) dissections.

We pose the interesting and challenging problem of identifying general techniques to produce swing-hingeable dissections. We introduce and prove conditions for which the superposition of tessellations and the crossposition of T-strips produce hingeable dissections. We show how to exploit the structure of regular polygons and stars for these dissections. We adopt the natural goal of minimizing the number of pieces, subject to the dissection being hingeable.

The second way to hinge dissections seems to be relatively recent. A *twist hinge* has a point of rotation on the interior of the line segment along which two pieces touch edge-to-edge. This allows one piece to be flipped over relative to the other, using rotation by 180° through the third dimension. One of the earliest such dissections is by William Esser, III, who was awarded a U.S. patent in 1985 for what was essentially the dissection of an ellipsoid to a heart [8]. A small open circle represents a twist hinge, which is typically positioned at the midpoint of a common edge. Each piece that ends up being flipped an odd number of times is marked with an "∗" in the ellipse, and with a "⋆" in the heart. Ernst Lurker discovered a similar dissection around the same time.

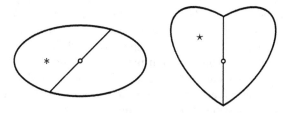

**Fig. 2.** Twist-hingeable dissection of an ellipse to a heart

Such movement is remarkably limited. It comes as somewhat of a surprise that there are a relatively large number of twist-hingeable dissections. What we might earlier have characterized as improbable we now pose as:

**Open Problem II**: For any two figures of equal area and bounded by straight line segments, is a twist-hingeable dissection possible?

We pose the interesting and challenging problem of creating general techniques to produce twist-hingeable dissections. We introduce two techniques to convert swing-hingeable dissections to be twist-hingeable, a technique to change the length (and thus the height) of a parallelogram, and a technique that produces two intriguing (infinite) families of dissections.

Works that have either identified dissections as swing-hingeable or come close to doing so include [1, 3, 5, 11, 15, 16, 19, 27, 31]. Recently, [1, 10] have focused exclusively on hingeable dissections. This paper is excerpted from [10].

## 2   Definitions

We consider dissections of regular polygons and regular star polygons. Let $\{p\}$ be a regular polygon with $p$ sides. Let $\{p/q\}$ be a star polygon with $p$ points (vertices), where each point is connected to the $q$-th points clockwise and counterclockwise from it.

We assume that a figure such as a polygon or a star is an open set, so that its boundary is not part of the figure itself. When we cut the figure along a sequence of line segments, we effectively remove all points on those line segments, resulting in pieces that are open sets. When we assemble two pieces, we "glue" them together by adding their common boundary, minus the endpoints of the common boundary.

Three (or more) pieces are also allowed to share a swing hinge, but the clockwise order of the pieces may not change. Also, two different swing hinges are allowed to abut up against each other.

## 3   Swing-hingeable dissections

Given a plane figure, a *tessellation of the plane* is a covering of the plane with copies of the figure without gaps and without overlap [14]. The figure that we use to tile the plane is a *tessellation element* and consists of one or more pieces. The technique of *superposing tessellations* is the following [11, 20]: Take two tessellations with the same pattern of repetition and overlay them so that the combined figure preserves this common pattern of repetition. The line segments in one tessellation induce cuts in the figure of the other, and vice versa.

One more restriction makes the dissections hingeable. A tessellation has *rotational symmetry* if rotating it by some angle smaller than $2\pi$ radians leaves it coinciding in every detail with the original. It possesses *n-fold rotational symmetry* if the angle of rotation is $2\pi/n$. Call a point about which there is rotational symmetry a *symmetry point*. Let $\mathcal{T}_1$ and $\mathcal{T}_2$ be two tessellations of hinged elements with the same pattern of repetition. Let $\mathcal{T}_1$ and $\mathcal{T}_2$ be superposed so that points of intersection between line segments are at symmetry points. If $\mathcal{T}_1$ and $\mathcal{T}_2$ share no line segments of positive length in the superposition, then we call their superposition *proper intersecting*.

**Theorem 1.** *Let $\mathcal{T}_1$ and $\mathcal{T}_2$ be two tessellations of hinged elements that have a superposition that is proper intersecting. Then the induced dissection is hingeable.*

*Proof (Idea).* Let $A$ be a symmetry point of $k$-fold rotational symmetry in both $\mathcal{T}_1$ and $\mathcal{T}_2$. Let $P$ and $P'$ be pieces in $\mathcal{T}_1$ incident on $A$, such that $P'$ is the image of $P$ after a rotation of $2\pi/k$ radians. Let $Q$ and $Q'$ be pieces in $\mathcal{T}_2$ incident on $A$, such that $Q \cap P \neq \emptyset$, at most 2 pieces in $\mathcal{T}_1$ incident on $A$ have nonempty intersection with $Q$, and $Q'$ is the image of $Q$ after a rotation of $2\pi/k$ radians. Then the pieces $P' \cap Q'$ and $P - Q$ resulting from the superposition of $\mathcal{T}_1$ and $\mathcal{T}_2$ can hinge together. The more general case, in which there are more than 2 pieces in $\mathcal{T}_1$ incident on $A$ that have nonempty intersection with $Q$, can be similarly be seen to lead to pieces that can be hinged. □

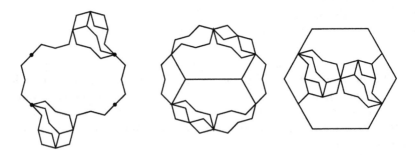

**Fig. 3.** {12/2} Element        **Fig. 4.** Hingeable {12/2} to a hexagon

An example is the dissection of a {12/2} to a hexagon. Start with the tessellation element in Figure 3, which is formed from a {12/2} using hinges. Superposition as in Figure 6 gives the 12-piece hingeable dissection in Figure 4. Small dots indicate the points of 2-fold symmetry. The hinged pieces are in Figure 5.

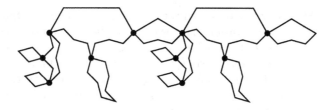

**Fig. 5.** Hinges for a cross of a {12/2} to a hexagon

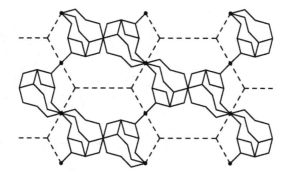

**Fig. 6.** Superposition of tessellations for a {12/2} to a hexagon

The T-strip technique is as follows [11, 20]: Cut a figure into pieces that form a strip element. Then fit copies of this element together to form a strip, rotating every second element in the strip by 180°. Thus every two consecutive elements in the strip share a point of 2-fold rotational symmetry, called an *anchor point*. Similarly create a T-strip for the other figure. Then crosspose the two T-strips, forcing an anchor point in one strip either to overlay an anchor point in the other strip or to fall on a boundary edge of the other strip. We adapt the T-strip technique by further requiring that nonboundary edges of the strips cross only at a common anchor point.

Consider dissecting a pentagon to a square. There is a 4-piece dissection of a pentagon into a hinged assemblage (Figure 7), where all the angles are multiples of $\pi/5$ radians. It folds into a T-strip element as shown in the crossposition of Figure 8. The small dots indicate the anchor points. Note that the right boundary for the strip of squares passes (just barely) to the right of a point where several pieces meet in the pentagon strip, and symmetrically for the left boundary. The resulting 7-piece dissection is in Figure 9.

**Theorem 2.** *Let $S_1$ and $S_2$ be two T-strips of hinged elements. If $S_1$ and $S_2$ are crossposed so that points of intersection between line segments are where two anchor points coincide, or where an anchor point falls on a strip boundary, or where two strip boundaries cross, then the induced dissection is hingeable.*

*Proof (Idea).* Macaulay [23] observed that the strip technique is a type of tessellation method. The crossposition induces two tessellations and their corresponding superposition. The critical points mentioned above then become symmetry points. You can line up multiple copies of the same strip to fill out the plane. If you shift the strips relative to each other by an appropriate offset, and do this for both of the strips that you are crossposing, then the resulting superposed tessellations produce the desired dissection.  □

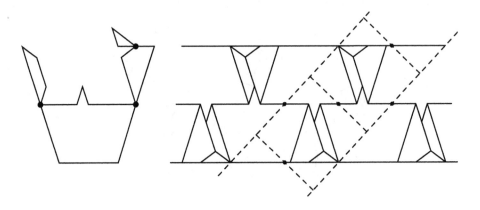

**Fig. 7.** Pentagon unfolded      **Fig. 8.** Crossposition: pentagons and squares

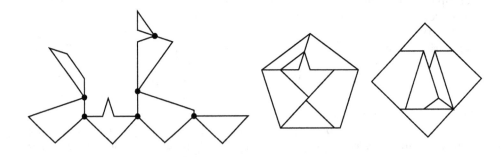

**Fig. 9.** Hinged dissection of a pentagon to a square

Akiyama and Nakamura [1] also recognized the importance of crossing tessellations at the midpoints of certain line segments in order to find hingeable dissections.

Lindgren introduced the *quadrilateral slide*, or *Q-slide*, in [18, 19]. It transforms one quadrilateral to another with the same angles. To emphasize that the dissection is swing-hingeable, we rename the technique the *Q-swing*.      Hanegraaf [15] introduced a technique, called the *trapezoid slide* or *T-slide* in [11], to convert one rectangle to another. It transforms a trapezoid to another with an equal angle. Since this dissection is swing-hingeable, we rename the technique the *T-swing*. Readers are referred to [10] for a description of these swings and a discussion of their properties.

Regular polygons have an internal structure of rhombuses and half-rhombuses that can be exploited in various dissections [11, 20]. For example, a hexagon decomposes into three 60°-rhombuses, and a hexagram into six. Thus there is a simple hinged dissection (Figure 10).

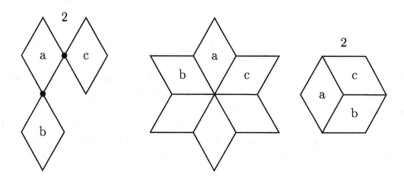

**Fig. 10.** Hinged dissection of a hexagram to two hexagons

## 4    Twist-hingeable dissections

There are two general techniques for converting many of the swing-hingeable dissections to be twist-hingeable. Two pieces that are connected by a hinge are *hinge-snug* if they are adjacent along different line segments in each of the figures formed, and each such line segment has one endpoint at the hinge.

**Theorem 3.** *Let $\mathcal{D}$ be a swing-hingeable dissection such that each pair of pieces connected by a hinge is hinge-snug. We can then replace each swing hinge with a new piece and two twist hinges, so that the resulting dissection $\mathcal{D}'$ is twist-hingeable.*

*Proof (Idea).* Consider the line segments along which a pair of pieces is hinge-snug in the two different figures formed. Take $\delta$ to be one half of the minimum length of the line segments. Identify an isosceles triangle in each of the two pieces, with apex at the hinge point and two equal sides of length $\delta$ coincident with the line segments along which the pieces are hinge-snug. Cut the isosceles triangles out of the pieces and merge them together to give the new piece. □

The conversion for the dissection of a triangle to a square (Figure 1) is shown in Figure 11, where dashed edges indicate the bases of isosceles triangles adjacent

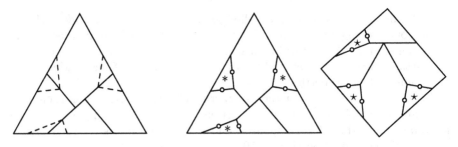

**Fig. 11.** Add isosceles          **Fig. 12.** Triangle to square

to each hinge point. In the 7-piece twist-hingeable dissection (Figure 12) these isosceles triangles are merged together.

The second technique converts a swing hinge to a single twist hinge with no increase in the number of pieces. It applies when the swing hinge connects two pieces that are hinge-snug, and the hinged assemblage on one of the sides of the hinge is "hinge-reflective". A hinged assemblage is *hinge-reflective* if when we flip all pieces in this hinged assemblage on to their other side, then there is no effective change to the whole hinged assemblage.

**Theorem 4.** *Let hinged dissection $\mathcal{D}$ have two hinge-snug pieces, such that the hinged assemblage on one side of the swing hinge is hinge-reflective. Then we can modify the two pieces and replace the swing hinge with a twist hinge.*

*Proof. (Idea)* For each such hinge, cut an isosceles triangle out of one piece and attach it to the piece that is part of the hinge-reflective hinged assemblage. □

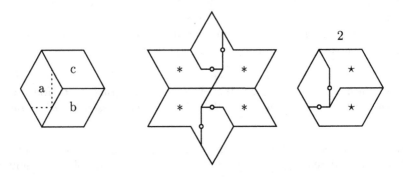

**Fig. 13.** Add isosceles          **Fig. 14.** Hexagram to two hexagons

We convert Figure 10 to Figure 14. One of the hexagons from Figure 10 is on the left, with dotted lines indicating the bases of the isosceles triangles.

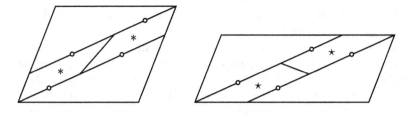

**Fig. 15.** Twist-hingeable parallelogram to same-angled parallelogram

Another general technique transforms a parallelogram to another parallelogram with the same angles (Figure 15). We call this the *parallelogram twist*, or *P-twist*. A beautiful feature of the P-twist is that the pieces are cyclicly hinged. Since rectangles are parallelograms, the P-twist can transform one rectangle to another.

**Theorem 5.** *The P-twist can convert a parallelogram with sides $a$ and $b \leq a$ and nonacute angle $\theta$ to any parallelogram with the same nonacute angle and a side from $a$ up to, but not including, $a + \sqrt{a^2 + b^2 - 2ab\cos\theta}$.*

*Proof (Idea).* To convert one parallelogram into another that is shorter, make the height of each of the two triangles be the height of the desired parallelogram. To produce the correct cut between the resulting trapezoids, place these triangles in the desired parallelogram and measure the angles that the trapezoids make with the top and bottom of the desired parallelogram.

The second term in the above expression represents the length of the longer diagonal in the parallelogram and is derived using the law of cosines. □

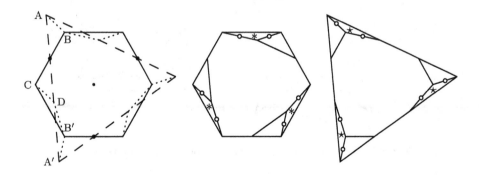

**Fig. 16.** Twist-hingeable dissection of a hexagon to a triangle

Next is a wonderful family of dissections. For any $p > 2$ there is a $(2p + 1)$-piece twist-hingeable dissection of a $\{2p\}$ to a $\{p\}$. The dissection exhibits $p$-fold rotational symmetry. An example is the twist-hingeable dissection of a hexagon to a triangle is in Figure 16.

The dissection technique is: Overlay the $\{2p\}$ and the $\{p\}$ so that their centers coincide and each side of the $\{p\}$ goes through the midpoint of a side of the $\{2p\}$. For each vertex A of the $\{p\}$, do the following: Draw a line segment to the nearest vertex B of the $\{2p\}$. Identify the vertex C of the $\{2p\}$ such that B and C are the endpoints of a side and the side of the $\{p\}$ goes through the midpoint of this side. Draw a line segment from C to the nearest side of the $\{p\}$, with the line segment parallel to the line segment from A to B and meeting the side at D. From D draw a line segment to the next nearest vertex B' of the $\{2p\}$ other than C. In conjunction with the sides of $\{2p\}$ and $\{p\}$, these line segments identify the appropriate cuts to make.

Because the method is related to the completing the tessellation method (See [11, 20]) and yet we have tessellations only when $p = 3$ and $p = 4$, we call it *completing the pseudo-tessellation*.

**Theorem 6.** *Completing the pseudo-tessellation gives a $(2p + 1)$-piece twist-hingeable dissection of a $\{2p\}$ to a $\{p\}$ of equal area.*

*Proof (Idea).* We prove that the length of the line segment from D to A' equals the length of the side of the $\{2p\}$, using formulas for the side lengths of both figures along with a formula for the cotangent of a double angle. □

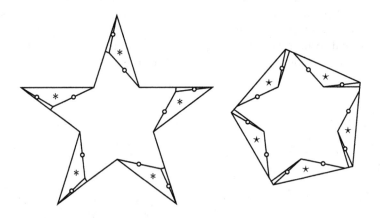

**Fig. 17.** Twist-hingeable dissection of a pentagram to a pentagon

Remarkably, there is another family of dissections of a similar nature. Consider any integers $p > 4$ and $2 \leq q \leq (p + 1)/3$. Then there is a $(2p + 1)$-piece twist-hingeable dissection of the $\{p/q\}$ to the $\{p\}$. The dissection exhibits $p$-fold rotational symmetry. As an example, the twist-hingeable dissection of a pentagram to a pentagon is in Figure 17.

The approach is the same as what we used for the previous family of dissections, if we treat the reflex angles of the $\{p/q\}$ as vertices too and force vertex B to be a reflex angle.

**Theorem 7.** *Completing the pseudo-tessellation gives a $(2p + 1)$-piece twist-hingeable dissection of a $\{p/q\}$ to a $\{p\}$ of equal area whenever $p \geq 3q - 1$.*

*Proof (Sketch).* We prove that the length of the line segment from D to A′ equals the length of the side of the $\{p/q\}$ using formulas for the side lengths of both figures, the law of sines, the law of cosines, and the formula for the cosine of a double angle. We also determine the values of $q$ and $p$ for which the sides of $\{p\}$ will go through the midpoints of the sides of $\{p/q\}$, applying the following lemma. □

**Lemma 1.** *The condition $4\cos(q\pi/p)\cos((q-1)\pi/p) \geq \cos(\pi/p)$ is equivalent to $p \geq 3q - 1$, for positive integers $q > 1$ and $p \geq 2q + 1$.*

*Proof (Idea).* If for any real $q \geq 1$, $4\cos(q\pi/(3q-1))\cos((q-1)\pi/(3q-1)) \geq \cos(\pi/(3q-1))$ and $4\cos(q\pi/(3q-2))\cos((q-1)\pi/(3q-2)) < \cos(\pi/(3q-2))$, the lemma follows. To prove these, we rewrite the angles on the left sides of the inequalities as the sum of $\pi/3$ and what remains, apply the rule for the cosine of a sum of two angles, and simplify using the law of sines and the formulas for sines and cosines of double and triple angles. □

# References

1. Jin Akiyama and Gisaku Nakamura. Dudeney dissection of polygons. In Jin Akiyama, Mikio Kano, and Masatsugu Urabe, editors, *Discrete and Computational Geometry, Japanese Conference, JCDCG'98, Lecture Notes in Computer Science*, volume 1763, pages 14–29. Springer Verlag, 2000.
2. Farkas Bolyai. *Tentamen juventutem.* Typis Collegii Reformatorum per Josephum et Simeonem Kali, Maros Vasarhelyini, 1832.
3. Donald L. Bruyr. *Geometrical Models and Demonstrations.* J. Weston Walch, Portland, Maine, 1963.
4. H. Martyn Cundy and A. P. Rollett. *Mathematical Models.* Oxford, 1952.
5. Erik D. Demaine, Martin L. Demaine, David Eppstein, and Erich Friedman. Hinged dissection of polyominoes and polyiamonds. In *Proceedings of the 11th Canadian Conf. on Computational Geometry*, Vancouver, 1999.
6. Henry E. Dudeney. Puzzles and prizes. Column in *Weekly Dispatch*, April 19, 1896–Dec. 26, 1903.
7. Henry Ernest Dudeney. *The Canterbury Puzzles and Other Curious Problems.* W. Heinemann, London, 1907.
8. William L. Esser, III. Jewelry and the like adapted to define a plurality of objects or shapes. U.S. Patent 4,542,631, 1985. Filed 1983.
9. Howard Eves. *A Survey of Geometry*, volume I. Allyn and Bacon, Boston, 1963.
10. Greg N. Frederickson. Dissections Too! Swingin'. Manuscript, 320 pages, August, 2000.
11. Greg N. Frederickson. *Dissections Plane & Fancy.* Cambridge University Press, New York, 1997.

12. Martin Gardner. *The 2nd Scientific American Book of Mathematical Puzzles & Diversions*. Simon and Schuster, New York, 1961.
13. P. Gerwien. Zerschneidung jeder beliebigen Anzahl von gleichen geradlinigen Figuren in dieselben Stücke. *Journal für die reine und angewandte Mathematik (Crelle's Journal)*, 10:228–234 and Taf. III, 1833.
14. Branko Grünbaum and G. C. Shephard. *Tilings and Patterns*. W. H. Freeman and Company, New York, 1987.
15. Anton Hanegraaf. The Delian altar dissection. Elst, the Netherlands, 1989. First booklet in his projected series *Polyhedral Dissections*.
16. Philip Kelland. On superposition. Part II. *Transactions of the Royal Society of Edinburgh*, 33:471–473 and plate XX, 1864.
17. Glenn A. Kramer. *Solving Geometric Constraint Systems*. MIT Press, Cambridge, MA, 1992.
18. H. Lindgren. Problem E1210: A dissection of a pair of equilateral triangles: Solution. *American Mathematical Monthly*, 63:667–668, 1956.
19. H. Lindgren. A quadrilateral dissection. *Australian Mathematics Teacher*, 16:64–65, 1960.
20. Harry Lindgren. *Geometric Dissections*. D. Van Nostrand Company, Princeton, New Jersey, 1964.
21. Sam Loyd. Mental Gymnastics. Puzzle column in Sunday edition of *Philadelphia Inquirer*, October 23, 1898–1901.
22. Ernst Lurker. Heart pill. 7 inch tall model in nickel-plated aluminum, limited edition of 80 produced by Bayer, in Germany, 1984.
23. W. H. Macaulay. The dissection of rectilineal figures (continued). *Messenger of Mathematics*, 52:53–56, 1922.
24. I. J. Schoenberg. *Mathematical Time Exposures*. Mathematical Association of America, Washington, DC, 1982.
25. Hugo Steinhaus. *Mathematical Snapshots, 3rd edition*. Oxford University Press, New York, 1969.
26. Ian Stewart. *The Problems of Mathematics*. Oxford University Press, Oxford, 1987.
27. H. M. Taylor. On some geometrical dissections. *Messenger of Mathematics*, 35:81–101, 1905.
28. William Wallace, editor. *Elements of Geometry*. Bell & Bradfute, Edinburgh, eighth edition, 1831. First six books of Euclid, with a supplement by John Playfair.
29. Eric W. Weisstein. *CRC Concise Encyclopedia of Mathematics*. CRC Press, 1998.
30. David Wells. *The Penguin Dictionary of Curious and Interesting Geometry*. Penguin Books, London, 1991.
31. Robert C. Yates. *Geometrical Tools, a Mathematical Sketch and Model Book*. Educational Publishers, St. Louis, 1949.

# On Convex Decompositions of Points

Kiyoshi Hosono[1], David Rappaport[2], and Masatsugu Urabe[1]

[1] Department of Mathematics
Tokai University 3-20-1, Orido, Shimizu, Shizuoka 424-8610, JAPAN
[2] Department of Computing and Information Science
Queen's University, Kingston, Ontario, K7L 3N6, CANADA

**Abstract.** Given a planar point set in general position, $S$, we seek a partition of the points into convex cells, such that the union of the cells forms a simple polygon, $P$, and every point from $S$ is on the boundary of $P$. Let $f(S)$ denote the minimum number of cells in such a partition of $S$. Let $F(n)$ be defined as the maximum value of $f(S)$ when $S$ has $n$ points. In this paper we show that $\lceil (n-1)/4 \rceil \leq F(n) \leq \lfloor (3n-2)/5 \rfloor$.

## 1 Introduction

Partitions of point sets into convex subsets is a ubiquitous problem in discrete geometry. The domain is a finite set of points in the plane, which we will usually denote by $S$. The points are assumed to be in general position, that is, no three points on a line. A subset of $S$ that are the vertices of a convex $k$-gon is a *convex subset*. A convex subset of $S$ with no points of $S$ in its open interior is called an *empty convex subset*. The landmark paper of Erdős and Szekeres [3] asks for the value of the smallest integer $A(k)$ such that any set of $A(k)$ points contains a convex subset of size $k$. Subsequently a similar question is asked by Erdős in [2] for the value of the smallest integer $B(k)$ such that any set of $B(k)$ points contains an empty convex subset of size $k$. Values for $B(k)$ are known for all values of $k$ except $k = 6$. In [2] it is shown that $B(3) = 4$, and $B(4) = 5$. In [4] it is shown that $B(5) = 10$, in fact, Figure 1 shows a 9 point set with no empty convex pentagons. Horton [5] gives a construction showing that $B(7)$ is not finite, that is, there are arbitrarily many points with no empty convex 7-gons. The value of $B(6)$ is not known, and this remains a tantalizing long outstanding open problem. Some experimental results showing that $B(6) > 20$ as well as an algorithm for computing maximum empty convex subsets were first shown in [1]. A 26 point set with no empty hexagons is given in [6]. Some combinatorial results on partitions of points sets in two and three dimensions, are presented in [7] and [8].

In this paper we consider the following variation on the convex partition theme. Given a set of points $S$ we want to partition $S$ into empty convex subsets such that the union of the subsets form a single simple polygon $P$, and every point in $S$ is on the boundary of $P$. Here, we call such an empty convex subset of $k$-gon in $P$ a *k-cell*. Given $S$, let $f(S)$ represent the minimum number of cells

J. Akiyama, M. Kano, and M. Urabe (Eds.): JCDCG 2000, LNCS 2098, pp. 149–155, 2001.
© Springer-Verlag Berlin Heidelberg 2001

obtained in such a partition of $S$. Let $F(n)$ denote the maximum value of $f(S)$ over all sets $S$ with n points.

For example, a 9 point set $S$ in Figure 1 gives $f(S) = 4$. For any simple polygon with order $n$, since we can always triangulate its interior, the trivial upper bound of $F(n)$ is $n - 2$.

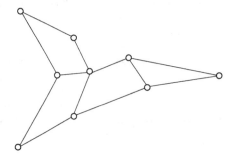

**Fig. 1.** The construction with 9 points, partitioned into three 4-cells and one 3-cell.

In the next section we prove the following theorem.

**Theorem 1.** $\left\lceil \dfrac{n - 1}{4} \right\rceil \leq F(n) \leq \left\lfloor \dfrac{3n - 2}{5} \right\rfloor$

## 2   Upper and lower bounds

As was shown in [5] there exists sets $n = 2^k$ with no empty convex hexagons giving the lower bound $n/4$. We obtain here the lower bound for any integer $n$.

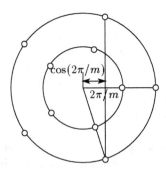

**Fig. 2.** An example to illustrate the lower bound

**Lemma 1.** $F(n) \geq \left\lceil \frac{n-1}{4} \right\rceil$.

*Proof.* We first construct a set of $n = 2m$ points, such that we have $m$ points equally space on a circle of radius 1, and the other $m$ points on a circle of radius $\cos(2\pi/m) + \epsilon$, where $\epsilon$ is a small positive value. If necessary we perturb the points slightly so that no 3 points are on the same line. See Figure 2. If $n$ is odd, place an additional point close to the center. This ensures that there are no empty triangles made up of points on the outer circle. Thus no cell uses more than two points from the outer circle, achieving the desired bound.          □

For the upper bound we present an iterative construction. Let $CH(S)$ denote the anti-clockwise circular sequence of points when traversing the vertices of the convex hull of $S$. The input to the construction are the points of $S$ and two consecutive points on $CH(S)$ where the line segment between the two points is called a *starting edge*. We induce a permutation of the points $\Pi(S) = (1, 2, 3, \ldots n)$. Thus let points 1 and 2 be a fixed pair of consecutive points on $CH(S)$. The rest of the sequence is obtained by sweeping a ray anchored at 1, and passing through 2, 3, .., $n$, which is implicitly the other neighbour of 1 on $CH(S)$. See Figure 3. We assume that $n \geq 3$. It is easy to see that $F(3) = 1$ and $F(4) \leq 2$. Using the fact that every 5 points has an empty convex 4-gon, that is, $B(4) = 5$, we deduce that $F(5) \leq 2$ and $F(6) \leq 3$. For values of $n \geq 7$ we use an iterative approach.

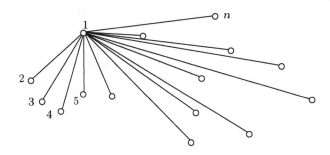

**Fig. 3.** The points in $S$ are ranked according their order in an angular sweep of a ray anchored at 1.

After each iteration there are two basic outcomes.

In the first instance we look for a point $w$ so that $123w$ is a 4-cell. Then the points 3 and $w$ are neighbours in $CH(S \setminus \{1, 2\})$, and the edge $3w$ is the starting edge for the next iteration. Observe that 3 is always in $CH(S \setminus \{1, 2\})$. If $\triangle 13n$ is empty, then set $w$ to $n$, otherwise set $w$ to the neighbour of 3 on $CH(S \setminus \{1, 2\})$, contained in $\triangle 13n$. If $123w$ is convex, then it is also empty and we are done. See Figure 4.

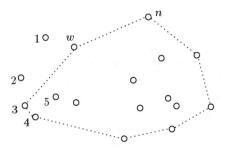

**Fig. 4.** The case where $123w$ is convex.

Suppose $123w$ is not convex. The second possibility operates on a 7 point subset. Then we will find either 6 or 7 point subset $S'$ such that $S'$ can be decomposed into 2 or 3 cells, respectively, so that one of the cells is $\triangle 123$, and another cell uses the edge $uv$ so that $u$ and $v$ are consecutive on $CH(S \setminus S' \cup \{u, v\})$. Thus we can use $uv$ as the starting edge for the next iteration. We will detail the decomposition process on this 7 point subset. Observe that the maximum number of cells produce by this iterative process is $\lfloor (3n - 2)/5 \rfloor$.

Consider the lines passing through the points 1,3 and 2,3. This partitions the plane into 4 regions plus the lines themselves. An important consequence of the fact that there is no fourth point that makes an empty convex 4-gon with 1,2,3 is that one of these 4 regions contains all of $S \setminus \{1, 2, 3\}$. See Figure 5.

Let $S_0 = \{4, 5, 6\}$ and $S_1 = S \setminus \{1, 2, 3, 4, 5, 6\}$. Consider the inner common tangent line $L$ of $CH(S_0)$ and $CH(S_1)$ such that one half plane bounded by $L$ contains $\{1\}$ and $S_0$. Let $p$ be a point in $L$ and $S_0$ and $q$ be a point in $L$ and $S_1$. Let the two points in $S_0 \setminus \{p\}$ be called $s$ and $t$. See also Figure 5. There are two primary cases to consider.

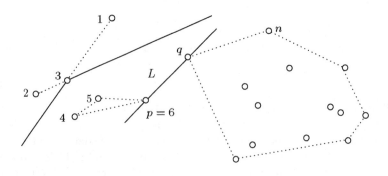

**Fig. 5.** We find a pair of points $p$ and $q$ that are in $CH(S \setminus S' \cup \{p, q\})$.

Case A) The line through $s$ and $t$ does not intersect the line segment between $p$ and $q$. Then $pqst$ is an empty convex 4-gon. For convenience we rename the points $q_0, q_1, q_2, q_3$ as they appear in anti-clockwise order on the boundary of the 4-cell with $q = q_0$, where $q_3 = p$. See Figure 6. If a line through $q_0, q_1$ does not intersect the edge 13, then we obtain a 4-cell $q_0 13 q_1$ to add to $pqst$ and $\triangle 123$.The starting edge for the next iteration is $q_0 q_3$. Otherwise, let $i$ be the largest integer such that the directed ray anchored at $q_{i-1}$ and passing through $q_i$, intersects the edge 13. If $i \neq 3$ we set $j = i + 1$. If the line through $q_i, q_j$ also intersects with the edge 13, then we use three cells $q_i 32 q_j, q_0 q_1 q_2 q_3$ and $\triangle 123$, and $q_0 q_3$ is the starting edge for the next iteration. If $q_i, q_j$ does not intersect with the edge 13 we use $q_i 13 q_j, q_0 q_1 q_2 q_3$ and $\triangle 123$ with the starting edge $q_0 q_3$. If $i = 3$, the line through $q_0, q_3$ necessarily intersects the edge 13. In this case we obtain two cells, a 5-cell $q_1 q_2 q_3 32$ and $\triangle 123$, and use $2q_1$ as the starting edge for the next iteration.

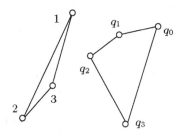

**Fig. 6.**

Case B) The line through $s$ and $t$ intersects the line segment between $p$ and $q$. Let $t$ be the point closest to the segment $pq$. Denote the anti-clockwise angle of $(x, y, z)$ for the points $x, y, z$ by $\angle(x, y, z)$. If $\angle(3, s, t)$ is convex, then we obtain three cells $3stq1, ptq$ and $\triangle 123$ with the starting edge $pq$ for the next iteration. Otherwise, we examine the pentagon $2pts3$, and if it is convex we use it with $\triangle 123$ and $tpq$. The starting edge for the next iteration is $pq$. However, $2pts3$ may not be convex, it may not even be simple. Thus we need to examine a few more sub-cases.

We enumerate the possible cases according to which of $p, s, t$ is equal to 6.

1. $p = 6$: This implies that $2pts3$ is convex so nothing more needs to be done.
2. $s = 6$: Let $r \in S_1$ such that the angle $\angle(r, s, 1)$ is maximized. Since the line through $s$ and $t$ intersects the line segment between $p$ and $q$ we deduce that $p < t$. See Figure 7. We obtain cells $rts$, and $\triangle 123$, and we use $sr$ as the starting edge for the next iteration. If $\angle(t, p, 3)$ is convex then we also use the 5-cell $13pts$. On the other hand if $\angle(3, p, t)$ is convex we use the 5-cell $2rtp3$.

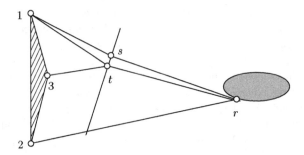

**Fig. 7.** Either $\angle(3, p, t)$ is convex or $\angle(t, p, 3)$ convex.

3. $t = 6$: Let $r, r' \in S_1$ such that $\angle(r, t, 1)$ and $\angle(r', t, 1)$ are maximum and minimum, respectively. We consider the situation according to the position of $p$. Note that $p \notin \triangle 1st$. See Figure 8.
   (a) $p \in \triangle 13s$. We obtain the two cells $\triangle 123$ and $32tsp$, and use the starting edge $2t$ for the next iteration.
   (b) $p \in \triangle 23s$. Two of the three desired cells are $\triangle 123$ and $13ps$. At least one of $tspr$ or $1str'$, must be convex. If the 4-gon $tspr$ is convex, choose this as the third cell and $tr$ as the starting edge for the next iteration. Otherwise, choose $1str'$ as the third cell and $tr'$ for the starting edge.
   (c) $p \in \triangle 2st$. We obtain three cells $123, 32ps$ and $2rtp$, and the edge $tr$ is the starting edge for the next iteration.

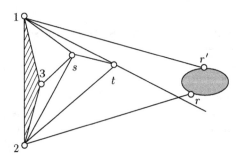

**Fig. 8.** We can put $p$ in $\triangle 13s$ or $\triangle 23s$ or $\triangle 2st$.

Thus this construction proves the following lemma.

**Lemma 2.** $F(n) \leq \left\lfloor \dfrac{3n-2}{5} \right\rfloor$.

Combining lemmas 1 and 2 proves our theorem.

# 3 Discussion

We give upper and lower bounds for a variation of convex partitions. There is still a substantial spread between the upper and lower bound. From our point of view a more complicated algorithm may be able to improve the upper bound somewhat, but we conjecture that the correct bound is $n/2$. The lower bound although easy to prove seems to be more difficult to improve upon, as there seems to be a connection with settling the issue of whether $B(6)$ is finite or not. Observe that, $F(n) < (n-2)/3$ for all but finitely many $n$, implies the finiteness of $B(6)$.

# Acknowledgement

We would like to thank the referee for the careful reading of the manuscript and the many helpful comments.

# References

1. D. Avis and D. Rappaport. Computing the largest empty convex subset of a set of points. In *Proc. 1st Annu. ACM Sympos. Comput. Geom.*, 161–167, 1985.
2. P. Erdős. Some combinatorial problems in geometry. *Lecture Notes in Mathematics*, 792:46–53, 1980.
3. P. Erdős and G. Szekeres. A combinatorial problem in geometry. *Compositio Math.*, 2:463–470, 1935.
4. H. Harborth. Konvexe Fünfecke in ebenen Punktmengen. *Elem. Math.*, 33:116–118, 1978.
5. J. D. Horton. Sets with no empty convex 7-gons. *Canad. Math. Bull.*, 26:482–484, 1983.
6. M. H. Overmars, B. Scholten, and I. Vincent. Sets without empty convex 6-gons. *Bull. EATCS.*, 37:160–168, 1989.
7. M. Urabe. On a partition into convex polygons. *Discrete Appl. Math.*, 64:179–191, 1996.
8. M. Urabe. Partitioning point sets into disjoint convex polytopes. *Comput. Geom. Theory Appl.*, 13:173–178, 1999.

# Volume Queries in Polyhedra

John Iacono* and Stefan Langerman**

Department of Computer Science
Rutgers University
New Brunswick, NJ, USA

**Abstract.** We present a simple and practical data structure for storing a (not necessarily convex) polyhedron $P$ which can, given a query surface $S$ cutting the polyhedron, determine the volume and the area of the portion of the polyhedron above $S$. The queries are answered in a time linear in the size (complexity) of $S$. The space and preprocessing time for this data structure are linear in the size of $P$. We also present an intermediary data structure for planar graphs which is of use in other application domains.

## 1 Introduction: Polygons

This paper studies a very natural generalization of the following problem in $R^2$: Given a simple polygon $P$, construct a data structure which, given a query chord $c$ (a line segment cutting $P$ into exactly two pieces, whose boundaries are on edges of $P$), returns the area of the polygon above $c$. All geometric objects in this paper are assumed to have nonnegative coordinates.

In [2] (also [3]), a solution is proposed for convex polygons: consider the polygon $P$ with edges $e_1, ..., e_n$ oriented in clockwise order starting at the lowest vertex of $P$ (smallest $y$ coordinate). Let $a(e)$ be the signed area of the trapezoid defined by the edge $e$ and its projection on the $x$ axis. The area $a(e)$ is negative iff the edge $e_i$ is a bottom edge, where the interior of the polygon lies above (higher $x$ coordinate) the edge. We call $a(e)$ the projective area of $e$. The area of $P$ is simply $\sum_{i=1}^{n} a(e_i)$. Suppose that the chord $c$ has its endpoints on the edges $e_j$ and $e_k$, where $j \leq k$.. Let $e'_j$ and $e'_k$ be the portions of $e_j$ and $e_k$ below $c$. The area of $P$ above $c$ is $\sum_{i=j}^{k} a(e_i) - a(e'_j) - a(e'_k) + a(c)$.

By storing $s_t = \sum_{i=1}^{t} a(e_i)$ for $t = 1, ..., n$, we can compute $\sum_{i=j}^{k} a_i = s_k - s_{j-1}$ in constant time, and so we get:

**Theorem 1** *Given a convex polygon $P$ with $n$ vertices, there is a data structure that after $O(n)$ preprocessing time can return the area of $P$ above a query chord $c$ in constant time.*

For non convex polygons, [3] describes a data structure that requires $O(n \log n)$ preprocessing time and that give the area of the polygon above a query chord in $O(\log n)$ time. Recently, in [1], the authors notice that if the query chord doesn't cross any edge of the polygon, the algorithm used for Theorem 1 works without modification. They also show other applications of the structure.

---

\* `iacono@cs.rutgers.edu`, Research supported by NSF grant CCR-9732689
\*\* `lfalse@cs.rutgers.edu`

J. Akiyama, M. Kano, and M. Urabe (Eds.): JCDCG 2000, LNCS 2098, pp. 156-159, 2001.

## 2   Polyhedra

We now discuss a three-dimensional generalization to the polygon problem. It is interesting to note that the ideas presented here could be used to extend the data structure to higher dimensions, although we do not know of any applications of such a structure.

In order to generalize the problem to 3 dimensions, we first need to generalize the concept of a chord. In our setting, we have a polyhedron $P$ with $n$ vertices. A *query surface* $S$ is a polyhedral surface separating $P$ into exactly two pieces, and the boundary of $S$ lies on the surface of $P$. Note that this implies that the description of $S$ will be at least as large as the number of edges of $P$ crossed by the boundary of $S$. On the other side, this is usually likely to be much smaller than $n$. A volume query on a surface $s$ will return the volume of the portion of $P$ lying above $s$. We will show the following:

**Theorem 2** *Given a polyhedron of size $n$, there is a structure that with $O(n)$ preprocessing time will answer volume queries in time linear in the size of the query surface.*

The idea will be similar to the 2-dimensional case. First we define the (signed) *projective volume* $v(f)$ of a polygon $f$ to be the volume of the polyhedron defined by $f$ and its projection on the plane $z = 0$. For a facet $f$ of $P$, $v(f)$ will be negated if the inside of $P$ is above $f$. The volume of $P$ can then be expressed as the sum for all facets $f$ of $P$ of $v(f)$.

So, in order to answer a volume query for a surface $s$, we need to

(i) compute the sum of $v(f)$ for all facets $f$ strictly above $s$,

(ii) add the projective volume of all the upper portions of the facets that are cut by $s$ and

(iii) subtract the projective volume of all the polygons forming $s$.

Let $k$ be the size of the description of $s$. We can easily compute the value for (iii) in $O(k)$ time.

By storing a variant of the 2-dimensional data structure of Theorem 1 in each facet of $P$, it is possible to find the projective volume of a cut facet in $O(1)$ time. This can be achieved by replacing the projective area of an edge with the volume of a prism defined by the edge, the edge's projection onto the $z = 0$ plane and any fixed point on the intersection of the $z = 0$ plane and the plane containing the facet. Since the number of cut facets is less than $k$, we can compute (ii) in $O(k)$ time.

So the only task remaining is to compute (i). In order to do that, we need some preprocessing: compute the projective volume of each face. Label each face with this value. from this point on we will treat the polyhedra as a planar graph, with each face labeled with its projective volume. We can now reduce our problem to a problem for embedded planar graphs.

## 3   Planar Graphs

We define a query cut of a planar graph to be a set of directed edges that when removed from a planar graph separate a connected component, the query component, from the rest of the graph. The edges of a query cut are to be presented in clockwise order about the query component.

**Theorem 3** *Given an embedding of a planar graph $G$, where weights are assigned to every face, there is a structure that with $O(n)$ preprocessing time will return the sum of the weights of the faces in a query component given a query cut in time linear in the number of edges in the cut.*

The boundary of a query surface for the polyhedron clearly defines a query cut for the planar graph, and the answer returned by the planar graph structure will answer to the point (i) of the previous section.

Note that this data structure is of independent interest and could be useful in areas such as cartography and computer networks. One possible application would be in geometric routing, where the user circles an area on a network to broadcast a message to, it may be useful to first compute the size or other statistic of the selected range to see if it lies within acceptable parameters. See

We will first solve a slightly simpler problem: define a query circuit to be a circular vertex-disjoint alternating sequence of vertices and edges in an embedding of a planar graph.

**Theorem 4** *Given an embedding of a planar graph $G$, where weights are assigned to every face, there is a structure that with $O(n)$ preprocessing time will return the sum of the weights of the faces inside a query circuit in time linear in the size of the circuit.*

**Preprocessing:** First, construct a directed path $Q$ on the plane that visits every face at least once, and starts and ends in the outside face. This can be done quickly by constructing the Eulerian path of a spanning tree for the dual graph of $G$. For each face $f$, assign the weight of that face to some point on $Q$ in that face. For any point $p$ on $Q$, define $t(p)$ to be the sum of the weight points from the beginning of the path to $p$.

Consider every cell as a clockwise oriented cycle (i.e. every edge of $G$ becomes two directed edges). Start with all edges weights at 0, and follow the path $Q$. For each edge $Q$ crosses (with, say, intersection $p$), add $t(p)$ to its weight if it is a right turn from $Q$, and $-t(p)$ if it is a left turn.

**Query:** Traverse the edges of the query circuit in clockwise order. Sum the weights of the directed edges traversed. This will actually compute the sum of all weight points on $Q$ (and weights of the corresponding faces) that are inside the query circuit. This can be done in time linear in the size of the query circuit.

We now describe the modification for obtaining Theorem 3:

**Preprocessing:** The same preprocessing is done as for the query cycle based structure. In addition for each face, defined by edges $e_1 \ldots e_m$ in clockwise order, the structure used in Theorem 1 is used to store $w(e_1) \ldots w(e_m)$.

**Query:** The edge cut set defines a query circuit. However, as there may be many edges on the query circuit between each pair of cut edges, it is undesirable to directly sum the weights of each edge on the query circuit as is done in the previous structure. However, by querying the structure of Theorem 1 for every face between adjacent query edges, one can compute the sum of the weights of all edges on a face that are on the query circuit in constant time. Thus the entire query may be carried out in time linear in the size of the edge cut set.

# References

1. R. Boland and J. Urrutia. Polygon Area Problems. *Proc. of the 12th Canadian Conf. on Computational Geometry*, Fredericton, NB, Canada, 2000.
2. F. Contreras-Alcalá. Cutting polygons and a problem on illumination of stages. *Masters thesis, Dept. Comp. Sci. University of Ottawa*, Ottawa, ON, Canada, 1998. http://www.csi.uottawa.ca/~fhca/thesis/
3. J. Czyzowicz, F. Contreras-Alcalá and J. Urrutia. On measuring areas of polygons. *Proc. of the 10th Canadian Conf. on Computational Geometry*, Montréal, QC, Canada, 1998.
4. Julio C. Navas. *Geometric Routing in a Datagram Network*. Ph.D. Thesis. Rutgers University, 2001.

# Sum of Edge Lengths of a Graph Drawn on a Convex Polygon

Hiro Ito, Hideyuki Uehara, and Mitsuo Yokoyama

Department of Information and Computer Sciences
Toyohashi University of Technology, Toyohashi, 441-8580, Japan
{ito,uehara,yokoyama}@tutics.tut.ac.jp

**Abstract.** Let $x_0, x_1, ..., x_{n-1}$ be vertices of a convex $n$-gon $P$ in the plane, where, $x_0x_1$, $x_1x_2$, $...$, $x_{n-2}x_{n-1}$, and $x_{n-1}x_0$ are edges of $P$. Let $G = (N, E)$ be a graph, such that $N = \{0, 1, ..., n-1\}$. Consider a graph drawing of $G$ such that each vertex $i \in N$ is represented by $x_i$ and each edge $(i, j) \in E$ is drawn by a straight line segment. Denote the sum of lengths of graph edges in such drawing by $S_P(G)$. If $S_P(G) \le S_P(G')$ for any convex $n$-gon $P$, then we write as $G \preceq_l G'$. This paper shows two necessary and sufficient conditions of $G \preceq_l G'$. Moreover, these conditions can be calculated in polynomial time for any given $G$ and $G'$.

## 1 Introduction

Let $x_0, x_1, ..., x_{n-1}$ be vertices of a convex $n$-gon in the plane (each internal angle may be equal to $\pi$), where, $x_0x_1$, $x_1x_2$, $...$, $x_{n-2}x_{n-1}$, and $x_{n-1}x_0$ are edges of the $n$-gon. Denote the length of the line segment $x_ix_j$ by $d(i, j)$. $i \bmod n$ denotes $i'$ such that $i \equiv i' \pmod n$ and $0 \le i' \le n-1$.

Let $G = (N, E)$ be a graph with a vertex set $N = \{0, 1, ..., n-1\}$ and an edge set $E$. Parallel edges and self-loops are permitted in $G$. In $G = (N, E)$, $E$ may be denoted by $E(G)$. In this paper, a vertex set of each graph is fixed to $N = \{0, 1, ..., n-1\}$. Define a length of $G$ with respect to an $n$-gon $P$ as

$$S_P(G) := \sum_{(i,j) \in E(G)} d(i, j).$$

$S_P(G)$ can be regarded as a sum of edge length of a graph $G$ drawn in the plane such that each vertex of $G$ is equal to a corresponding vertex of $P$ and each edge of $G$ is written by a straight line segment. Graph drawing has recently become a very important research area and the sum of edge lengths is one of the crucial criteria for evaluating drawing methods[1].

**A Partial-Order "$\preceq_l$" Based on $S_P(G)$.**
We introduce a partial-order "$\preceq_l$" as follows. Let $G$ and $G'$ be two graphs. If $S_P(G) \le S_P(G')$ for any convex polygon $P$, then $G \preceq_l G'$ ("l" means length).

J. Akiyama, M. Kano, and M. Urabe (Eds.): JCDCG 2000, LNCS 2098, pp. 160–166, 2001.
© Springer-Verlag Berlin Heidelberg 2001

If $G \preceq_l G'$ and $G \neq G'$, then $G \prec_l G'$. $\preceq_l$ is clearly a partial-order.

For any two subsets $X, Y \subseteq V$, $E(X, Y; G)$ denotes an edge set between $X$ and $Y$, i.e.,

$$E(X, Y; G) := \{(i, j) \in E \mid i \in X, \ j \in Y\}.$$

For $i, j \in N$, define

$$N[i, j] := \begin{cases} \{i, \ i+1, \ \ldots, \ j\}, & \text{if } i \leq j, \\ \{i, \ i+1, \ \ldots, \ n-1, \ 0, \ 1, \ \ldots, \ j\}, & \text{if } i > j. \end{cases}$$

If $N[i, j]$ is a proper subset of $N$, $N[i, j]$ is called a *neighbor-cut*.

If there is no neighbor-cut $N[i, j]$ such that $E(N[i, j]; G) = \emptyset$, then $G$ is called *neighbor-connected*. We define

$$E_q := \{(i, \ i+q \bmod n) \mid i \in N\}$$

for each integer $0 \leq q \leq \lfloor n/2 \rfloor$. $G_q := (N, \ E_q)$. $G_q$ is a 2-regular graph.

The authors have already presented the following properties[2,3].

**Theorem A**

*(1)* $S_P(G_q) \prec_l S_P(G_{q+1})$ *for* $q = 0, \ 1, \ \ldots, \ \lfloor n/2 \rfloor - 1$.

*(2)* *For any 2-regular graph* $G(\neq G_{\lfloor n/2 \rfloor})$, $G \prec_l G_{\lfloor n/2 \rfloor}$.

*(3)* *If* $G(\neq G_1)$ *is a neighbor-connected 2-regular graph,* $G_1 \prec_l G$. □

In this paper, we present a general rule on $S_P(G)$, which includes Theorem A. For explaining this rule, we give some notations.

$E(X, V-X; G)$ can be also represented as $E(X; G)$ for notational simplicity. $|E(X, Y; G)|$ and $|E(X; G)|$ may be written as $c(X, Y; G)$ and $c(X; G)$, respectively. A singleton set $\{x\}$ may be simply written as $x$. $N(i, j) := N[i, j] - \{i, j\}$, $N(i, j] := N[i, j] - \{i\}$, $N[i, j) := N[i, j] - \{j\}$.

**Cross-Operation and a Partial-Order "$\preceq_o$".**

If two distinct edges $(i, j), (h, k) \in E$ satisfy that $h, k \in N(i, j)$ or $h, k \in N(j, i)$, then we say that $(i, j)$ and $(h, k)$ are *separated*. If "$h \in N(i, j)$ and $k \in N(j, i)$" or "$k \in N(i, j)$ and $h \in N(j, i)$," then we say that $(i, j)$ and $(h, k)$ are *crossing*.

Let $(i, j)$ and $(h, k)$ be a separated pair of edges. Without loss of generality, we assume that $i \in N(h, j]$ and $k \in N(j, h]$ (see Figure 1 (a)). By deleting $(i, j)$ and $(h, k)$ from $E(G)$ and putting $(i, k)$ and $(h, j)$ in $E(G)$, a new graph $G' = (N, E')$ is obtained (see Figure 1). This operation is called a *cross-operation*.

If $G'$ can be obtained from $G$ by applying a sequence of cross-operations, then $G \preceq_o G'$ ("o" means operation). $G \prec_o G'$ means $G \preceq_o G'$ and $G \neq G'$.

**A Partial-Order '$\preceq_c$' Based on the Size of Neighbor-Cuts.**

If $c(N[i, j]; G) \leq c(N[i, j]; G')$ for every neighbor-cut $N[i, j]$, then $G \preceq_c G'$ ("c" means cut). We will show in Corollary 1 that if $G \preceq_c G'$ and $G' \preceq_c G$, then

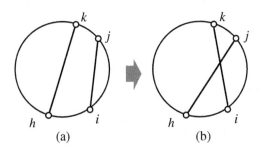

**Fig. 1.** Cross-operation.

$G = G'$. Thus we can say $G \prec_c G'$ if $G \preceq_c G'$ and $G \neq G'$.

**Main Theorem**    *Three partial orders $\preceq_l$, $\preceq_o$, and $\preceq_c$ are equivalent for any pair of graphs $G = (N, E)$ and $G' = (N, E')$ with $|E| = |E'|$. That is, each one of $G \preceq_l G'$, $G \preceq_o G'$, and $G \preceq_c G'$ implies the others.*

Theorem A is a corollary of the main theorem.

## 2    Proof

**Lemma 1.** *If $G \preceq_o G'$, then $G \preceq_l G'$.*

*Proof.* It is clear from the triangle inequality.    □

**Lemma 2.** *If $G \preceq_l G'$, then $G \preceq_c G'$.*

*Proof.* Suppose that $G \preceq_c G'$ does not hold, i.e., there are $i, j \in N$ such that $c(N[i, j]; G) > c(N[i, j]; G')$. We construct a polygon $P$ satisfying $S_P(G) > S_P(G')$ as follows. $X = \{x_k \mid k \in N[i, j]\}$ and $Y = \{x_k \mid k \in N(j, i)\}$. Put all vertices $x \in X$ in a circle whose center is $(0, 0)$ and radius is $r$. Let $p > 0$ be a real number. Put all vertices $x \in Y$ in a circle whose center is $(p, 0)$ and radius is $r$. We can locate all vertices satisfying the above conditions and convexity for each $r$ and $p$. By letting $p$ be far larger than $r$, $S_P(G) > S_P(G')$.    □

**Lemma 3.** *For any pair of graphs $G = (N, E)$ and $G' = (N, E')$ with $|E(G)| = |E(G')|$, $G \preceq_c G'$ implies $G \preceq_o G'$.*

For proving this lemma, we need some lemmas as follows.

**Lemma 4.** *Let $G = (N, E)$ and $G' = (N, E')$ be two graphs. For any neighbor-cut $N[i, j]$, $c(N[i, j]; G) - c(N[i, j]; G')$ is even if and only if $\sum \{c(k; G) - c(k; G') \mid k \in N[i, j]\}$ is even.*

*Proof.*

$$\sum_{k \in N[i,j]} c(k;\, G) = c(N[i,j];\, G) + 2c(N[i,j],\, N[i,j];\, G) \tag{1}$$

$$\sum_{k \in N[i,j]} c(k;\, G') = c(N[i,j];\, G') + 2c(N[i,j],\, N[i,j];\, G') \tag{2}$$

From (1)−(2), the statement is assured.     □

For $i, j \in N$, a graph obtained from $G = (N, E)$ by contracting $N(j, i)$ to a vertex is denoted by $G[i, j]$. For an integer $k$ $(0 \leq k \leq n)$,

$$N_{=k} := \{(i,j) \mid i, j \in N,\ |N[i,j]| = k\},$$
$$N_{<k} := \{(i,j) \mid i, j \in N,\ |N[i,j]| < k\},$$
$$N_{\leq k} := \{(i,j) \mid i, j \in N,\ |N[i,j]| \leq k\}.$$

**Lemma 5.** *Let $0 \leq k \leq n - 1$ be an integer. If $c(N[i,j]; G) = c(N[i,j]; G')$ for all $(i,j) \in N_{\leq k}$, then $G[i,j] = G'[i,j]$ for all $(i,j) \in N_{\leq k}$.*

*Proof.* We use induction. If $k = 0$ or $1$, it is clear. Assume that for an $h \geq 2$ if $k < h$, the statement is correct. Further assume that $c(N[i,j]; G) = c(N[i,j]; G')$ for all $(i,j) \in N_{\leq h}$. From these assumptions, we derive that $G[i,j] = G'[i,j]$ for all $(i,j) \in N_{\leq h}$.

Consider $(i,j) \in N_{=h}$. If $c(i,j; G) = c(i,j; G')$, then $G[i,j] = G'[i,j]$. Then we assume that $c(i,j; G) > c(i,j; G')$ without loss of generality. From $c(i; G) = c(i; G')$ and $c(i, N(i,j); G) = c(i, N(i,j); G')$,

$$c(i, N(j,i); G) < c(i, N(j,i); G'). \tag{3}$$

Similarly,

$$c(j, N(j,i); G) < c(j, N(j,i); G'). \tag{4}$$

From the assumption,

$$c(N(i,j), N[j,i]; G) = c(N(i,j), N[j,i]; G'),$$
$$c(N(i,j), i; G) = c(N(i,j), i; G'), \text{ and}$$
$$c(N(i,j), j; G) = c(N(i,j), j; G'),$$

hence

$$c(N(i,j), N(j,i); G) = c(N(i,j), N(j,i); G'). \tag{5}$$

From (3), (4), and (5), $c(N[i,j]; G) < c(N[i,j]; G')$, contradicting the assumption.     □

**Corollary 1.** *If $G \preceq_c G'$ and $G' \preceq_c G$, then $G = G'$.*

*Proof:* By letting $k = n - 1$ in Lemma 5, the statement can be obtained.     □

**Lemma 6.** *Let $k$ ($2 \leq k \leq n - 1$) be an integer. Let $i, j$ be integers such as $(i, j) \in N_{=k}$. If $c(N[i'\ j']; G) = c(N[i', j']; G')$ for all $(i', j') \in N_{<k}$ and $c(N[i, j]; G) < c(N[i, j]; G')$, then $c(i, j; G) > c(i, j; G')$.*

*Proof:* From Lemma 5, we obtain

$$c(i, N(i, j); G) = c(i, N(i, j); G') \text{ and}$$
$$c(j, N(i, j); G) = c(j, N(i, j); G').$$

From the assumption,

$$c(N(i, j); G) = c(N(i, j); G').$$

Thus,

$$c(N(i, j), N(j, i); G) = c(N(i, j), N(j, i); G').$$

By considering $c(N[i, j]; G) < c(N[i, j]; G')$,

$$c(i, N(j, i); G) < c(i, N(j, i); G') \text{ or } c(j, N(j, i); G) < c(j, N(j, i); G')$$

hold. Without loss of generality,

$$c(i, N(j, i); G) < c(i, N(j, i); G').$$

We have

$$c(i; G) = c(i; G') \text{ and } c(i, N(i, j); G) = c(i, N(i, j); G'),$$

thus $c(i, j; G) > c(i, j; G')$ is obtained.     □

Now, we can prove Lemma 3.

*Proof of Lemma 3:* Assume that $G \preceq_c G'$. Let $k$ ($0 \leq k \leq n - 1$) be a largest integer satisfying that

$$c(N[i, j]; G) = c(N[i, j]; G') \text{ for all } (i, j) \in N_{\leq k}.$$

From Lemma 5,
$$G[i, j] = G'[i, j] \text{ for all } (i, j) \in N_{\leq k}.$$

Then if $k = n - 1$, the statement is trivial.

Otherwise, let $i_0, j_0 \in \{N \mid |N[i_0, j_0]| = k + 1\}$ be a pair such that $c(N[i_0, j_0]; G) \neq c(N[i_0, j_0]; G')$ (see Figure 2).

From $G \preceq_c G'$,
$$c(N[i_0, j_0]; G) < c(N[i_0, j_0]; G').$$

From Lemma 6, $c(i_0, j_0; G) > c(i_0, j_0; G')$, hence $(i_0, j_0) \in E(G)$.

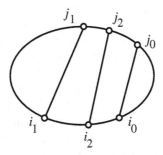

**Fig. 2.** Relation of $i_0$, $j_0$, $i_1$, $j_1$, $i_2$, and $j_2$.

$c(N[i_0, j_0]; G) < c(N[i_0, j_0]; G')$ is equivalent to $c(N(j_0, i_0); G) < c(N(j_0, i_0); G')$. Let $i_1$ be an integer such that

$$c(N(j_0, i); G) < c(N(j_0, i); G') \text{ for } i \in N[i_1, i_0] \text{ and}$$

$$c(N(j_0, i_1); G) = c(N(j_0, i_1); G').$$

(From the assumption, there exists such an $i_1$.) It follows that there exists an edge $(i_1, j_1) \in E(G)$ such that $j_1 \in N(j_0, i_1)$. Let $i_2 \in N[i_1, i_0]$ and $j_2 \in N(j_0, j_1]$ be integers such that $(i_2, j_2) \in E(G)$,

$$c(N[i_1, i_0], N(j_0, j_2); G) = 0 \tag{6}$$

and $c(N(i_2, i_0), j_2; G) = 0$. Note that $i_2$ and $j_2$ may be equal to $i_1$ and $j_1$, respectively.

We will show that $c(N[j, i]; G) < c(N[j, i]; G')$ for any $i \in N[i_2, i_0]$ and $j \in N(j_0, j_2]$ as follows. Assume that there are $i' \in N[i_2, i_0]$ and $j' \in N(j_0, j_2]$ such that $c(N[j', i']; G) = c(N[j', i']; G')$. By considering

$$c(N(j_0, i_1); G) = c(N(j_0, i_1); G'),$$
$$c(N(j_0, i'); G) < c(N(j_0, i'); G'), \text{ and}$$
$$c(N[j', i_1); G) \leq c(N[j', i_1); G'),$$

we obtain that $c(N[i_1, i'], N(j_0, j'); G) > 0$, contradicting (6). Therefore,

$$c(N[j, i]; G) < c(N[j, i]; G') \text{ for any } i \in N[i_2, i_0] \text{ and } j \in N(j_0, j_2]. \tag{7}$$

From the assumption, $c(i; G) = c(i; G')$ for all $i \in N$. Thus from Lemma 4, expression (7) can be rewritten as

$$c(N[j, i]; G) \leq c(N[j, i]; G') - 2 \text{ for any } i \in N[i_2, i_0] \text{ and } j \in N(j_0, j_2]. \tag{8}$$

Denote a graph obtained by applying cross-operation $(i_0, j_0; j_2, i_2)$ to $G$ by $G''$. Clearly, $G \prec_c G''$. From (8), $G'' \preceq_c G'$. Thus by applying the above operation recursively, we can get a sequence of cross-operations for modifying $G$ to $G'$, i.e., $G \preceq_o G'$. □

*Proof of Main Theorem*: It is clear from Lemmas 1, 2, and 3. □

## 3   Concluding Remarks

This paper shows that three partial-orders $\preceq_l$, $\preceq_o$, and $\preceq_c$ are equivalent. For investigating $G \preceq_c G'$, only neighbor-cuts are tested, thus it can be determined in polynomial time. Therefore, we can solve a problem of determining whether or not $S_P(G) \leq S_P(G')$ for any convex polygon $P$ for given two labeled graphs $G$ and $G'$ with $|E(G)| = |E(G')|$ in polynomial time. Moreover, if $G \preceq_c G'$, we can find a sequence of cross-operations for modifying $G$ to $G'$ by using the discussion of the proof of Lemma 3 in polynomial time.

In this paper, Euclidean distance is used. However, for any distance (for example, $L_k$ distance) in which the triangle inequality holds, the same results can be obtained.

**Acknowledgments**

We would like to express our appreciation to Professor Jorge Urrutia of Universidad Nacional Autónoma de México who gave us the motivation for this research. We also wish to thank to Professor Nagamochi Hiroshi of Toyohasi University of Technology for his valuable suggestions.

## References

1. Battista, G. D., Eades, P., Tamassia, R., and Tollis, I. G., Algorithms for drawing graphs: An annotated bibliography,
   `http://www.cs.brown.edu/people/rt/gd-biblio.html`, 1994.
2. Ito, H., Uehara, H., and Yokoyama, M, A consideration on lengths of permutations on a vertex set of a convex polygon, Extended Abstracts of JCDCG'99, Tokai Univ., Nov. 26–27, pp. 40–41, 1999.
3. Ito, H., Uehara, H., and Yokoyama, M, Lengths of tours and permutations on a vertex set of a convex polygon, Discrete Applied Mathematics, (to appear).

# On double bound graphs with respect to graph operations

Shin-ich Iwai[1], Kenjiro Ogawa[2] and Morimasa Tsuchiya[3]

[1] Department of Mathematical Sciences, Tokai University
Hiratsuka 259-1292, JAPAN
[2] Department of Mathematical Sciences, Tokai University
Hiratsuka 259-1292, JAPAN
[3] Department of Mathematical Sciences, Tokai University
Hiratsuka 259-1292, JAPAN
tsuchiya@ss.u-tokai.ac.jp

**Abstract.** We consider upper bound graphs with respect to operations on graphs, for example, the sum, the Cartesian product, the corona and the middle graphs of graphs, etc. According to the characterization of double bound graphs, we deal with characterizations of double bound graphs obtained by graph operations. For example, The Cartesian product $G \times H$ of two graphs $G$ and $H$ is a DB-graph if and only if both $G$ and $H$ are bipartite graphs, the corona $G \circ H$ of two graphs $G$ and $H$ is a DB-graph if and only if $G$ is a bipartite graph and $H$ is a UB-graph, and the middle graph $M(G)$ of a graph $G$ is a DB-graph if and only if $G$ is an even cycle or a path, etc.

## 1 Introduction.

In this paper, we consider finite undirected simple graphs. For a vertex $v$ in $G$, the *neighborhood* of $v$ is the set of vertices which are adjacent to $v$, and denoted by $N_G(v)$. For $S \subseteq V(G)$, the induced subgraph of $S$ is denoted by $< S >_G$.

We know some kinds of graphs related to posets, that is, upper bound graphs and double bound graphs, as follows. For a poset $P = (X, \leq)$, the *upper bound graph* (UB-graph) of $P$ is the graph $U = (X, E_U)$ where $xy \in E_U$ if and only if $x \neq y$ and there exists $m \in X$ such that $x, y \leq m$. The *double bound graph* (DB-graph) of $P = (X, \leq)$ is the graph $D = (X, E_D)$ where $xy \in E_D$ if and only if $x \neq y$ and there exist $m, n \in X$ such that $n \leq u, v \leq m$. McMorris and Zaslavsky [9] introduced these concepts. Upper bound graphs are concerned with maximal order ideals of posets and double bound graphs are concerned with maximal intervals of posets. Properties of upper bound graphs and double bound graphs play important roles in geometrical properties of posets.

McMorris and Zaslavsky [9] gives a characterization of upper bound graphs. A *clique* in the graph $G$ is the vertex set of a maximal complete subgraph. In some cases we consider that a clique is a maximal complete subgraph. A family $\mathcal{C}$ of complete subgraphs *edge covers* $G$ if and only if for each $uv \in E(G)$ there exists $C \in \mathcal{C}$ such that $u, v \in C$.

J. Akiyama, M. Kano, and M. Urabe (Eds.): JCDCG 2000, LNCS 2098, pp. 167-175, 2001.
© Springer-Verlag Berlin Heidelberg 2001

**Theorem 1 (McMorris  and  Zaslavsky [9]).** *A graph $G$ is a UB-graph if and only if there exists a family $\mathcal{C} = \{C_1, ..., C_n\}$ of complete subgraphs of $G$ such that*

*(i) $\mathcal{C}$ edge covers $G$, and*

*(ii) for each $C_i$, there is a vertex $v_i \in C_i - (\bigcup_{j \neq i} C_j)$.*

   *Furthermore, such a family $\mathcal{C}$ must consist of cliques of $G$ and is the only such family if $G$ has no isolated vertices.*

   For an edge clique cover $\mathcal{C} = \{C_1, C_2, ..., C_n\}$ satisfying the conditions of Theorem 1, a *representation vertex set* $R(\mathcal{C})$ on $\mathcal{C}$ is a vertex subset $\{v_1, v_2, ..., v_n\}$ such that $v_i \in C_i - (\bigcup_{j \neq i} C_j)$ for each $i = 1, ..., n$. A vertex $v$ is called a *simplicial vertex* if $N_G(v)$ is a complete subgraph. Each vertex $v_i \in R(\mathcal{C})$ is a simplicial vertex.

   For a graph $G$ with two disjoint independent subsets $M$ and $N$ of $V(G)$ and $v \in V(G) - (M \cup N)$, define the sets $U_G(v) = \{u \in M \; ; \; uv \in E(G)\}$, $L_G(v) = \{u \in N \; ; \; uv \in E(G)\}$. Diny [2] gives a characterization of double bound graphs.

**Theorem 2 (Diny [2]).** *A graph $G$ is a DB-graph if and only if there exists a family $\mathcal{C} = \{C_1, ..., C_n\}$ of complete subgraphs of $G$ and disjoint independent subsets $M$ and $N$ such that*

*(1) $\mathcal{C}$ edge covers $G$,*

*(2) for each $C_i$, there exist $m_i \in M$, $n_i \in N$ such that $\{m_i, n_i\} \subseteq C_i$ and $\{m_i, n_i\} \nsubseteq C_j$ for all $i \neq j$, and*

*(3) for each $v \in V(G) - (M \cup N)$, $|U_G(v)| \times |L_G(v)|$ equals the number of cliques of $\mathcal{C}$ containing $v$.*

   *Furthermore, a family $\mathcal{C}$ is the unique, minimal edge covering family of cliques in $G$.*

   For a DB-graph $G$ and an edge clique cover $\mathcal{C} = \{C_1, C_2, ..., C_n\}$ satisfying the conditions of Theorem 2, which is said to be a *DB edge clique cover*, $M$ is called an *upper kernel* $\mathcal{U}(G)$ of $G$ and $N$ is called a *lower kernel* $\mathcal{L}(G)$ of $G$. We already know the fact that for a corresponding poset $P$ of a DB-graph G, $\mathcal{U}(G)$ corresponds to the set of all maximal elements of $P$ and $\mathcal{L}(G)$ corresponds to the set of all minimal elements of $P$.

   In [8] we obtained another characterization of double bound graphs as follows. In this theorem we deal with construction methods of double bound graphs.

**Theorem 3 ([8]).** *Let $G$ be a connected graph. Then $G$ is a DB-graph if and only if there exist two disjoint independent sets $M$ and $N$ such that the graph obtained by successive deletions of all vertices $v \in V(G) - (M \cup N)$ satisfying the following conditions is a bipartite graph with partite sets $M$ and $N$:*

   *(a) $< X \cup Y >_V$ is a complete bipartite graph.*

   *(b) $uv \in E(G)$ for $u \in V(G) - (M \cup N)$ if and only if $N_G(u) \cap X \neq \emptyset$ and $N_G(u) \cap Y \neq \emptyset$,*

*where $X = N_G(v) \cap M$, and $Y = N_G(v) \cap N$.*

In [3], [4] and [5] we obtained some properties on upper bound graphs and order ideals by graph operations. In this paper we consider double bound graphs in terms of graph operations.

## 2  Sum.

The *sum*  $G + H$  of two graphs $G$ and $H$ is the graph with the vertex set $V(G + H) = V(G) \cup V(H)$ and the edge set $E(G + H) = E(G) \cup E(H) \cup \{uv; u \in V(G), v \in V(H)\}$.

**Proposition 1 (Sum).** *For graphs $G$ and $H$, $G+H$ is a DB-graph if and only if (1) $G$ and $H$ are UB-graphs, or (2) $G$ is a DB-graph such that $< \mathcal{U}(G) \cup \mathcal{L}(G) >_G$ is a complete bipartite graph and $H$ is a complete graph.*

*Proof.* To show the sufficiency of the Proposition, first we consider the case (1). Let $G$ and $H$ be UB-graphs. Let $\mathcal{C}(G) = \{C_{G,1}, C_{G,2}, ..., C_{G,s}\}$ be a family of cliques of $G$ satisfying the conditions of Theorem 1 and $R(\mathcal{C}(G))$ be a representation vertex set of $\mathcal{C}(G)$. Let $\mathcal{C}(H) = \{C_{H,1}, C_{H,2}, ..., C_{H,t}\}$ be a family of cliques of $H$ satisfying the conditions of Theorem 1 and $R(\mathcal{C}(H))$ be a representation vertex set of $\mathcal{C}(H)$. Then $C_{G,i} \cup C_{H,j}$ is a complete subgraph of $G+H$, and $R(\mathcal{C}(G))$ and $R(\mathcal{C}(H))$ are independent sets of $G + H$. Since $< R(\mathcal{C}(G)) \cup R(\mathcal{C}(H)) >_{G+H}$ is a complete bipartite graph, $\{C_{G,i} \cup C_{H,j} ; \forall i, j\}$ satisfies the conditions of Theorem 2 and $G + H$ is a DB-graph.

Next we consider the case (2). Let $\mathcal{C}(G) = \{C_1, C_2, ..., C_n\}$ be a family of cliques of $G$ satisfying the conditions of Theorem 2. Since for $\forall i$, $C_i \cup V(H)$ is a complete subgraph of $G + H$, $\{C_i \cup V(H) ; \forall i\}$ satisfies the conditions of Theorem 2 and $G + H$ is a DB-graph.

Conversely we assume that $G+H$ is a DB-graph. Let $\mathcal{U}(G+H)$ be an upper kernel and $\mathcal{L}(G + H)$ be a lower kernel. Since $uv \in E(G + H)$ for $\forall u \in V(G)$ and $\forall v \in V(H)$, there exists no independent set $S$ such that $S \cap V(G) \neq \emptyset$ and $S \cap V(H) \neq \emptyset$. So we consider the following two cases.

*Case 1.* $\mathcal{U}(G + H) \subseteq V(G)$ and $\mathcal{L}(G + H) \subseteq V(H)$.

For $v_i \in \mathcal{L}(G + H)$, $< V(G) \cup \{v_i\} >_{G+H}$ is a DB-graph with an upper kernel $\mathcal{U}(G + H)$ and a lower kernel $\{v_i\}$. Then $G$ is a UB-graph. Similarly $H$ is a UB-graph.

*Case 2.* $\mathcal{U}(G + H), \mathcal{L}(G + H) \subseteq V(G)$.

Since for $\forall v \in V(H)$, $\forall m \in \mathcal{U}(G + H)$ and $\forall n \in \mathcal{L}(G + H)$, $mv, nv \in E(G+H)$, $U_G(v) = \mathcal{U}(G)$, $L_G(v) = \mathcal{L}(G)$ and $|U_G(v)| \times |L_G(v)| = |\mathcal{U}(G)| \times |\mathcal{L}(G)|$. Thus $< \mathcal{U}(G) \cup \mathcal{L}(G) >_{G+H}$ is a complete bipartite graph. For $\forall u, v \in V(H)$ such that $mv, nv \in E(G + H)$, $m \in \mathcal{U}(G + H)$ and $n \in \mathcal{L}(G + H)$, $u$ and $v$ belong to a clique containing $m$ and $n$. Thus $uv \in E(H)$ and $H$ is a complete graph. Since $V(H) \cap (\mathcal{U}(G) \cup \mathcal{L}(G)) = \emptyset$, $(G + H) - V(H) = G$ is a DB-graph by Theorem 3.  □

## 3  Corona

The *corona* $G \circ H$ of two graphs $G$ and $H$ is defined as the graph obtained by taking one copy of $G$ and $|V(G)|$ copies of $H$, and then joining the $i$th vertex of $G$ to every vertex in the $i$th copy of $H$.

**Proposition 2 (Corona).** *For connected graphs $G$ and $H$ with $|V(G)| \geq 2$, $G \circ H$ is a DB-graph if and only if $G$ is a bipartite graph and $H$ is a UB-graph.*

*Proof.* Let $G$ be a bipartite graph with partite sets $M$ and $N$. Then $G$ is a DB-graph with an upper kernel $M$ and a lower kernel $N$. Based on the fact that $H + v$ is a DB-graph for a UB-graph $H$, $G \circ H$ is a DB-graph.

Conversely let $G \circ H$ be a DB-graph with an upper kernel $\mathcal{U}(G \circ H)$ and a lower kernel $\mathcal{L}(G \circ H)$. Then we have the following claim.

**Claim 1.** *Let $C_{m,n}$ be a clique of $G \circ H$ such that $m \in \mathcal{U}(G \circ H)$, $n \in \mathcal{L}(G \circ H)$ and $m, n \in C_{m,n}$. For $\forall uv \in E(G)$, if $u, v \in C_{m,n}$, then $m, n \in V(G)$.*

*Proof (of Claim 1).* Since $u, v \in V(G)$, $N_{G \circ H}(v) \cap N_{G \circ H}(u) \subseteq V(G)$. Then $m, n \in N_{G \circ H}(v) \cap N_{G \circ H}(u) \subseteq V(G)$. $\qquad\square$

We assume that $V(G) - (\mathcal{U}(G \circ H) \cup \mathcal{L}(G \circ H)) \neq \emptyset$ and $v \in V(G) - (\mathcal{U}(G \circ H) \cup \mathcal{L}(G \circ H))$. Since $G$ is connected and $v$ is adjacent to another vertex $u$ of $G$, there exists a clique $C$ of $G$ such that $u, v \in C$, $m \in \mathcal{U}(G \circ H) \cap C$ and $n \in \mathcal{L}(G \circ H) \cap C$. Let $H_v$ be a copy of $H$ corresponding to $v$. Then for each vertex $x \in H_v$, $v$ is adjacent to $x$. Thus $U_{G \circ H}(v) \cap V(H_v) \neq \emptyset$ and $L_{G \circ H}(v) \cap V(H_v) \neq \emptyset$. Since $v$ is a cut vertex of $G \circ H$, $< U_{G \circ H}(v) \cup L_{G \circ H}(v) >_{G \circ H}$ is not a complete bipartite graph, which contradicts to the conditions of Theorem 3. Therefore $V(G) - (\mathcal{U}(G \circ H) \cup \mathcal{L}(G \circ H)) = \emptyset$ and $G$ is a bipartite graph. Since $v$ is a cut vertex of $G \circ H$ and $v \in \mathcal{U}(G \circ H) \cup \mathcal{L}(G \circ H)$, $H_v + v \cong H + v$ is a DB-graph and $H$ is a UB-graph. $\qquad\square$

## 4  Cartesian product

The *Cartesian product* $G \times H$ of two graphs $G$ and $H$ is the graph with vertex set $V(G \times H) = V(G) \times V(H)$, where the second $\times$ is the set Cartesian product, and edges defined as follows: $(u_1, v_1)$ is adjacent to $(u_2, v_2)$ if either $u_1 = u_2$ and $v_1$ is adjacent to $v_2$ in $H$, or $u_1$ is adjacent to $u_2$ in $G$ and $v_1 = v_2$. If $E(G) = \emptyset$, then $G \times H$ is the union of $|V(G)|$ copies of $H$. We know the following facts by the definition.

**Fact (1)** For graphs $G$ and $H$ with $|V(G)| \geq 2$ and $|V(H)| \geq 2$, $G \times H$ is a bipartite graph if and only if both $G$ and $H$ are bipartite graphs.

**Fact (2)** Let $C$ be a clique of $G \times H$. If $(u, v) \in C$, then for all $(x, y) \in C$, $x = u$, or for all $(x, y) \in C$, $y = v$.

**Proposition 3 (Cartesian product).** *For connected graphs $G$ and $H$ with $|V(G)| \geq 2$ and $|V(H)| \geq 2$, $G \times H$ is a DB-graph if and only if both $G$ and $H$ are bipartite graphs.*

*Proof.* If $G$ and $H$ are bipartite graphs, then $G \times H$ is a bipartite graph and a DB-graph.

Conversely let $G \times H$ be a DB-graph with an upper kernel $\mathcal{U}(G \times H)$ and a lower kernel $\mathcal{L}(G \times H)$. We assume that $V(G \times H) - \mathcal{U}(G \times H) \cup \mathcal{L}(G \times H) \neq \emptyset$ and $(u, v) \in \mathcal{U}(G \times H) \cup \mathcal{L}(G \times H)$. Since $|V(G)| \geq 2$ and $|V(H)| \geq 2$, there exist edges $(u, v)(w_G, v)$ and $(u, v)(u, w_H)$. By Fact (2), there exist $(m_G, v) \in \mathcal{U}(G \times H)$, $(n_G, v) \in \mathcal{L}(G \times H)$, $(u, m_H) \in \mathcal{U}(G \times H)$ and $(u, n_H) \in \mathcal{L}(G \times H)$. $(m_G, v)$ and $(n_G, v)$ are adjacent to $(u, v)$ and $(w_G, v)$, and $(u, m_H)$ and $(u, n_H)$ are adjacent to $(u, v)$ and $(u, w_H)$. Thus $(m_G, v)$ is adjacent to $(u, n_H)$ which contradicts the definition of $G \times H$. Therefore $V(G \times H) = \mathcal{U}(G \times H) \cup \mathcal{L}(G \times H)$ and $G \times H$ is a bipartite graph. By Fact (1), $G$ and $H$ are bipartite graphs.     □

## 5   Composition

The *composition* $G[H]$ of two graphs $G$ and $H$ is the graph with the vertex set $V(G[H]) = V(G) \times V(H)$, and edges defined as follows: $(u_1, v_1)$ is adjacent to $(u_2, v_2)$ if either $u_1$ is adjacent to $u_2$ in $G$, or $u_1 = u_2$ and $v_1$ is adjacent to $v_2$ in $H$.

**Proposition 4 (Composition).** *If $G$ is a DB-graph and $H$ is a complete graph, then $G[H]$ is a DB-graph.*

*Proof.* Let $\mathcal{C} = \{C_1, C_2, ..., C_n\}$ be a DB edge clique cover of $G$. Since $H$ is a complete graph, $C_i \times V(H)$ is a complete subgraph. We fix $\alpha, \beta \in V(H)$. For an upper kernel $\mathcal{U}(G) = \{m_1, m_2, ..., m_s\}$ of $G$ and a lower kernel $\mathcal{L}(G) = \{n_1, n_2, ..., n_t\}$ of $G$, $M_{G[H]} = \{(m_i, \alpha) \; ; \; m_i \in \mathcal{U}(G)\}$ and $N_{G[H]} = \{(n_i, \beta) \; ; \; m_i \in \mathcal{L}(G)\}$. $\mathcal{C} \times V(H) = \{C_1 \times V(H), C_2 \times V(H), ..., C_n \times V(H)\}$ is an edge covers of $G[H]$. Furthermore for each $C_i \times V(H)$, there exist $(m_i, \alpha) \in M_{G[H]}$ and $(n_i, \beta) \in N_{G[H]}$ such that $\{(m_i, \alpha), (n_i, \beta)\} \subseteq C_i \times V(H)$ and $\{(m_i, \alpha), (n_i, \beta)\} \not\subseteq C_j \times V(H)$ for all $i \neq j$. Since $|U_G(u)| \times |L_G(u)|$ equals the number of cliques of $\mathcal{C}$ containing $u$, for each $(u, v) \in V(G[H]) - (M_{G[H]} \cup N_{G[H]})$, $|U_{G[H]}((u, v))| \times |L_{G[H]}((u, v))|$ equals to the number of cliques of $\mathcal{C} \times V(H)$ containing $(u, v)$. Thus $\mathcal{C} \times V(H)$ is a DB edge clique cover of $G[H]$.     □

## 6   Normal product

The *normal product* $G \star H$ of two graphs $G$ and $H$ is the graph with the vertex set $V(G \star H) = V(G) \times V(H)$, and edges defined as follows: $(u_1, v_1)$ is adjacent to $(u_2, v_2)$ if one of the following holds: (1) $u_1$ is adjacent to $u_2$ in $G$ and $v_1$ is adjacent to $v_2$ in $H$, (2) $u_1 = u_2$ and $v_1$ is adjacent to $v_2$ in $H$, (3) $u_1$ is adjacent to $u_2$ in $G$ and $v_1 = v_2$. For posets $P$ and $Q$, the *product* of $P$ and $Q$ is the poset $P \times Q$ on the Cartesian product of the element set of $P$ and the element set of $Q$ such that $(u, x) \leq_{P \times Q} (v, y)$ if $u \leq_P v$ and $x \leq_Q y$. By these definitions we obtain the following fact. For DB-graphs $G$ and $H$, $G \star H$ is a DB-graph of a poset $P_G \times P_H$, where $P_G$ and $P_H$ are posets whose DB-graphs are $G$ and $H$. So we have the following result.

**Proposition 5 (Normal product).** *For DB-graphs $G$ and $H$, $G \star H$ is a DB-graph.*

## 7   Conjunction.

The *conjunction* $G \wedge H$ of two graphs $G$ and $H$ is the graph with the vertex set $V(G \wedge H) = V(G) \times V(H)$, and edges defined as follows: $(u_1, v_1)$ is adjacent to $(u_2, v_2)$ if $u_1$ is adjacent to $u_2$ in $G$ and $v_1$ is adjacent to $v_2$ in $H$.

**Lemma 1.** *For connected graphs $G$ and $H$, $G \wedge H$ is a bipartite graph if and only if $G$ or $H$ is a bipartite graph.*

*Proof.* We assume that both $G$ and $H$ are not bipartite graphs. Let $C : x, v_1, ...,$ $v_{2n}, x$ be an odd cycle of $G$ and $D : y, u_1, ..., u_{2m}, y$ be an odd cycle of $H$ ($n \geq m$). Then in the case $n = m$, $(x, y), (v_1, u_1), (v_2, u_2), ..., (v_i, u_i), ..., (v_{2n}, u_{2m}),$ $(x, y)$ is an odd cycle of $G \wedge H$. In the case $n \neq m$, $(x, y), (v_1, u_1), (v_2, y),$ $(v_3, u_1), (v_4, y), ..., (v_{2n-2m-1}, u_1), (v_{2n-2m}, y), (v_{2n-2m+1}, u_1), (v_{2n-2m+2}, u_2)$ $, ..., (v_{2n-2m+i}, u_i), ..., (v_{2n}, u_{2m}), (x, y)$ is an odd cycle of $G \wedge H$. Thus $G \wedge H$ is not a bipartite graph.

Let $G$ be a bipartite graph with partite sets $M$ and $N$. Then $V(G \wedge H) = (M \times V(H)) \cup (N \times V(H))$ and $G \wedge H$ is a bipartite graph with partite sets $M \times V(H)$ and $N \times V(H)$. □

**Lemma 2.** *Let $G \wedge H$ be a DB-graph with an upper kernel $\mathcal{U}(G \wedge H)$ and a lower kernel $\mathcal{L}(G \wedge H)$. If $(a, b), (c, d) \notin \mathcal{U}(G \wedge H) \cup \mathcal{L}(G \wedge H)$, then $(a, b)(c, d) \notin E(G \wedge H)$.*

*Proof.* We assume that $(a, b)(c, d) \in E(G \wedge H)$ and $(a, b), (c, d) \notin \mathcal{U}(G \wedge H) \cup \mathcal{L}(G \wedge H)$. Then there exist $(x, y) \in \mathcal{U}(G \wedge H)$ and $(z, w) \in \mathcal{L}(G \wedge H)$ such that $(a, b)(x, y), (c, d)(x, y), (a, b)(z, w), (c, d)(z, w) \in E(G \wedge H)$. Thus $ax, az \in E(G)$, $dy, dw \in E(H)$, and $(a, d)$ is adjacent to $(x, y)$ and $(z, w)$. Since $G \wedge H$ is a DB-graph, $< N_{G \wedge H}((z, w)) \cap N_{G \wedge H}((x, y)) >_{G \wedge H}$ is a complete subgraph. So $(a, d)$ is adjacent to $(a, b)$, which contradicts to the definition of conjunctions. □

**Lemma 3.** *Let $G$ and $H$ be connected graphs. $G \wedge H$ is a DB-graph if and only if $G \wedge H$ is a bipartite graph.*

*Proof.* If $G \wedge H$ is a bipartite graph, then $G \wedge H$ is a DB-graph.

Conversely let $G \wedge H$ be a DB-graph with an upper kernel $\mathcal{U}(G \wedge H)$ and a lower kernel $\mathcal{L}(G \wedge H)$. We assume that $V(G \wedge H) - (\mathcal{U}(G \wedge H) \cup \mathcal{L}(G \wedge H)) \neq \emptyset$. Let $(a, b) \in V(G \wedge H) - (\mathcal{U}(G \wedge H) \cup \mathcal{L}(G \wedge H))$, $(x, y) \in \mathcal{U}(G \wedge H)$ such that $(x, y)$

is adjacent to $(a, b)$, and $(z, w) \in \mathcal{L}(G \wedge H)$ such that $(z, w)$ is adjacent to $(a, b)$. Then $(x, w)$ is adjacent to $(a, b)$ and $(x, w) \in (\mathcal{U}(G \wedge H) \cup \mathcal{L}(G \wedge H))$ by Lemma 2. Since $(x, w)$ is not adjacent to $(x, y)$ and $(z, w)$, we have a contradiction. □

By Lemma 1, 2 and 3, we obtain the following result.

**Proposition 6 (Conjunction).** *For connected graphs $G$ and $H$, $G \wedge H$ is a DB-graph if and only if $G$ or $H$ is a bipartite graph.*

## 8  Middle graphs.

The *middle graph* $M(G)$ of a graph $G$ is the intersection graph of $\mathcal{F} = \{\{v_1\}, \{v_2\}, ..., \{v_p\}\} \cup E(G)$, where $V(G) = \{v_1, v_2, ..., v_p\}$ is the vertex set of $G$ and $E(G)$ is the edge set of $G$. That is, $V(M(G)) = V(G) \cup E(G)$ and two vertices $\alpha$ and $\beta$ in $M(G)$ is adjacent if and only if (1) $\alpha, \beta \in E(G)$ and $\alpha$ is adjacent to $\beta$, or (2) $\alpha \in V(G), \beta \in E(G)$ and $\alpha$ is incident to $\beta$.

**Lemma 4.** *Let $G$ be a connected graph and $M(G)$ is a DB-graph with an upper kernel $\mathcal{U}(M(G))$ and a lower kernel $\mathcal{L}(M(G))$. Let $v$ be a vertex of $G$. If $v \in \mathcal{U}(M(G)) \cup \mathcal{L}(M(G))$, then $deg_G(v) = 1$.*

*Proof.* We assume that $v \in \mathcal{U}(M(G))$, and $deg_G(v) \geq 2$. Since $< \{vw \; ; \; w \in N_G(v)\} >_{M(G)} \cup \{v\}$ forms a clique of $M(G)$ and $|\{vw \; ; \; w \in N_G(v)\}| \geq 2$, we can assume that $vx$ belongs to $\mathcal{L}(M(G))$ and $vy \notin \mathcal{U}(M(G)) \cup \mathcal{L}(M(G))$. $y$ is adjacent to $vy$ in $M(G)$ and $y$ is not adjacent to $vx$ and $v$ in $M(G)$. Then there exist $y\beta \in \mathcal{U}(M(G))$ and $y\alpha \in \mathcal{L}(M(G))$. Since $y\alpha$ is adjacent to $vy$ and is not adjacent to $v$, we have a contradiction. □

**Lemma 5.** *Let $G$ be a connected graph and $M(G)$ is a DB-graph with an upper kernel $\mathcal{U}(M(G))$ and a lower kernel $\mathcal{L}(M(G))$. If $uv \in E(G)$, then $uv \in \mathcal{U}(M(G)) \cup \mathcal{L}(M(G))$.*

*Proof.* We assume that $uv \in E(G)$ and $uv \notin \mathcal{U}(M(G)) \cup \mathcal{L}(M(G))$. If $deg_G(u) = 1$, then $u$ is only adjacent to $uv$ in $M(G)$. Thus $uv \in \mathcal{U}(M(G)) \cup \mathcal{L}(M(G))$. So $deg_G(u) \geq 2$. Similarly $deg_G(v) \geq 2$. By Lemma 4 $u, v \notin \mathcal{U}(M(G)) \cup \mathcal{L}(M(G))$. Then there exist $u\alpha \in \mathcal{U}(M(G))$, $v\beta \in \mathcal{L}(M(G))$. Since $uv$ adjacent to $u\alpha, v\beta$, $u\alpha$ is adjacent to $v\beta$, and $\alpha = \beta$. Then $\alpha$ is adjacent to $uv$ in $M(G)$, which is a contradiction. □

**Proposition 7 (Middle graph).** *For a connected graph $G$, $M(G)$ is a DB-graph if and only if $G$ is an even cycle or a path.*

*Proof.* Let $G$ be an even cycle or a path. Then $M(G)$ is a DB-graph.

Conversely we assume that $M(G)$ is a DB-graph. If $G$ has a vertex $v$ with $deg_G(v) \geq 3$, then $< \{uv \; ; \; u \text{ is adjacent to } v\} >_{M(G)}$ is a complete subgraph of $M(G)$ with at least 3 vertices, which contradicts to the fact that $< E(G) >_{M(G)}$ is a bipartite graph by Lemma 5. Thus the maximum degree of $G$ is at most 2, and $G$ is a cycle or a path. If $G$ is an odd cycle, then $< E(G) >_{M(G)}$ has an odd cycle. Thus $G$ is an even cycle or a path. □

## 9   Total graphs.

The *total graph* $T(G)$ of a graph $G$ is the graph with the vertex set $V(T(G)) = V(G) \cup E(G)$ and two vertices $\alpha$ and $\beta$ in $T(G)$ is adjacent if and only if (1) $\alpha, \beta \in V(G)$ and $\alpha$ is adjacent to $\beta$, or (2) $\alpha, \beta \in E(G)$ and $\alpha$ is adjacent to $\beta$, or (3) $\alpha \in V(G)$, $\beta \in E(G)$ and $\alpha$ is incident to $\beta$.

**Lemma 6.** *Let $G$ be a connected graph and $T(G)$ is a DB-graph with an upper kernel $\mathcal{U}(T(G))$ and a lower kernel $\mathcal{L}(T(G))$. Let $v$ be a vertex of $G$. If $deg_G(v) \geq 2$, then $uv \notin \mathcal{U}(T(G)) \cup \mathcal{L}(T(G))$ for all $u \in N_G(v)$.*

*Proof.* We assume that $deg_G(v) \geq 2$ and $uv \in \mathcal{U}(M(G))$. Since $u$ is adjacent to $v$, $u, v \notin \mathcal{U}(T(G)) \cup \mathcal{L}(T(G))$, or $u \in \mathcal{L}(T(G))$ and $v \notin \mathcal{U}(T(G)) \cup \mathcal{L}(T(G))$, or $v \in \mathcal{L}(T(G))$ and $u \notin \mathcal{U}(T(G)) \cup \mathcal{L}(T(G))$. In the case that $u, v \notin \mathcal{U}(T(G)) \cup \mathcal{L}(T(G))$, there exist no vertices of $\mathcal{L}(T(G))$ adjacent to $u$, $v$ and $uv$. In the case that $u \in \mathcal{L}(T(G))$ and $v \notin \mathcal{U}(T(G)) \cup \mathcal{L}(T(G))$, $uv$ and $v$ are adjacent to $vw_v$ and $uv$ is not adjacent to $w_v$, where $w_v$ is another adjacent vertex of $v$. Thus $vw_v \in \mathcal{L}(T(G))$, $w_v \in \mathcal{U}(T(G))$, and $u$ is adjacent to $w_v$. Then $uv$ and $u$ are adjacent to $uw_v$. However $uw_v$ is not adjacent to $v$. In the case $v \in \mathcal{L}(T(G))$ and $u \notin \mathcal{U}(T(G)) \cup \mathcal{L}(T(G))$, since $uv$ and $v$ are adjacent to $vw_v$ in $T(G)$, where $w_v$ is another adjacent vertex of $v$, $vw_v$ is adjacent to $u$. In every case we have a contradiction. So $uv \notin \mathcal{U}(T(G))$ for all $u \in N_G(v)$ if $deg_G(v) \geq 2$.     □

**Proposition 8 (Total graph).** *For a connected graph $G$ with $p \geq 3$ vertices, $T(G)$ is not a DB-graph.*

*Proof.* There exists a vertex $v \in V(G)$ with $deg_G(v) \geq 2$. Let $x, y$ be adjacent to $v$ in $G$. Then $vx$ is adjacent to $vy$ in $T(G)$ and $vx, vy \notin \mathcal{U}(T(G)) \cup \mathcal{L}(T(G))$ by Lemma 6. Thus there exist vertices of $\mathcal{U}(T(G)) \cup \mathcal{L}(T(G))$ adjacent to $vx$ and $vy$ in $T(G)$. By Lemma 6 $vz \notin \mathcal{U}(T(G)) \cup \mathcal{L}(T(G))$ for all vertex $z$ adjacent to $v$. Therefore we have only two cases: (1)$v \in \mathcal{U}(T(G))$ and $xy \in \mathcal{L}(T(G))$, (2)$v \in \mathcal{L}(T(G))$ and $xy \in \mathcal{U}(T(G))$. However $v$ is not adjacent to $xy$. So there exists no clique containing the edge $vx$. Hence $T(G)$ is not a DB-graph.     □

## References

1. M.Behzad and G.Chartrand, Total graphs and traversability, Proc. Edinburgh Math. Soc., **15** (1966)117-120.
2. D.Diny, The double bound graph of a partially ordered set, Journal of Combinatorics, Infomation & System Sciences, **10**(1985),52-56.
3. H.Era, K.Ogawa, and M.Tsuchiya, On upper bound graphs with respect to unary operations on graphs, (preprint).
4. H.Era, K.Ogawa, and M.Tsuchiya, On upper bound graphs with respect to operations on graphs, Theoretical Computer Science, **235**(2)(2000),219-223.
5. H.Era and M.Tsuchiya, On upper bound graphs whose complements are also upper bound graphs, Discrete Mathematics, **197**(1998),103-109.
6. T.Hamada and I.Yoshimura, Traversability and connectivity of the middle graph of a graph, Discrete Mathematics, **14**(1976)247-255.

7. F.Harary, Graph Theory, ( Addison-Wesley, 1969).

8. S.-i.Iwai, K.Ogawa and M.Tsuchiya, A note on construction of double bound graphs, (preprint).

9. F.R.McMorris and T.Zaslavsky, Bound graphs of a partially ordered set, Journal of Combinatorics, Infomation & System Sciences, **7**(1982),134-138.

10. K.Ogawa and M.Tsuchiya, On upper bound graphs with respect to line graphs, Southeast Asian Bulletin of Mathematics, **23**(1999),265-269.

# Generalized Balanced Partitions of Two Sets of Points in the Plane

Atsushi Kaneko[1] and M.Kano[2]

[1] Department of Computer Science and Communication Engineering
Kogakuin University, Shinjuku-ku, Tokyo 1563-8677, Japan
[2] Department of Computer and Information Sciences
Ibaraki University, Hitachi 316-8511, Japan

**Abstract.** We consider the following problem. Let $n \geq 2$, $b \geq 1$ and $q \geq 2$ be integers. Let $R$ and $B$ be two disjoint sets of $n$ red points and $bn$ blue points in the plane, respectively, such that no three points of $R \cup B$ lie on the same line. Let $n = n_1 + n_2 + \cdots + n_q$ be an integer-partition of $n$ such that $1 \leq n_i$ for every $1 \leq i \leq q$. Then we want to partition $R \cup B$ into $q$ disjoint subsets $P_1 \cup P_2 \cup \cdots \cup P_q$ that satisfy the following two conditions: $(i)$ conv $(P_i) \cap$ conv $(P_j) = \emptyset$ for all $1 \leq i < j \leq q$, where conv$(P_i)$ denotes the convex hull of $P_i$; and $(ii)$ each $P_i$ contains exactly $n_i$ red points and $bn_i$ blue points for every $1 \leq i \leq q$.
We shall prove that the above partition exists in the case where (i) $2 \leq n \leq 8$ and $1 \leq n_i \leq n/2$ for every $1 \leq i \leq q$, and (ii) $n_1 = n_2 = \cdots = n_{q-1} = 2$ and $n_q = 1$.

## 1   Introduction

For a set $X$ of points in the plane, we denote by conv$(X)$ the convex hull of $X$, which is the smallest convex set containing $X$. We shall consider the following problem:

*Problem 1.* Let $R$ and $B$ be any two disjoint sets of $n$ red points and $bn$ blue points in the plane, respectively, such that no three points of $R \cup B$ lie on the same line. Find positive integer-partitions $n = n_1 + n_2 + \cdots + n_q$ for which $R \cup B$ can be partitioned into $q$ disjoint subsets $P_1 \cup P_2 \cup \cdots \cup P_q$ that satisfy the following two conditions:

(1) conv $(P_i) \cap$ conv $(P_j) = \emptyset$ for all $1 \leq i < j \leq q$; and
(2) each $P_i$ contains exactly $n_i$ red points and $bn_i$ blue points.

If $R \cup B$ can be partitioned into $q$ subsets in the above way, then we say that $R \cup B$ is partitioned into $(n_1, n_2, \ldots, n_k)$-*balanced subsets*, or $R \cup B$ has a $(n_1, n_2, \ldots, n_k)$-*balanced partition*.

Figure 1 (a) gives a $(3, 2, 2, 1)$-balanced partition. Figure 1 (b) shows configurations $R \cup B$ having no $(2, 1)$-balanced partition or no $(3, 2)$-balanced partition, respectively. Non-existence of such balanced partitions can be shown by the fact that if such balanced partitions exist, then there exist lines that partition $R \cup B$

J. Akiyama, M. Kano, and M. Urabe (Eds.): JCDCG 2000, LNCS 2098, pp. 176–186, 2001.

into two subsets $P_1 \cup P_2$ satisfying the above conditions (1) and (2). However there exist no such lines.

By a similar argument given above, if $n/2 < n_1 < n$, then we can easily give configurations $R \cup B$ having no $(n_1, n_2)$-balanced partitions as follows: Let $C_1$ and $C_2$ be two circles with the same center in the plane such that the radius of $C_1$ is much smaller than that of $C_2$, and $b$ be a sufficient large integer. Then we uniformly place $n$ red points and $bn$ blue points on the boundaries of $C_1$ and $C_2$, respectively (see Figure 1(c)).

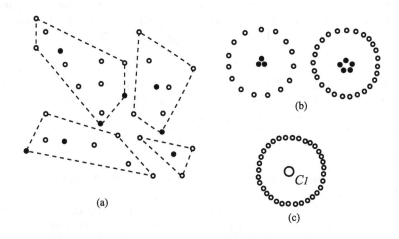

(b)

(a)

$C_1$

(c)

**Fig. 1.** (a) A $(3, 2, 2, 1)$-balanced partition. (b) Configurations having no $(2, 1)$-balanced partition and no $(3, 2)$-balanced partition, respectively. (c) A configuration having no $(n_1, n_2)$-balanced partition with $n/2 < n_1 < n$.

Throughout this paper, let $R$ and $B$ denote two disjoint sets of red points and blue points in the plane, respectively, such that no three points of $R \cup B$ lie on the same line. We shall give some parts of the following theorems since their complete proofs are quite long and rather tedious.

**Theorem 1.** Let $2 \le n \le 8$, $1 \le b$ and $2 \le q$ be integers. Let $R$ and $B$ be disjoint sets of $n$ red points and $bn$ blue points in the plane respectively. Then for every integer-partition $n = n_1 + n_2 + \ldots + n_q$ such that $1 \le n_i \le n/2$ for every $1 \le i \le q$, $R \cup B$ has an $(n_1, n_2, \ldots, n_q)$-balanced partition.

**Theorem 2.** Let $5 \le n$ be an odd integer, and $b \ge 1$ an integer. Let $R$ and $B$ be disjoint sets of $n$ red points and $bn$ blue points in the plane respectively. Then $R \cup B$ has a $(2, 2, \ldots, 2, 1)$-balanced partition.

By the above results and by an example given latter, we propose the following conjecture.

*Conjecture 1.* Let $n \geq 3$, $a \geq 1$, $b \geq 1$ and $q \geq 2$ be integers. Let $R$ and $B$ be disjoint sets of $an$ red points and $bn$ blue points in the plane respectively. Let $n = n_1 + n_2 + \cdots + n_k$ be an integer-partition such that $1 \leq n_i \leq n/3$ for every $1 \leq i \leq q$. Then $R \cup B$ can be partitioned into $q$ disjoint subsets $P_1 \cup P_2 \cup \cdots \cup P_q$ so that (i) conv $(P_i) \cap$ conv $(P_j) = \emptyset$ for all $1 \leq i < j \leq q$; and (ii) each $P_i$ contains exactly $an_i$ red points and $bn_i$ blue points.

The conjecture is true if either $a = 1$ and $n \leq 8$ or $a = 1$, $n_1 = \cdots = n_{q-1} = 2$ and $n_q = 1$ by Theorems 1 and 2. Moreover the conjecture is true in the case where $n_1 = \cdots = n_q = 1$, that is, the following Theorem 3 holds. This theorem was partially proved by [5] and [6], and completely proved by Bespamyatnikh, Kirkpatrick and Snoeyink [2], Ito, Uehara and Yokoyama [4] and Sakai [8] independently.

**Theorem 3.** *Let $R$ and $B$ be two disjoint sets of $an$ red points and $bn$ blue points in the plane respectively, where $a \geq 1$ and $b \geq 1$. Then $R \cup B$ can be partitioned into $n$ disjoint subsets $P_1 \cup P_2 \cup \ldots \cup P_n$ so that (i) conv$(P_i) \cap$ conv$(P_j) = \emptyset$ for all $1 \leq i < j \leq n$, and (ii) every $P_i$ contains exactly $a$ red points and $b$ blue points.*

Other related results can be found in [1] and [7], which deal with balanced partition of convex sets in the plane.

We conclude this section with an example which shows that the condition that $1 \leq n_i \leq n/3$ for every $1 \leq i \leq q$ in Conjecture 1 or other similar condition is necessary, and that Conjecture 1 does not hold under the condition that $1 \leq n_i \leq n/2$ for every $1 \leq i \leq q$. Namely, the configuration of $R \cup B$ whose rough sketch is given in Figure 2 has no $(8, 8, 4)$-balanced partition, and there might exist a similar configuration having no $(7, 7, 6)$-balanced partition.

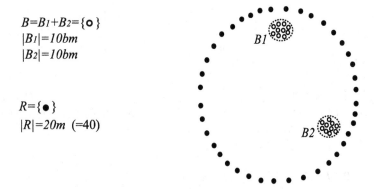

$B = B_1 + B_2 = \{\circ\}$
$|B_1| = 10bm$
$|B_2| = 10bm$

$R = \{\bullet\}$
$|R| = 20m \ (=40)$

**Fig. 2.** A configuration having no $(8, 8, 4)$-balanced partition.

## 2   Proofs of Theorems

In this section we shall give some parts of proofs of Theorems as we mentioned before. We deal only with *directed lines* in order to define the right side of a line and the left side of it. Thus a *line* means a directed line. A line $l$ dissects the plane into three pieces: $l$ and two open half-planes $R(l)$ and $L(l)$, where $R(l)$ and $L(l)$ denote the *open half-planes* which are on the right side and on the left side of $l$, respectively (see Figure 3). Let $r_1$ and $r_2$ be two rays emanating from the same point $p$. Then we denote by $R(r_1) \cap L(r_2)$ the open region that is swept by the ray being rotated clockwise around $p$ from $r_1$ to $r_2$, and does not contain the point $p$ (see Figure 3). Similarly we define the open region $L(r_1) \cap R(r_2)$, which is swept by the ray being rotated counterclockwise around $p$ from $r_1$ to $r_2$, and does not contain the point $p$. Then $r_1 \cup r_2$ dissects the plane into three pieces: $r_1 \cup r_2$ and two open regions $R(r_1) \cap L(r_2)$ and $L(r_1) \cap R(r_2)$.

If the internal angle $\angle r_1 p r_2 = \angle r_1 r_2$ of $R(r_1) \cap L(r_2)$ is less than $\pi$, then we call $R(r_1) \cap L(r_2)$ the *wedge* defined by $r_1$ and $r_2$, and denote it by $wdg(r_1 r_2)$ or $wdg(r_2 r_1)$.

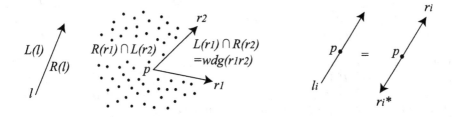

**Fig. 3.** Open regions $R(l)$, $L(l)$ and $L(r_1) \cap R(r_2)$ and a wedge $wdg(r_1 p r_2) = R(r_1) \cap L(r_2)$.

Let $l_i$ be a line with suffix $i$, and $p$ a point on $l_i$. We define $l_i^*$ as the line lying on $l_i$ and having the opposite direction of $l_i$. We next define the two rays $r_i$ and $r_i^*$ lying on the line $l_i$ and having the same starting point $p$ such that $r_i$ has the same direction as $l_i$ and $r_i^*$ has the opposite direction of $l_i$. In particular, $l_i = r_i \cup r_i^*$ (see Figure 3). Conversely, given a ray $r_i$, we can similarly define the ray $r_i^*$, whose direction is opposite to $r_i$, and the line $l_i = r_i \cup r_i^*$, which has the same direction as $r_i$.

For a line $l$ which passes through some points in $R \cup B$, there exist two lines $l_1$ and $l_2$ obtained from $l$ by very small translations that pass through no points in $R \cup B$, and satisfy $L(l_1) \cap (R \cup B) = L(l) \cap (R \cup B)$, and $R(l_2) \cap (R \cup B) = R(l) \cap (R \cup B)$. We often use this fact without mentioning.

**Lemma 1 (Ham-sandwich Theorem [3]).** *Let $R$ and $B$ be disjoint sets of red points and blue points in the plane respectively. Then there exists a line $l$ such that $|L(l) \cap R| = |R(l) \cap R|$, $|l \cap R| \leq 1$, $|L(l) \cap B| = |R(l) \cap B|$ and $|l \cap B| \leq 1$.*

If $l \cap R = \emptyset$ and $l \cap B = \emptyset$, that is, if $l$ passes through no red point and no blue point, then we say that $R \cup B$ is partitioned into two balanced subsets by $l$.

The following Lemma 2 is known, and its proof can be found in [5] and [2].

**Lemma 2.** *Let $R$ and $B$ be disjoint sets of red points and blue points in the plane respectively. If there exist two lines $l_1$ and $l_2$ such that $|L(l_1) \cap R| = |L(l_2) \cap R|$, $|L(l_1) \cap B| < |L(l_2) \cap B|$, and they might pass through some points in $R \cup B$, then for every integer $i$, $|L(l_1) \cap B| \leq i \leq |L(l_2) \cap B|$, there exists a line $l_3$ such that $|L(l_3) \cap R| = |L(l_1) \cap R|$, $|L(l_3) \cap B| = i$ and $l_3$ passes through no point in $R \cup B$.*

The next Lemma 3 can be proved by the same argument as in the proof of the above Lemma 2, that is, we can continuously move a line $l$ from $l_1$ to $l_2$ in such a way that $|L(l) \cap R| = |L(l_1) \cap R|$, $|L(l) \cap B|$ changes $\pm 1$, and $l$ passes through exactly one red point.

**Lemma 3.** *Let $R$ and $B$ be disjoint sets of red points and blue points in the plane respectively. If there exist two lines $l_1$ and $l_2$ such that $|L(l_1) \cap R| = |L(l_2) \cap R|$, $|L(l_1) \cap B| < |L(l_2) \cap B|$ and both $l_1$ and $l_2$ pass through exactly one red point, respectively, and might pass through blue points, then for every integer $i$, $|L(l_1) \cap B| \leq i \leq |L(l_2) \cap B|$, there exists a line $l_3$ such that $|L(l_3) \cap R| = |L(l_1) \cap R|$, $|L(l_3) \cap B| = i$ and $l_3$ passes through exactly one red point and no blue point.*

**Lemma 4.** *Let $R$ and $B$ be disjoint sets of $n$ red points and $bn$ blue points in the plane respectively. If $3 \leq |R| \leq 4$ and there exists a line $l_1$ such that $|L(l_1) \cap R| = 1$ and $|L(l_1) \cap B| \leq b$, then there exists a line $l_2$ such that $|L(l_2) \cap R| = 1$, $|L(l_2) \cap B| = b$ and $l_2$ passes through no point in $R \cup B$.*

*Proof.* Suppose first $|R| = 3$. Let $l_3$ be a line that passes through exactly one red point, and satisfies $|L(l_3) \cap R| = |R(l_3) \cap R| = 1$. Then at least one of $L(l_3) \cap B$ and $R(l_3) \cap B = L(l_3^*) \cap B$ contains at least $b$ blue points. By applying Lemma 2 to $l_1$ and to either $l_3$ or $l_3^*$, we can obtain the desired line $l_2$.

We next consider the case $|R| = 4$. Let $l_4$ be a line that passes through two red points and satisfies $|L(l_4) \cap R| = |L(l_4) \cap R| = 1$. Then at least one of $L(l_4) \cap B$ and $R(l_4) \cap B$ contains at least $b$ blue points, and so the lemma holds by Lemma 2.

**Lemma 5.** *Let $R$ and $B$ be disjoint sets of 5 red points and $5b$ blue points respectively. If there exists a line $l_1$ such that $|L(l_1) \cap R| = 2$ and $|L(l_1) \cap B| \leq 2b$, then there exists a line $l_2$ such that $|L(l_2) \cap R| = 2$, $|L(l_2) \cap B| = 2b$ and $l_2$ passes through no point in $R \cup B$.*

*Proof.* Consider a line $l_3$ passing through exactly one red point and satisfying $|L(l_3) \cap R| = 2$ and $|R(l_3) \cap R| = 2$. Then $|L(l_3) \cap B| \geq 2b$ or $|R(l_3) \cap B| \geq 2b$. Thus by Lemma 2, there exists the desired line $l_2$.

**Proposition 1.** *Let $R$ and $B$ be disjoint sets of 5 red points and $5b$ blue points in the plane respectively. Then $R \cup B$ has a $(2, 2, 1)$-balanced partition.*

*Proof.* Unless otherwise stated, except when it moves, we always consider a line that passes through no point in $R \cup B$. We begin with two Claims.

**Claim 1.** If there exists a line $l_1$ such that $|L(l_1) \cap R| = 1$ and $|L(l_2) \cap B| \le b$, then the proposition holds. Thus we may assume that for every line $l$ with $|L(l) \cap R| = 1$, it follows that $|L(l) \cap B| > b$. By considering $l^*$, we may also assume that for every line $l$ with $|R(l) \cap R| = |L(l^*) \cap R| = 1$, it follows that $|R(l) \cap B| = |L(l^*) \cap B| > b$.

Suppose that there exists a line $l_1$ such that $|L(l_1) \cap R| = 1$ and $|L(l_1) \cap B| \le b$. If there exists a line $l_2$ such that $|L(l_2) \cap R| = 1$ and $|L(l_2) \cap B| \ge b$, then there exists a line $l_3$ satisfying $|L(l_3) \cap R| = 1$ and $|L(l_3) \cap B| = b$ by Lemma 2. By Ham-sandwich Theorem, $R(l_3) \cap (R \cup B)$ can be partitioned into two balanced subsets. Thus the proposition holds. Therefore, we may assume that for every line $l$ with $|L(l) \cap R| = 1$, it follows that $|L(l) \cap B| < b$. Similarly, we may assume that for every line $l$ with $|R(l) \cap R| = 1$, it follows that $|R(l) \cap B| < b$.

Let $l_4$ and $l_5$ be two parallel lines with the same direction that pass through exactly one red point respectively, and satisfy $|L(l_4) \cap R| = 1$ and $|R(l_5) \cap R| = 1$. Then $|L(l_4) \cap B| < b$, $|R(l_5) \cap B| < b$ and $R(l_4) \cap L(l_5)$ contains exatly one red point. By the symmetry, we may assume that there exists a line $l_6$ that is parallel to $l_4$, lies between $l_4$ and $l_5$, and satisfies $|L(l_6) \cap R| = 2$ and $|L(l_6) \cap B| = 2b$.

By Lemma 4 and $l_5^*$, $R(l_6) \cap (R \cup B)$ has a $(2,1)$-balanced partition. Hence the proposition holds. Consequently Claim 1 is proved.

**Claim 2.** If there exists a line $l_1$ such that $|L(l_1) \cap R| = 2$ and $|L(l_2) \cap B| \ge 3b$, then the proposition follows. Thus we may assume that for every line $l$ with $|L(l_1) \cap R| = 2$, it follows that $|L(l) \cap B| < 3b$.

Suppose that there exists a line $l_1$ such that $|L(l_1) \cap R| = 2$ and $|L(l_2) \cap B| \ge 3b$. Then there exists a line $l_2$ in $R(l_1)$ such that $l_2$ passes through exactly one red point and no blue point, $|L(l_2) \cap R| = |R(l_2) \cap R| = 2$, $|L(l_2) \cap B| \ge 3b$ and $|L(l_2^*) \cap B| = |R(l_2) \cap B| \le 2b$. By applying Lemma 3 to $l_2$ and $l_2^*$, there exists a line $l_3$ that passes through exactly one red point, say $x$, and no blue point, and satisfies $|L(l_3) \cap R| = 2$ and $|L(l_3) \cap B| = 3b$.

We now consider $(L(l_3) \cap (R \cup B)) \cup \{x\}$, which contains three red points and $3b$ blue points. By Lemma 4, $(L(l_3) \cap (R \cup B)) \cup \{x\}$ has a $(2,1)$-balanced partition, and thus $R \cup B$ has the desired $(2,2,1)$-balanced partition. Therefore Claim 2 is proved.

Let $l_1$ be a line that passes through two red points, say $x$ and $y$, and satisfies $|L(l_1) \cap R| = 1$ (see Figure 4). Put $L(l_1) \cap R = \{z_1\}$ and $R(l_1) \cap R = \{z_2, z_3\}$. By considering a line $l_1'$ very closed to $l_1$ such that $L(l_1') \cap R = \{x, z_1\}$, $y \in R(l_1')$ and $L(l_1') \cap B = L(l_1) \cap B$, it follows from Claim 2 that $|L(l_1) \cap B| = |L(l_1') \cap B| < 3b$. By Claim 2, we have $|L(l_1^*) \cap B| = |R(l_1) \cap B| < 3b$, which implies $|L(l_1) \cap B| > 2b$. Therefore

$$2b < |L(l_1) \cap B| < 3b \quad \text{and} \quad |R(l_1) \cap B| < 3b. \qquad (3)$$

Let $p$ be any point on $l_1$, and $r_2$ a ray emanating from $p$ such that $R(r_1^*) \cap L(r_2)$ contains exactly $2b$ blue points, where $r_1^*$ denotes the ray on $l_1$ emanating from $p$ and having the opposite direction of $l_1$. By (3), $r_2$ is contained in $L(l_1)$, and so $R(r_1^*) \cap L(r_2)$ is a wedge $wdg(r_1^*r_2)$. We next take a ray $r_3$ emanating from $p$ such that $R(r_2) \cap L(r_3)$ contains exactly $b$ blue points. By (3), $r_3$ is contained in $R(l_1)$, and if $R(l_2) \cup l_2$ contains a red point, then $r_3$ must be contained in $R(l_2)$ by Claim 1, and thus $R(r_2) \cap L(r_3)$ is a wedge $wdg(r_2r_3)$.

We first consider the case of $p = x$, that is, we take a point $p$ on $x$. In this case, $R(r_2) \cap L(r_3)$ is a wedge $wdg(r_2r_3)$. If $wdg(r_2r_3)$ contains no red point, then $z_1$ is contained in $wdg(r_1^*r_2)$, and we get the desired balanced partition $wdg(r_1^*r_2) \cup \{y\}$, $wdg(r_2r_3) \cup \{x\}$ and $wdg(r_1^*r_3)$, where $wdg(r_1^*r_3)$ contains the two red points $z_2$ and $z_3$. Hence we may assume that $wdg(r_2r_3)$ contains at least one red point when $p = x$.

By moving a point $p$ along $l_1$ in its direction from $x$ to a point very far from $x$, we can find either (i) a point $p_1$ for which $r_2$ passes through the red point $z_1$ and $wdg(r_2r_3)$ contains no red points, or (ii) a point $p_2$ for which $r_3$ passes through one red point, say $z_2$, and $wdg(r_2r_3)$ contains no red point (see Figure 4). Since $wdg(r_1^*r_3)$ contains exactly $2b$ blue points, in each case we can easily obtain the desired $(2, 2, 1)$-balanced partition.

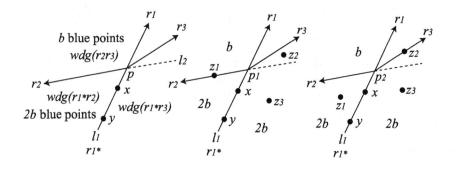

**Fig. 4.** A line $l_1$ and rays $r_1$, $r_1^*$, $r_2$ and $r_3$.

**Proposition 2.** *Let $R$ and $B$ be disjoint sets of 6 red points and $6b$ blue points respectively. Then $R \cup B$ has a $(3, 2, 1)$-balanced partition.*

*Proof.* Unless otherwise stated, we always consider a line that passes through no point in $R \cup B$. We begin with some Claims.

**Claim 1.** If there exists a line $l_1$ such that $|L(l_1) \cap R| = 2$ and $|L(l_1) \cap B| \geq 3b$, then the proposition holds. Thus we may assume that for every line $l$ with $|L(l) \cap R| = 2$, it follows that $|L(l) \cap B| < 3b$. By considering $l^*$, we may also assume that for every line $l$ with $|R(l) \cap R| = 2$, it follows that $|R(l) \cap B| < 3b$.

Suppose that there exists a line $l_1$ such that $|L(l_1) \cap R| = 2$ and $|L(l_2) \cap B| \geq 3b$. By a suitable very small rotation of $l_1$, we may assume that every line parallel to $l_1$ passes through at most one point in $R \cup B$. Then there exist two lines $l_2$ and $l_3$ in $R(l_1)$ such that they are parallel to $l_1$ and have the same direction as $l_1$, $|L(l_2) \cap R| = 2$, $|L(l_2) \cap B| \geq 3b$, $|R(l_3) \cap R| = 2$, $|R(l_3) \cap B| \leq 3b$ and such that each of $l_2$ and $l_3$ passes through exactly one red point, in particular, $R(l_2) \cap L(l_3)$ contains no red points.

By applying Lemma 3 to $l_2$ and $l_3^*$, there exists a line $l_4$ that passes through exactly one red point, say $x$, and satisfies $|L(l_4) \cap R| = 2$ and $|L(l_4) \cap B| = 3b$. Then $R(l_4)$ contains three red points and $3b$ blue points, and $Q = (L(l_4) \cap (R \cup B)) \cup \{x\}$ also contains three red points and $3b$ blue points. By Lemma 4, $Q$ has a $(2, 1)$-balanced partition. Hence $R \cup B$ has the desired $(3, 2, 1)$-balanced partition. Therefore the claim is proved.

**Claim 2.** If there exists a line $l_1$ such that $|L(l_1) \cap R| = 1$ and $|L(l_2) \cap B| \leq b$, then the proposition holds. Thus we may assume that for every line $l$ with $|L(l) \cap R| = 1$, it follows that $|L(l) \cap B| > b$.

Suppose that there exists a line $l_1$ such that $|L(l_1) \cap R| = 1$ and $|L(l_1) \cap B| \leq b$. We now assume that there exists a line $l_2$ such that $|L(l_2) \cap R| = 1$ and $|L(l_2) \cap B| \geq b$. Then by Lemma 2, there exists a line $l_3$ such that $|L(l_3) \cap R| = 1$ and $|L(l_3) \cap B| = b$. By considering a very small rotation of $l_3$ if necessary, we may assume that every line parallel to $l_3$ passes through at most one point in $R \cup B$.

Let $l_4$ be a line that is parallel to $l_3$, pass through exactly one red point, and satisfy $|L(l_4) \cap R| = 2$. Then by Claim 1, $R(l_3) \cap L(l_4)$ contains at most $2b - 1$ blue points, which implies that a line $l_5$ very closed to $l_4$ and lying to the right of $l_4$ satisfies $|(L(l_5) \cap R(l_3)) \cap R| = 2$ and $|(L(l_5) \cap R(l_3)) \cap B| < 2b$. By applying Lemma 5 to $R(l_3) \cap (R \cup B)$ and $l_5$, we obtain that $R(l_3) \cap (R \cup B)$ has a $(3, 2)$-balanced partition, which implies that $R \cup B$ has the desired $(3, 2, 1)$-balanced partition.

Hence we may assume that for every line $l$ with $|L(l) \cap R| = 1$, it follows that $|L(l) \cap B| < b$.

By Ham-sandwich Theorem, $R \cup B$ can be partitioned into two balanced subsets $Q_1 \cup Q_2$, each of which contains exactly three red points and $3b$ blue points. By Lemma 4 and by the above statement on a line $l$, the subset $Q_1$ has a $(2, 1)$-balanced partition. Hence $R \cup B$ has the desired $(3, 2, 1)$-balanced partition. Therefore the claim is proved.

**Claim 3.** If there exists a line $l_1$ such that $|L(l_1) \cap R| = 2$ and $|L(l_2) \cap B| \leq 2b$, then the proposition holds. Thus we may assume that for every line $l$ with $|L(l) \cap R| = 2$, it follows that $|L(l) \cap B| > 2b$.

Suppose that there exists a line $l_1$ such that $|L(l_1) \cap R| = 2$ and $|L(l_1) \cap B| \leq 2b$. By considering a line $l_2$ passing through two red points and satisfying $|L(l_2) \cap R| = |R(l_2) \cap R| = 2$, we may assume that $|L(l_2) \cap B| \geq 2b$ by symmetry. Thus there exists a line $l_3$ such that $|L(l_3) \cap R| = 2$ and $|L(l_3) \cap B| = 2b$. Let $l_4$ be a line that is parallel to $l_3$, passes through exactly one red point, and satisfies $|L(l_4) \cap R| = 2$. Then by Claim 1, the number of blue points lying between $l_3$ and $l_4$ is at most $b - 1$. Hence by Lemma 4, $R(l_3) \cap (R \cup B)$ has a $(3, 1)$-balanced partition. Thus $R \cup B$ has a $(3, 2, 1)$-balanced partition, and hence the claim is proved.

Let $l_1$ be a line that passes through two red points, say $x$ and $y$, and satisfies $|L(l_1) \cap R| = 1$. Put $L(l_1) \cap R = \{z_1\}$ and $R(l_1) \cap R = \{z_2, z_3, z_4\}$. By considering a line $l_1'$ very closed to $l_1$ such that $L(l_1') \cap R = \{x, z_1\}$ and $L(l_1') \cap B = L(l_1) \cap B$, it follows from Claims 1 and 2 that $2b < |L(l_1) \cap B| = |L(l_1') \cap B| < 3b$, that is, the following inequality holds:

$$2b < |L(l_1) \cap B| < 3b \quad \text{and} \quad |R(l_1) \cap B| < 4b. \tag{4}$$

Let $p$ be any point on $l_1$, and $r_2$ a ray emanating from $p$ such that $R(r_1^*) \cap L(r_2)$ contains exactly $2b$ blue points, where $r_1^*$ denotes the ray on $l_1$ emanating from $p$ and having the opposite direction of $l_1$. By (4), $r_2$ is contained in $L(l_1)$, and so $R(r_1^*) \cap L(r_2)$ is a wedge $wdg(r_1^* r_2)$. We next take a ray $r_3$ emanating from $p$ such that $R(r_2) \cap L(r_3)$ contains exactly $b$ blue points. By (4), $r_3$ is contained in $R(l_1)$, and if $R(l_2) \cup l_2$ contains a red point, then $r_3$ must be contained in $R(l_2)$ by Claim 2, and so $R(r_2) \cap L(r_3)$ is a wedge $wdg(r_2 r_3)$.

Hereafter, by similar arguments in the proof of Proposition 1, we can prove Proposition 2.

We omit the proof of the following Proposition 3, but in each case of Proposition 3, we can individually prove it by similar arguments in the proofs of the above two propositions.

**Proposition 3.**  *Let $R$ and $B$ be two disjoint sets of $n$ red points and $bn$ blue points respectively. If $n = 7$, then $R \cup B$ has both a $(3, 3, 1)$-balanced partition and a $(3, 2, 2)$-balanced partition. If $n = 8$, then $R \cup B$ has both a $(4, 3, 1)$-balanced partition and a $(3, 3, 2)$-balanced partition.*

Theorem 1 can be easily proved by using Propositions 1, 2, and 3, Ham-sandwich Theorem and Theorem 3. For example, the existence of a $(2, 2, 2)$-balanced partition is guaranteed by Theorem 3, and a $(3, 1, 1, 1)$-balanced partition is obtained from a $(3, 2, 1)$-balanced partition by applying Ham-sandwich Theorem to its subset containing two red points and $2b$ blue points.

We now give a sketch of the proof of Theorem 2 since it is quite long, but it is similar to the proof given in [5].

Let $R$ and $B$ be two disjoint sets of $n$ red points and $bn$ blue points respectively. Put $n = 2k + 1$. We prove the theorem by induction on $n$.

**Claim 1.** If $n = 5, 7, 9$, then $R \cup B$ has a $(2, 2, \ldots, 2, 1)$-balanced partition. Thus we may assume that $n \geq 11$.

**Claim 2.** If there exists a line $l$ such that $5 \leq |L(l) \cap R| = t \leq n - 5$ and $|L(l) \cap B| = bt$, then $R \cup B$ has a $(2, 2, \cdots, 2, 1)$-balanced partition.

We give a proof of Claim 2 since it is short. If the integer $t$ defined in the claim is odd, then $L(l) \cap (R \cup B)$ has a $(2, 2, \cdots, 2, 1)$-balanced partition by the induction hypothesis. Since $n-t$ is even, $R(l) \cap (R \cup B)$ has a $(2, 2, \ldots, 2)$-balanced partition by Ham-sandwich Theorem 3. Hence $R \cup B$ has the desired balanced partition. If $t$ is even, then we apply the inductive hypothesis to $R(l) \cap (R \cup B)$, and we can similarly obtain the desired partition as above.

**Claim 3.** Let $i$ be an integer such that $1 \le i \le k$ and $i \ne 3$. Then for every line $l$ with $|L(l) \cap R| = i$, we may assume that $|L(l) \cap B| > bi$ since otherwise the theorem holds. Moreove if $|L(l) \cap R| = k$, then we may assume that $km < |L(l) \cap B| < k(m+1)$.

Let $l_1$ be a line passing through one red point, say $x$, such that $R(l_1)$ contains no red point, and $l_2$ be a line passes through $x$ and one more red point, say $y$, such that $|R(l_2) \cap R| = k$ and $|L(l_2) \cap R| = k - 1$. Without loss of generality, we may assume that the direction of $l_2$ is downward. Let $r_3$ be a ray emanating from $x$ so that $R(r_2) \cap L(r_3)$ contains exactly $2m$ blue points.

**Claim 4.** We may assume that $R(r_2) \cap L(r_3)$ contains at least two red points since otherwise the theorem holds.

Let $r_4$ be a ray emanating from $x$ such that $R(r_2) \cap L(r_4)$ contains as many as points in $R \cup B$ subject to $R(r_2) \cap L(r_4)$ contains exactly $2i$ red points and at most $2ib$ blue points, and lies in $R(l_2)$.

The existence of the above $r_4$ is guaranteed by Claim 3. By applying similar arguments given in [5] to these two lines $l_1$ and $l_2$, rays $r_3$ and $r_4$, and other rays emanating from $x$, we can prove Theorem 2.

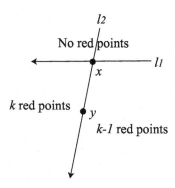

**Fig. 5.** Two lines $l_1$ and $l_2$.

# References

1. J.Akiyma, A.Kaneko, M.Kano, G.Nakamura, E.Rivera-Campo, S.Tokunaga and J.Urrutia, Radical Perfect partitions of convex sets in the plane. *Discrete and Computational Geometry* (Lecture notes in computer science No.1763) 1-13, (2000).

2. S.Bespamyatnikh, D.Kirkpatrick and J.Snoeyink,  Generalizing ham sandwich cuts to equitable subdivisions,  preprint.
3. J. Goodman and J. O'Rourke, *Handbook of Discrete and Computational Geometry*, CRC Press,  p.211 (1997)
4. H.Ito, H.Uehara, and M.Yokoyama,  2-dimentinal ham-sandwich theorem for partitioning into three convex pieces,  *Discrete and Computational Geometry* (Lecture notes in computer science No.1763),  129-157,  (2000).
5. A.Kaneko, and M.Kano,  Balanced partitions of two sets of points in the plane. Computational Geometry: Theory and Applications, **13**, 253-261 (1999).
6. A.Kaneko, and M.Kano,  A balanced partition of points in the plane and tree embedding problems,  preprint
   Computatinal Geometry: Theory and Applications, **13**, 253-261 (1999).
7. A.Kaneko, and M.Kano,  Perfect partitions of convex sets in the plane,   preprint.
8. T.Sakai, Balanced Convex Partitions of Measures in $R^2$,  to appear in *Graphs and Combinatorics*.

# On Paths in a Complete Bipartite Geometric Graph

Atsushi Kaneko[1] and M.Kano[2]

[1] Department of Computer Science and Communication Engineering,
Kogakuin University, Shinjuku-ku, Tokyo 1563-8677, Japan
[2] Department of Computer and Information Sciences,
Ibaraki University, Hitachi 316-8511, Japan

**Abstract.** Let $A$ and $B$ be two disjoint sets of points in the plane such
that no three points of $A \cup B$ are collinear, and let $n$ be the number of
points in $A$. A geometric complete bipartite graph $K(A, B)$ is a complete
bipartite graph with partite sets $A$ and $B$ which is drawn in the plane
such that each edge of $K(A, B)$ is a straight-line segment. We prove that
(i) If $|B| \geq (n + 1)(2n - 4) + 1$, then the geometric complete bipartite
graph $K(A, B)$ contains a path that passes through all the points in $A$
and has no crossings; and (ii) There exists a configuration of $A \cup B$ with
$|B| = \frac{n^2}{16} + \frac{n}{2} - 1$ such that in $K(A, B)$ every path containing the set $A$
has at least one crossing.

## 1 Introduction

Let $G$ be a finite graph without loops or multiple edges. We denote by $V(G)$ and
$E(G)$ the set of vertices and the set of edges of $G$, respectively. For a vertex $v$
of $G$, we denote by $\deg_G(v)$ the degree of $v$ in $G$. For a set $X$, we denote by $|X|$
the cardinality of $X$. A *geometric graph* $G = (V(G), E(G))$ is a graph drawn in
the plane such that $V(G)$ is a set of points in the plane, no three of which are
collinear, and $E(G)$ is a set of (possibly crossing) straight-line segments whose
endpoints belong to $V(G)$. If a geometric graph $G$ is a complete bipartite graph
with partite sets $A$ and $B$, i.e., $V(G) = A \cup B$, then $G$ is denoted by $K(A, B)$,
which may be called *a geometric complete bipartite graph.*

In 1996, M. Abellanas, J. García, G. Hernández, M. Noy and P. Ramos [1]
showed the following result.

**Theorem A ( Abellanas et al. [1])**    *Let $A$ and $B$ be two disjoint sets of
points in the plane such that $|A| = |B|$ and no three points of $A \cup B$ are collinear.
Then the geometric complete bipartite graph $K(A, B)$ contains a spanning tree
$T$ without crossings such that the maximum degree of $T$ is $O(\log |A|)$.*

In 1999, Kaneko [3] improved their result and proved the following theorem.

**Theorem B ( Kaneko [3])**    *Let $A$ and $B$ be two disjoint sets of points in the
plane such that $|A| = |B|$ and no three points of $A \cup B$ are collinear. Then the
geometric complete bipartite graph $K(A, B)$ contains a spanning tree $T$ without
crossings such that the maximum degree of $T$ is at most 3.*

J. Akiyama, M. Kano, and M. Urabe (Eds.): JCDCG 2000, LNCS 2098, pp. 187-191, 2001.

It is well-known that under the same condition in Theorem B, there are configurations of $A \cup B$ such that $K(A, B)$ does not contain a hamiltonian path without crossings [2]. Note that the upper bound of the number of crossings of hamiltonian cycles in $K(A, B)$ is given in [4]. So we are led to the following problem. Given two disjoint sets $A$ and $B$ of points in the plane such that no three points of $A \cup B$ are collinear, if $|B|$ is large compared with $|A|$, then does $K(A, B)$ contain a path $P$ without crossings such that $V(P)$ contains the set $A$? The answer to the above question is in the affirmative, as we shall see now. We prove the following theorem.

**Theorem 1.** *Let $A$ and $B$ be two disjoint sets of points in the plane such that no three points of $A \cup B$ are collinear, and let $n$ be the number of points in $A$.*
*(i) If $|B| \geq (n + 1)(2n - 4) + 1$, then the geometric complete bipartite graph $K(A, B)$ contains a path $P$ without crossings such that $V(P)$ contains the set $A$.*
*(ii) There exists a configuration of $A \cup B$ with $|B| = \frac{n^2}{16} + \frac{n}{2} - 1$ such that in $K(A, B)$ every path containing the set $A$ has at least one crossing.*

In order to prove Theorem 1, we need some notation and definitions. For a set $X$ of points in the plane, we denote by $\mathrm{conv}(X)$ the convex hull of $X$. The boundary of $\mathrm{conv}(X)$ is a polygon whose segments and extremes are called *the edges and the vertices* of $\mathrm{conv}(X)$, respectively. For two points $x$ and $y$ in the plane, we denote by $xy$ the straight line segment joining $x$ to $y$, which may be an edge of a geometric graph containing both $x$ and $y$ as it vertices. Let $A$ be a set of point in the plane, let $y$ be a vertex of $\mathrm{conv}(A)$ and let $x$ be a point exterior to $\mathrm{conv}(A)$. Then we say that $x$ *sees* $y$ on $\mathrm{conv}(A)$ if the line segment $xy$ intersects $\mathrm{conv}(A)$ only at $y$.

**Lemma 1.** *Let $R$ and $S$ be disjoint sets of points in the plane with $|R| \geq |S|$ such that no three points of $R \cup S$ are collinear. Suppose that there exists a line in the plane that separates $R$ and $S$. Let $x$ and $y$ be two vertices of $\mathrm{conv}(R \cup S)$ such that $x \in S$, $y \in R$, and $xy$ is an edge of $\mathrm{conv}(R \cup S)$. Then in $K(R, S)$, there exists a path $P$ without crossings such that*
*(i) the vertex $x$ is an end of $P$, and*
*(ii) $P$ passes through all the points in $A$.*

*Proof.* We prove the lemma by induction on $|R \cup S|$. If $|S| = 1$ or $|S| = 2$, then the lemma follows immediately, and so we may assume $|R| \geq |S| \geq 3$.

Let $x_1$ be the vertex of $\mathrm{conv}(R \cup S)$ such that $x_1 \in S$ and $xx_1$ is an edge of $\mathrm{conv}(R \cup S)$ (see Figure 1(b)).

Then we can find two points $z_1 \in S - \{x\}$ and $z \in R$ such that $x$ can see both $z_1$ and $z$, and $z_1z$ is an edge of $\mathrm{conv}(R \cup S - \{x\})$ (see Figure 1(b)). Note that it may occur that $z_1 = x_1$ and/or $z = y$. Similarly, we can find two more points $w_1 \in S - \{x\}$ and $w \in R - \{z\}$ such that $z$ can see both $w_1$ and $w$, and $w_1w$ is an edge of $\mathrm{conv}(R \cup S - \{x, z\})$ (see Figure 1(b)). Note that it may occur that $w_1 = z_1$ (and/or $w = y$ if $z \neq y$).

We now apply the inductive hypothesis to $S - \{x\}$, $R - \{z\}$, $w_1$ and $w$. Then there exists a path $P'$ in $K(S - \{x\}, R - \{z\})$ without crossings that starts with

$w_1$ and contains $S - \{x\}$. By adding two edges $w_1z$ and $zx$ to $P'$, we obtain the desired path in the lemma.

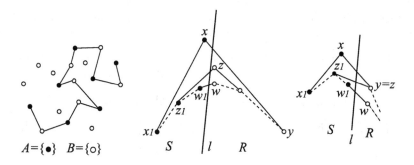

**Fig. 1.** (a) A path given in Theorem 1.    (b) Figure of proof of Lemma 1.

Now we proceed to prove part (i) of Theorem 1. We may assume that no two points of $A \cup B$ have the same $x$-coordinate. Let $a_1, a_2, \ldots a_n$ be points of $S$ sorted by their $x$-coordinate and let $l_i$ be the vertical line which passes through the point $a_i$, $1 \le i \le n$. These $n$ lines separate the plane into $n + 1$ regions and hence they separate the set $B$ into $n + 1$ disjoint subsets. Assume that these lines are directed upward. By the assumption, at least one subset contains at least $2n - 3$ points of $B$. We may assume that one of the regions which contains at least $2n - 3$ points of $B$ is bounded by the lines $l_j$ and $l_{j+1}$, $1 \le j \le n - 1$. (The leftmost and rightmost unbounded regions can be treated similarly.) Let $B_j$ be the subset of $B$ between $l_j$ and $l_{j+1}$, i.e., $|B_j| \ge 2n - 3$. Let $l_0$ be the line between $l_j$ and $l_{j+1}$ satisfying the following conditions:

(i) $l_0$ passes through a point $b_0$ of $B_j$ and is directed upward,
(ii) The number of points in $B_j - \{b_0\}$ to the left of $l_0$ is $2j - 2$.

Let $B_l$ be the subset of $B_j - \{b_0\}$ to the left of $l_0$, and $B_r = B_j - \{b_0\} - B_l$. Then $|B_l| = 2j - 2$ and $|B_r| \ge 2n - 2j - 2$. Let $A_l$ be the subset of $A$ to the left of $l_0$ and let $A_r$ be the subset of $A$ to the right of $l_0$. Trivially $|A_l| = j$ and $|A_r| = n - j$. Let $t_1$ and $t_2$ be the two rays emanating from $b_0$ such that $t_i$ is tangent to $\text{conv}(A_l)$ at $w_i$, $1 \le i \le 2$, and $t_1$ is above $t_2$. Also let $t_3$ and $t_4$ be the two rays emanating from $b_0$ such that $t_i$ is tangent to $\text{conv}(A_r)$ at $w_i$, $3 \le i \le 4$, and $t_3$ is above $t_4$. (Notice that since no three points of $A \cup B$ are collinear, each ray contains no point of $B_l \cup B_r$.) Let $B_l^+$ be the subset of $B_l$ above the ray $t_2$ and $B_l^-$ the subset of $B_l$ under the ray $t_1$. Also let $B_r^+$ be the subset of $B_r$ above the ray $t_4$ and $B_r^-$ the subset of $B_r$ under the ray $t_3$. Since $|B_l| = 2j - 2$, we have either $|B_l^+| \ge j - 1$ or $|B_l^-| \ge j - 1$, say $|B_l^+| \ge j - 1$. Similarly we have either $|B_r^+| \ge n - j - 1$ or $|B_r^-| \ge n - j - 1$, say $|B_r^+| \ge n - j - 1$. Consider now $K(B_l^+ \cup \{b_0\}, A_l)$. Since $|B_l^+ \cup \{b_0\}| \ge j = |A_l|$, applying Lemma 1 and letting

$x = b_0$, we can find a path $R_l$ in $K(B_l^+ \cup \{b_0\}, A_l)$ without crossings such that (i) the vertex $b_0$ is an end of $R_l$ and (ii) $V(R_l)$ contains $A_l$. In a similar manner, we can find a path $R_r$ in $K(B_r^+ \cup \{b_0\}, A_r)$ without crossings such that (i) the vertex $b_0$ is an end of $R_r$ and (ii) $V(R_r)$ contains $A_r$. Set $P = R_l \cup R_r$. Clearly $P$ is a path in $K(A, B)$ without crossings such that $V(P)$ contains the set $A$.

In order to show part (ii) of Theorem 1, suppose that $n = 4k$ and all points of $A = \{a_j^i\}$ and $B = \{b_j^i\}$ lie on a cycle in the following order:

$$a_1^0,\ a_2^0,\ \ldots,\ a_{k+2}^0,\ b_1^0,\ b_2^0,\ \ldots,\ b_k^0,\ a_1^1,\ a_2^1,\ b_1^1,\ b_2^1,\ \ldots,\ b_k^1,$$

$$a_1^2,\ a_2^2,\ b_1^2,\ b_2^2,\ \ldots,\ b_k^2,\ \ldots\ \ldots\ \ldots,\ a_1^{k-2},\ a_2^{k-2},\ b_1^{k-2},\ b_2^{k-2},\ \ldots,\ b_k^{k-2},$$

$$a_1^{k-1},\ a_2^{k-1},\ \ldots,\ a_{k+2}^{k-1},\ b_1^{k-1},\ b_2^{k-1},\ \ldots,\ b_{3k-1}^{k-1} \text{ (see Figure 2)}.$$

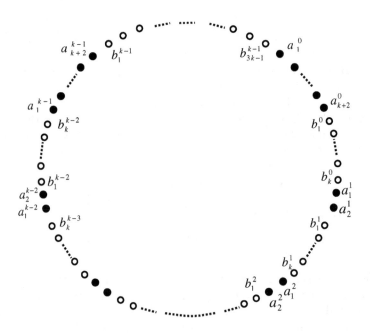

**Fig. 2.**

It is not difficult to show that $|A| = n$ and $|B| = \frac{n^2}{16} + \frac{n}{2} - 1$ and that in $K(A, B)$ every path containing the set $A$ has at least one crossing.

This completes the proof of Theorem 1.

# ACKNOWLEDGMENT

We would like to thank Professor H. Enomoto for showing us the configuration.

# References

1. M. Abellanas, J. García, G. Hernández, M. Noy and P. Ramos, Bipartite embeddings of trees in the plane, *Graph Drawing'96* , Springer Verlag LNCS **1190**, (1996), 1-10
2. J. Akiyama and J. Urrutia, Simple alternating path problem, Discrete Mathematics **84** no. 1, (1990), 101-103
3. A. Kaneko, On the maximum degree of bipartite embeddings of trees in the plane, *Discrete and Computational Geometry* Springer Verlag LNCS **1763**, (2000), 166-171.
4. A. Kaneko, M. Kano and K. Yoshimoto, Alternating hamilton cycles with minimum number of crossings in the plane, International Journal of Computational Geometry & Applications, **10** no.1, (2000), 73-78

# Approximating Uniform Triangular Meshes for Spheres

Naoki Katoh[1], Hiromichi Kojima[1], Ryo Taniguchi[2]

[1] Department of Architecture and Architectural Systems, Kyoto University
Yoshida-Honmachi, Sakyo-ku, Kyoto, 606-8501 Japan
naoki@archi.kyoto-u.ac.jp, kojima@is-mj.archi.kyoto-u.ac.jp
[2] Railway Information Systems
Hikari-machi,Kokubunji, Tokyo, 185-8510
ryou_taniguchi@jrs.co.jp

**Abstract.** We consider the problem of triangulating a convex polygon on spheres using $n$ Steiner points that minimizes the overall edge length ratio. We establish a relation of this problem to a certain extreme packing problem. Based on this relationship, we develop a heuristic producing 6-approximation for spheres (provided $n$ is chosen sufficiently large). That is, the produced triangular mesh is *uniform* in this respect.
The method is easy to implement and runs in $O(n^3)$ time and $O(n)$ space.

## 1 Introduction

Given a convex polygon $P$ on the sphere (see Section 2 for the precise definition for a convex polyhedron used in this paper) and a positive integer $n$, we consider the problem of generating a length-uniform triangular mesh for $P$ using $n$ Steiner points. More specifically, we want to find a set $S_n$ of $n$ points in the closure of $P$, and a triangulation of $P$ using $S_n$ that minimizes the ratio of the maximum edge length to the minimum one.

In our recent work [2], we considered the case of a convex polygon in the plane and proposed a 6-approximation algorithm for this problem. In this paper, we extend the result to spheres.

Although the NP-hardness has not been settled yet, finding an optimal solution for the problem for the convex polyhedron on the plane or the sphere seems to be difficult, in view of the NP-completeness of packing problems in the plane, see e.g. Johnson [12], or in view of the intrinsic complexity of Heilbronn's triangle problem, see [18].

In order to help understand our problem, it would be noteworthy to consider one-dimensional analog of the problem. Namely, given an interval on a circle with specified points on it, place $n$ points in the interval so as to minimize the ratio of maximum and minimum interpoint distances. On the contrary to the intrinsic difficulty of the problem dealt with in this paper, one-dimensional problem can be exactly solved in a straightforward manner although the details are omitted here. Alternatively, you can imagine a simple greedy procedure that iteratively

J. Akiyama, M. Kano, and M. Urabe (Eds.): JCDCG 2000, LNCS 2098, pp. 192-204, 2001.

places a new point at the middle of the two consecutive points that are farthest apart. Let $l_{\min}$ denote the minimum interpoint distance for the original point set. After the minimum interpoint distance becomes smaller than $l_{\min}$, the ratio becomes less than two. Also, as $n$ goes to infinity, the ratio converges to one. If we are given a line segment instead of an interval of a circle, the problem essentially remains the same.

Extending this idea to 2-dimensional case, [2] developed a 6-approximation algorithm for the case of convex polyhedron in the plane. The algorithm by [2] uses a heuristic called *canonical Voronoi insertion* which approximately solves a certain extreme packing problem for point sets within $P$, i.e., the problem that asks for a packing of $n$ circles with centers in $P$ such that the smallest radius is maximum. It has been shown that this heuristic gives a 2-approximation algorithm for this problem, and that the Delaunay triangulation for the point set appropriately modified from the solution produced by the algorithm gives a 6-approximation.

The algorithm proposed in this paper is based on the same idea as the one developed in [2]. We use again a canonical Voronoi insertion for the spherical region to produce a point set, and output as an approximate solution the convex hull for the point set appropriately modified from the solution produced by the canonical Voronoi insertion. The major difference between the algorithm given by [2] and the one proposed in this paper lies in that (1) we need to carry out 3-D Voronoi diagram computation for implementing canonical Voronoi insertion, and (2) in order to establish the approximation result, we need nontrivial arguments in order to generalize several lemmas given in [2] to spherical surfaces.

Here you may wonder why we use 3-D Voronoi diagram instead of 2-D counterpart for spherical region. Actually, If we use 2-D Voronoi insertion algorithm for spherical region, we need to use the metric based on the geodesic distance of the sphere. After obtaining the Delaunay triangulation of the spherical region (each edge is not a straight line segment, but a minor arc of the great circle passing through the end points), curved edge is then replaced by a straight line segment to obtain the final solution. In this case, however, as shown by Taniguchi [20] the worst-case ratio of maximum and minimum edge lengths becomes $2\sqrt{10}$ which is larger than 6. On the other hand, we shall show that based on 3-D Voronoi diagram, the ratio of the maximum and minimum edge lengths is at most 6 which matches the bound for the planar case treated by [2]. It will be shown that the ratio can be improved to $4+\epsilon$ under certain restricted conditions. The overall running time is $O(n^3)$.

We have carried out computational experiments to see the effectiveness of the proposed method. The computational results show that for all problem instances that are tested the ratio of maximum and minimum edge lengths is at most 2. Thus, there is a big gap between theoretical and experimental results. To close the gap remains as a future research.

Our study of generating uniform triangular meshes on spheres is motivated by the necessity of designing structures such as triangular trusses of large-span structures (see Fig. 1), where it is required to determine the shape from aesthetic points of view under the constraints concerning stress and nodal displacement. The study of finding an optimal shape of space structures has been studied by several researchers [14, 15, 17]. The truss can be viewed as a triangulation of points on the curved surface by regarding truss members and nodes as edges and points, respectively. When focusing on the shape, edge lengths should be as equal as possible from the viewpoint of design, mechanics and manufacturing. In such applications, the locations of the points are usually not fixed, but can be viewed as decision variables. In view of this, it is quite natural to consider the problem. The paper by Nooshin et al. [14] studied the same problem as ours, and proposed a heuristic approach based on potential function defined for every pair of points. The potential is proportional to the inverse of squared distance. The method by [14] finds a configuration of points minimizing the total potential by a heuristic method. The approach developed therein is a so-called local improvement method.

From the practical viewpoints, several other factors have to be taken into consideration. For instance, (i) the number of distinct edge lengths should be as small as possible because this will decrease the production cost of members, (ii) the triangular truss to be designed should possess a certain symmetry property if the initial design has such property, and etc. Thus, our research can be viewed as a fundamental one in view of the practical design of optimal triangular trusses.

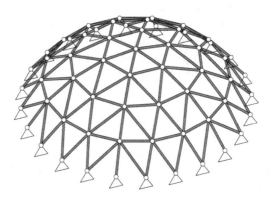

**Fig. 1.** Illustration of a triangular truss

The following notation will be used throughout. For two points $x$ and $y$ on the sphere, let $l(x, y)$ denote their Euclidean distance. The minimum (non-zero) distance between two point sets $X$ and $Y$ is defined as $l(X, Y) = \min\{l(x, y) \mid x \in X, y \in Y, x \neq y\}$. When $X$ is a singleton set $\{x\}$ we simply write $l(X, Y)$ as $l(x, Y)$. Note that $l(X, X)$ defines the minimum interpoint distance among the point set $X$.

## 2   Canonical Voronoi Insertion and Extreme Packing

A region on a sphere is *convex* if for any two points $u$ and $v$ in a region the shorter arc connecting $u$ and $v$ on a great circle passing through both $u$ and $v$ is inside the region. A convex region we are concerned with is assumed to be contained in a hemisphere; more specifically an upper hemisphere (i.e., north hemisphere). Together with a convex region on a sphere, a set of points (called vertices) on the boundary of the region is given as an input. Such a convex region together with boundary points is called a *convex polygon* on a sphere. Notice that the boundary edge of a convex polyhedron defined here is not a line segment in 3-D space but a curved segment. For a given point $p$ on a sphere and a positive number $r$, a *spherical disk* centered at $p$ with radius $r$ on a sphere is a region of the sphere such that for any point $x$ in the region the Euclidean distance from the point $p$ to $x$ is at most $r$. For our purpose, $r$ is assumed to be smaller than or equal to $\sqrt{2}$ times the radius of a given sphere.

In this section, we consider the following *extreme packing problem*. Let $\boldsymbol{P}$ be a (closed) convex polygon with vertex set $V$.

$$\text{Maximize } l(V \cup S_n, V \cup S_n)$$
$$\text{subject to a set } S_n \text{ of } n \text{ points within } \boldsymbol{P}.$$

In other words, the problem asks for a packing of $n$ disks with centers in $\boldsymbol{P}$ such that the smallest radius is maximum. We shall give a 2-approximation algorithm for this problem using *canonical Voronoi insertion*. In Section 3 we then show that the point set $S_n$ produced by this algorithm, as well as the triangulation induced by the upper hull of $S_n$, can be modified to give an approximate solution for the problem addressed in Section 1. The algorithm given here is a straightforward extension of the one proposed in [2] to the sphere, and thus we omit the proof of the lemma and the theorem given below.

The algorithm determines the location of the point set $S_n$ in a greedy manner. Namely, starting with an empty set $S$, it repeatedly places a new point inside $\boldsymbol{P}$ at the position which is farthest from the set $V \cup S$. The idea of the algorithm originates with Gonzalez [11] and Feder and Greene [9], and was developed for approximating minimax $k$-clusterings. Comparable insertion strategies are also used for mesh generation in 2D in Chew [8] and in Ruppert [19], there called *Delaunay refinement*. Their strategies aim at different quality measures, however, and insertion does not take place in a canonical manner. Various results on the size of circle packings are summarized in Fejes Tóth [10].

The algorithm is formally described below. It uses the Voronoi diagram of the current point set to select the next point to be inserted. We assume familiarity with the basic properties of a Voronoi diagram and its dual, the Delaunay triangulation, and refer to the survey paper [1]. In the following algorithm we use 3D Voronoi diagram consisting of Voronoi regions. A Voronoi region is a convex polyhedron in $\boldsymbol{R}^3$.

**Algorithm** INSERT

**Step 1:** Initialize $S := \emptyset$.

**Step 2:** Compute the Voronoi diagram (in 3D) $\text{Vor}(V \cup S)$ of $V \cup S$.

**Step 3:** Find the set $B$ of intersection points between edges of $\text{Vor}(V \cup S)$ and $P$. Find the set $C$ of intersection points between faces of $\text{Vor}(V \cup S)$ and the boundary of $P$. Among the points in $B \cup C$, choose the point $u$ which maximizes $l(u, V \cup S)$.

**Step 4:** Put $S := S \cup \{u\}$ and return to Step 2 if $|S| < n$.

Let $p_j$ and $S_j$, respectively, denote the point chosen in Step 3 and the set obtained in Step 4 at the $j$-th iteration of the algorithm. For an arbitrary point $x \in P$ define the *weight* of $x$ with respect to $S_j$ as $w_j(x) = l(x, S_j \cup V)$. That is, $w_j(x)$ is the radius of the largest disk centered at $x$ which does not enclose any point from $S_j \cup V$. By definition of a Voronoi diagram, the point $p_j$ maximizes $w_{j-1}(x)$ over all $x \in P$. Let

$$d_n = l(S_n \cup V, S_n \cup V) \tag{1}$$

be the minimum interpoint distance realized by $S_n \cup V$. Furthermore, denote by $S_n^*$ the optimal solution for the extreme packing problem for $P$ and let $d_n^*$ denote the corresponding objective value. Its proof is an adaptation of techniques in [11, 9] and contains observations that will be used in our further analysis.

**Theorem 1.** *The solution $S_n$ obtained by Algorithm INSERT is a 2-approximation of the extreme packing problem for $P$. That is, $d_n \geq d_n^*/2$.*

*Proof.* The proof is done by proving the claim that $p_n$ realizes the minimum (non-zero) distance from $S_n$ to $S_n \cup V$. Equivalently, the claim is

$$w_{n-1}(p_n) = l(S_n, S_n \cup V). \tag{2}$$

The remaining part of the proof is given in [2].

**Lemma 1.** *For any set $S \subset P$ of $n-1$ points there exists a point $x \in P$ with $l(x, S \cup V) \geq d_n^*/2$.*

We have assumed up to now that the vertex set of $P$ is given as an input. Besides such vertex set, the points inside the polygon are sometimes initially given. Notice that the above theorem and lemma also hold for this case.

Now we analyze the running time of the algorithm. The dominating part is the computation of 3D Voronoi diagram which requires $O(n^2)$ time ([7]). Since we have to compute 3D Voronoi diagram $O(n)$ times, the total time is $O(n^3)$.

## 3     Analysis of Approximation Ratio

For the planar case, the algorithm of [2] computes the Delaunay triangulation of a point set as the final output. On the other hand, for a set $S \cup V$ of points on $P$, we will output the upper part of the convex hull of the set as its triangulation (recall that $P$ is contained in the upper hemisphere). Since the convex hull of $S \cup V$ is a part of the Delaunay triangulation (DT) in $R^3$ of $S \cup V$, we shall denote the upper part of the convex hull of $S \cup V$ by DT$(S \cup V)$.

For $1 \leq j \leq n$, consider the triangulation DT$(S_j \cup V)$. For a triangle $\Delta$ of DT$(S_j \cup V)$, let $L$ denote the straight line which is the locus of the points with equal distance from three vertices of $\Delta$. If $L$ intersects $P$, $\Delta$ is called *non-critical*, otherwise *critical*.

**Lemma 2.** *No edge $e$ of a non-critical triangle $\Delta$ of $DT(S_j \cup V)$ is longer than* $2 \cdot w_{j-1}(p_j)$.

*Proof.* Let $e = (p, q)$ and denote with $x$ the intersection point of $L$ and $P$. As $x$ lies inside of $P$, we get $l(x, p) = l(x, q) = w_{j-1}(x)$. By the choice of point $p_j$ in Step 3 of Algorithm INSERT we have $w_{j-1}(x) \leq w_{j-1}(p_j)$. The triangle inequality now implies $l(p, q) \leq 2 \cdot w_{j-1}(p_j)$.

Our next observation is on critical triangles. Consider some edge $e$ of DT$(S_j \cup V)$ both of whose endpoints are on the boundary of $P$. Edge $e$ cuts off some part of the Voronoi diagram Vor$(S_j \cup V)$ that lies outside of $P$. If that part contains Voronoi vertices then we define the *critical region*, $R(e)$, for $e$ as the union of all the (critical) triangles that are dual to these vertices. As observed in [2], it is not hard to see that each critical triangle of DT$(S_j \cup V)$ belongs to a unique critical region.

**Lemma 3.** *No edge $f$ of a critical triangle in $R(e)$ is longer than $e$.*

*Proof.* Let $p$ be an endpoint of $f$. Then the Voronoi region of $p$ in Vor$(S_j \cup V)$ intersects $e$. We can choose point $x$ on the sphere which is inside this region but outside of $P$. There is some ball $B$ centered at $x$ which encloses $p$ but no endpoint of $e$. So edge $e$ cuts off a part of $B$ which, by $x \notin P$, is completely covered by the the ball $B(e)$ with diameter $e$. This implies that $p$ lies in $B(e)$. Therefore, there exists some ball $B$ centered at $x$ which encloses $p$ but no endpoint of $e$ such that $B$ is completely lies in the ball $B(e)$ with diameter $e$. Thus $p$ lies in $B(e)$. Similarly, the other end point $q$ of $f$ lies in $B(e)$ by considering ball centered at $x$ which encloses $q$. As the distance between any two points in $B(e)$ is at most $l(e)$, we get $l(f) \leq l(e)$.

Let us further distinguish between *independent* edges of DT$(S_j \cup V)$ and *dependent* ones, the latter type having both endpoints in $V$. The length of independent edges can be bounded as follows.

**Lemma 4.** *Each independent edge $e$ of $DT(S_j \cup V)$ has a length of at least* $w_{j-1}(p_j)$.

*Proof.* We have $l(e) \geq l(S_j, S_j \cup V)$ because $e$ has at most one endpoint in $V$. But from (2) we know $l(S_j, S_j \cup V) = w_{j-1}(p_j)$.

We are now ready to show how a triangulation with edge lengths related to $d_n$ can be computed. First, Algorithm INSERT is run on $P$, in order to compute the value $d_n$. We assume than $n$ is chosen sufficiently large to assure $d_n \leq l(V, V)/2$. This assumption is not unnatural as the shortest edge of the desired triangulation cannot be longer than the shortest edge of $P$. After having $d_n$ available, $k$ points $p'_1, \ldots, p'_k$ are placed on the boundary of $P$, with consecutive distances between $d_n$ and $3 \cdot d_n$, and such that $l(V', V') \geq d_n$ holds, for $V' = V \cup \{p'_1, \ldots, p'_k\}$. Notice that such a placement is always possible (the proof is omitted here). When the inner angle $\angle p_{j-1} p_j p_{j+1}$ for three consecutive vertices of $V$ is larger than or equal to $60°$, we can place points on the boundary so that the consecutive distance is between $d_n$ and $2 \cdot d_n$ and $l(V', V') \geq d_n$ holds. Finally, $n - k$ additional points $p'_{k+1}, \ldots, p'_n$ are produced by re-running Algorithm INSERT after this placement.

For $1 \leq j \leq n$, let $S'_j = \{p'_1, \ldots, p'_j\}$. Define $w(x) = l(x, S'_n \cup V)$ for a point $x \in P$. The value of $w(p'_n)$ will turn out to be crucial for analyzing the edge length behavior of the triangulation $DT(S'_n \cup V)$. Lemma 6 below asserts that $w(p'_n)$ gets small when $n$ exceeds a constant multiple of the number $k$ of prescribed points.

**Lemma 5.** *The area covered by a spherical disk with radius $r$ is equal to $\pi r^2$.*

*Proof.* Suppose the center of the disk is at the north pole, say $N$. The center of the ball defining the sphere and an arbitrary point on the boundary of the disk are denoted by $O$ and $p$, respectively. Let $R$ denote the radius of the ball, and let $\alpha$ denote the angle $\angle NOp$. Then $\cos \alpha = 1 - r^2/2R^2$. The area of the disk is computed by

$$\int_0^\alpha 2\pi R \cdot R \sin \theta d\theta = \pi r^2.$$

This lemma will be used in the proof of the following lemma.

**Lemma 6.** *For any positive constant $\alpha > 2$, suppose $n \geq \frac{\alpha^2}{\alpha^2 - 4} k$. Then $w(p'_n) \leq \cdot \alpha d_n$.*

*Proof.* The point set $S_n$ produced by Algorithm INSERT in the first run is large enough to ensure $d_n < l(V, V)/2$. So we get $d_n = w_{n-1}(p_n)$ from (2). As point $p_n$ maximizes $w_{n-1}(x)$ for all $x \in P$, the $n + |V|$ spherical disks centered at the points in $S_n \cup V$ and with radii $d_n$ completely cover the polygon $P$. Let $d_n = 1$ for the moment. We then have from Lemma 5

$$A(P) \leq \pi(n + |V|) - A' \tag{3}$$

where $A(P)$ is the area of $P$, and $A'$ denotes the area outside of $P$ which is covered by the spherical disks centered at $V$.

Assume now $w(p'_n) > \alpha \cdot d_n$. Draw a spherical disks with radius $\frac{\alpha}{2} d_n$ around each point in $S'_n \setminus S'_k$. Since $w(p'_n) = l(S'_n \setminus S'_k, S'_n \cup V)$ by (2), these disks are pairwise disjoint. By the same reason, and because boundary distances defined by

$V' = V \cup S'_k$ are at most $3 \cdot d_n$, these disks all lie completely inside $\boldsymbol{P}$. Obviously, these disks are also disjoint from the $|V|$ disks of radius $d_n$ centered at $V$. Finally, the latter disks are pairwise disjoint, since $d_n \leq l(V,V)/2$. Consequently,

$$A(\boldsymbol{P}) \geq \frac{\alpha^2}{4}\pi(n - k) + A'' \tag{4}$$

where $A''$ denotes the area inside of $\boldsymbol{P}$ which is covered by the disks centered at $V$. Combining (3) and (4), and observing $A' + A'' = \pi \cdot |V|$ now implies $n < \frac{\alpha^2}{\alpha^2 - 4}k$, a contradiction.

Typically, when $\alpha = 3$, the lemma says that $w(p'_n) \leq 3d_n$ if $n \geq 9k/5$. It has to be observed that the number $k$ depends on $n$. The following fact guarantees the assumption in Lemma 6, provided $n$ is sufficiently large. Let $B(\boldsymbol{P})$ denote the perimeter of $\boldsymbol{P}$.

**Lemma 7.** *The condition $d_n \leq (\alpha^2 - 4)A(\boldsymbol{P})/(\alpha^2\pi \cdot B(\boldsymbol{P}))$ implies $n \geq \frac{\alpha^2}{\alpha^2-4}k$.*

*Proof.* By (3) we have

$$n \geq \frac{A(\boldsymbol{P})}{\pi \cdot (d_n)^2} - |V|.$$

To get a bound on $k$, observe that at most $l(e)/d_n - 1$ points are placed on each edge $e$ of $\boldsymbol{P}$. This sums up to

$$k \leq \frac{B(\boldsymbol{P})}{d_n} - |V|.$$

Simple calculations now show that the condition on $d_n$ stated in the lemma implies $n \geq \frac{\alpha^2}{\alpha^2-4}k$.

The following is a main theorem of this paper.

**Theorem 2.** *Suppose $n$ is large enough to assure the conditions $d_n \leq l(V,V)/2$ and $d_n \leq A(\boldsymbol{P})/(2\pi \cdot B(\boldsymbol{P}))$. Then no edge in the triangulation $T^+ = DT(S'_n \cup V)$ is longer than $6 \cdot d_n$. Moreover, $T^+$ exhibits an edge length ratio of 6.*

*Proof.* The proof is essentially the same as the one given in [2], but for the completeness the proof is given. Two cases are distinguished, according to the value of $w(p'_n)$.

Case 1: $w(p'_n) < d_n$. Concerning upper bounds, Lemma 2 implies $l(e) \leq 2 \cdot w(p'_n) < 2 \cdot d_n$ for all edges $e$ belonging to non-critical triangles of $T^+$. If $e$ belongs to some critical triangle, Lemma 3 shows that $l(e)$ cannot be larger than the maximum edge length on the boundary of $\boldsymbol{P}$, which is at most $3 \cdot d_n$ by construction. Concerning lower bounds, Lemma 4 gives $l(e) \geq w(p'_n)$ for independent edges. (Independence is meant with respect to $V'$ now). We know $w(p'_n) \geq d_n^*/2$ from Lemma 1, which implies $l(e) \geq d_n/2$ because $d_n^* \geq d_n$. For dependent edges, that is, edges spanned by $V'$, we trivially obtain $l(e) \geq d_n$ as $l(V', V') \geq d_n$ by construction.

Case 2: $w(p'_n) \geq d_n$. The upper bound $2 \cdot w(p'_n)$ for non-critical triangles now gives $l(e) \leq 6 \cdot d_n$, due to Lemmas 6 and 7 with $\alpha = 3$. The lower bound for independent edges becomes $l(e) \geq w(p'_n) \geq d_n$. The remaining two bounds are the same as in the former case.

We do not believe that the ratio given by the theorem is tight as evidenced by the experimental results given in the following section. The following corollary states that under a certain condition the result can be strengthened.

**Corollary 1.** *Suppose that we can place the points on the boundary so that the consecutive distance is between $d_n$ and $2d_n$ and $l(V', V') \geq d_n$, and $d_n \leq A(\mathbf{P})/(\frac{\alpha^2}{\alpha^2-4}\pi \cdot B(\mathbf{P}))$ for some $\alpha$ with $2 < \alpha \leq 3$. Then no edge in the triangulation $T^+ = DT(S'_n \cup V)$ is longer than $\cdot d_n$. Moreover, $T^+$ exhibits an edge length ratio of $2\alpha$.*

*Proof.* The proof is done in a manner similar to the one for Theorem 2. For Case 1, the edge length is between $d_n/2$ and $2d_n$. Thus, the edge length ratio is at most 4. For Case 2, the edge length is at most $2\alpha d_n$ and thus the edge length ratio is at most $2\alpha$ because the edge length is at least $d_n$ in this case as in Case 2 of Theorem 2.

When the inner angle $\angle p_{j-1}p_jp_{j+1}$ for three consecutive vertices of $V$ is larger than or equal to $60°$, we can place points on the boundary so that the consecutive distance is between $d_n$ and $2 \cdot d_n$ in order to satisfy $d(V', V') \geq d_n$. In this case, we can get a bound $2\alpha$ which is better than 6 if $n$ is large enough so that $n \geq \frac{\alpha^2}{\alpha^2-4}k$ holds.

This result also holds for the planar case, and thus strengthens the one given by [2].

## 4    Experimental Results

We have implemented the proposed algorithm. We shall show experimental results obtained by applying the algorithm for the hemisphere with four initial points on the boundary as shown in Fig. 2. We have tested the cases of $n = 50, 100, 200, 300, 400, 500$. The ratio of the longest and shortest edge lengths for each case is given in Table 1. In the table, for each $n$, the values ratio1 and

**Table 1.** Edge length ratios for an example.

| $n$ | 50 | 100 | 200 | 300 | 400 | 500 |
|---|---|---|---|---|---|---|
| ratio1 | 2.6342 | 1.9213 | 1.9446 | 1.9567 | 1.9929 | 1.9755 |
| ratio2 | 1.9065 | 1.9331 | 1.9546 | 1.9241 | 1.9603 | 1.967 |

ratio2 represents the ratio obtained by the first run and the one by the second run of Voronoi insertion algorithm, respectively. It is observed from the table that the edge length ratio is close to 2 which is much better than the worst-case ratio of $2\alpha$ for $2 \le \alpha \le 3$ given by Corollary 1. In our experiments, $n$ is much larger than $k$ (typically $k$ is about 25 for $n = 100$ and for the hemisphere tested in our experiments, we can place points on the boundary so that the distances of consecutive points are almost the same (which is set to $1.5d_n$ in our experiments). The figures below show the solutions obtained by the algorithm for $n = 50, 100, 500$.

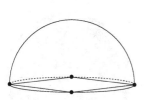

**Fig. 2.** Hemisphere and an initial point set

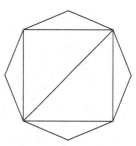

**Fig. 3.** Points placed after the first run

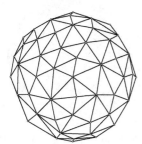

**Fig. 4.** Solution obtained by the first run ($n = 50$)

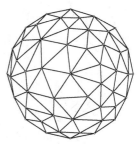

**Fig. 5.** Solution obtained by the second run ($n = 50$)

In order to see why the gap between the theoretical and experimental edge length ratios is large, the following observation might be helpful. In the proof of Theorem 2 and Corollary 1, we estimated the lower bound of $w(p'_n)$ as $d_n^*/2$. However, we believe that $w(p'_n)$ is close to $d_n^*$. Also the analysis made in the proof is loose, which might be strengthened, we hope. These facts may be an explanation for the big gap.

In applications to optimal triangular truss design, $n$, the number of points to be inserted, is not usually fixed. Alternatively, the lower bound of the edge length, say $\underline{d}$, may be given. Consider again the problem of minimizing the edge length ratio under such constraint. In this case, we can obtain the better edge

**Fig. 6.** Another view of Fig. 4 from a different angle

**Fig. 7.** Another view of Fig. 5 from a different angle

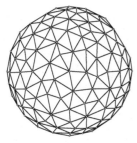

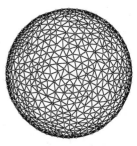

**Fig. 8.** Solution obtained by the first run ($n = 100$)

**Fig. 11.** Solution obtained by the first run ($n = 500$)

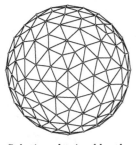

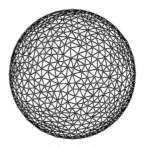

**Fig. 9.** Solution obtained by the second run ($n = 100$)

**Fig. 12.** Solution obtained by the first run ($n = 500$)

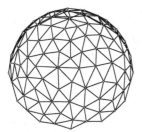

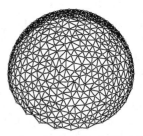

**Fig. 10.** Another view of Fig. 9 from a different angle

**Fig. 13.** Another view of Fig. 12 from a different angle

length ratio as follows. The algorithm first places the points on the boundary so that consecutive distances are between $\underline{d}$ and $\beta\underline{d}$ and $l(V', V') \geq \underline{d}$ holds ($\beta$ is chosen as small as possible, but is at most 3). Here $V'$ is the one defined in Section 3. Applying the Voronoi insertion algorithm, we then place the points in $\boldsymbol{P}$ as many as possible as long as the minimum interpoint distance is at least $\underline{d}$. On the contrary to what was described in the previous section, we execute the Voronoi insertion only once. Then we have the following theorem.

**Theorem 3.** *We assume $\underline{d} < d(V, V)/2$ and that the minimum edge length is at least $\underline{d}$. Suppose the points are placed on the boundary so that consecutive distances are between $\underline{d}$ and $\beta\underline{d}$. Then the edge length ratio obtained by the algorithm described above is at most $\max\{2, \beta\}$.*

*Proof.* Let $n$ denote the number of points that were inserted by the algorithm. Notice that when $(n + 1)$-st point is inserted, $w(p_{n+1})$ becomes less than or equal to $\beta\underline{d}$. In a manner similar to the proof of Lemma 2, we can obtain $l(e) \leq 2w(p_{n+1}) \leq 2\underline{d}$ for noncritical edge $e$. For critical edge, the edge length is at most $\beta\underline{d}$ from assumption. This completes the proof.

## 5    Extensions and Discussions

We have considered the problem of generating length-uniform triangular meshes for spheres. The method is easy to implement and seems to produce acceptably good triangular meshes as far as computational experiments are concerned. We are interested in obtaining a constant-approximation for the problem of minimizing the maximum edge length for which the previous paper [2] gave an $4\sqrt{3}$-approximation for the plane.

Viewed from the point of applications to the design of structures, it is also important to generate a uniform triangular mesh for more general surfaces such as a surface of revolution and a ruled surface. A difficulty of generalizing the result in this paper lies in that contrary to the spherical case (i.e., Lemma 5) we cannot evaluate good lower and upper bounds for the area of a disk for such surfaces. We are planning to further investigate the case of such general surfaces.

## Acknowledgements

This research was partially supported by the Grant-in-Aid for Scientific Research on Priority Areas (B) (No. 10205214) by the Ministry of Education, Science, Sports and Culture of Japan. The authors would like to thank the anonymous referee for helpful comments.

## References

1. F. Aurenhammer, "Voronoi diagrams – a survey of a fundamental geometric data structure", *ACM Computing Surveys* 23 (1991), 345-405.

2. F. Aurenhammer, N. Katoh, H. Kojima, M. Ohsaki and Y. Xu, Approximating Uniform Triangular Meshes in Polygons, Proc. of 6th Annual International Computing and Combinatorics Conference (COCOON 2000), LNCS 1858, Springer Verlag, pp. 23-33, 2000.

3. M. Bern, D. Dobkin and D. Eppstein, "Triangulating polygons without large angles", *Intl. J. Comput. Geom. and Appl.* 5 (1995), 171-192.

4. M. Bern and D. Eppstein, "Mesh generation and optimal triangulation", in D.-Z. Du (ed.), *Computing in Euclidean Geometry*, World Scientific Publishing, 1992, 47-123.

5. M. Bern, D. Eppstein and J.R. Gilbert, "Provably good mesh generation", *Journal of Computer and System Sciences* 48 (1994), 384-409.

6. M. Bern, S. Mitchell and J. Ruppert, "Linear-size nonobtuse triangulation of polygons", Proceedings of the 10th Ann. ACM Symposium on Computational Geometry (1994), 221-230.

7. J-D. Boissonnat and M. Yvinec, *Algorithmic Geometry*, Cambridge University Press, 1998.

8. P. Chew, "Guaranteed-Quality Mesh Generation for Curved Surfaces", Proceedings of the 9th Ann. ACM Symposium on Computational Geometry (1993), 274-280.

9. T. Feder and D.H. Greene, "Optimal Algorithms for Approximate Clustering", Proceedings of the 20th Ann. ACM Symposium STOC (1988), 434-444.

10. G. Fejes Tóth, "Packing and covering", in *Handbook of Discrete and Computational Geometry*, J.E. Goodman and J. O'Rourke (eds.), CRC Press Series on Discrete Mathematics and its Applications, 1997, 19-41.

11. T. Gonzalez, "Clustering to minimize the maximum intercluster distance", *Theoretical Computer Science* 38 (1985), 293-306.

12. D.S. Johnson, "The NP-completeness column: An ongoing guide", *Journal of Algorithms* 3 (1982), 182-195.

13. E. Melisseratos and D. Souvaine, "Coping with inconsistencies: A new approach to produce quality triangulations of polygonal domains with holes", Proceedings of the 8th Ann. ACM Symposium on Computational Geometry (1992),202-211.

14. H. Nooshin, K. Ishikawa, P.L. Disney and J.W. Butterworth, "The traviation process", *Journal of the International Association for Shell and Spatial Structures* 38 (1997), 165-175.

15. H. Nooshin, Y. Kuroiwa, P.L. Disney A genetic method for regularisation of structural configurations, Proc. of Int. Assoc. Space Structures - from recent past to next millennium, Madrid, 20-24 September 1999.

16. M. Ohsaki, T. Nakamura and M. Kohiyama, "Shape optimization of a double-layer space truss described by a parametric surface", *International Journal of Space Structures* 12 (1997), 109-119.

17. J. Roulier and T. Rando, Measures of fairness for curves and surfaces, *Designing Fair Curves and Surfaces*, N.S. Sapidis(Ed), SIAM, pp.75-122, 1994.

18. K.F. Roth, "On a problem of Heilbronn", *Proc. London Mathematical Society* 26 (1951), 198-204.

19. J. Ruppert, "A Delaunay Refinement Algorithm for Quality 2-Dimensional Mesh Generation", *Journal of Algorithms* 18 (1995), 548-585.

20. , R. Taniguchi, Triangulation of Curved Surfaces and Its Application to Optimal Shape Design (in Japanese), Master Thesis, Department of Architecture and Architectural Systems, Kyoto University, 2000.

# The construction of Delaunay diagrams by lob reduction

Pierre Kauffmann and Jean-Claude Spehner

Université de Haute Alsace – Laboratoire MAGE
6, rue des frères Lumière, F-68093 Mulhouse Cedex FRANCE
email : pika@ch.ibm.com and JC.Spehner@uha.fr
http://www.mage.uha.fr

**Abstract.** In this paper we present an algorithm that generalizes Lawson's algorithm for transforming any triangulation of a point set into the Delaunay triangulation. Lawson's algorithm flips illegal edges into legal edges. Our algorithm uses lob reductions instead of flips, in a lob reduction an arbitrary number (possibly zero) of illegal edges are replaced by a single edge. Our algorithm reaches the Delaunay triangulation of a set $S$ of $n$ sites in $O(n^2)$ time. If the elements of $S$ are the vertices of a convex polygon our algorithm is very efficient.

## 1 Introduction

The *Delaunay diagram* $Del(S)$ of a planar set $S$ of points (called *sites*) is a partition of the plane in regions whose circumcircles do not contain any site in their interior. This remarkable property leads to numerous applications of Delaunay diagrams in surface or volume meshing. If more than three sites are cocircular, the Delaunay diagram is not a triangulation but a more general inscribable diagram. A diagram $D$ is said to be *inscribable* if all its bounded regions are convex and inscribable in a circle, and if the complementary of the convex hull $conv(S)$ of the set $S$ of vertices of $D$ is the only unbounded region of $D$.

The dual diagram of $Del(S)$ is the well known Voronoi diagram $Vor(S)$. The first optimal construction (in $O(n log n)$) of $Vor(S)$ in the plane was given by Shamos and Hoey [SH75]; they used a divide and conquer algorithm. Fortune [For87] was the first to construct $Vor(S)$ by sweeping the plane with a straight line.

Lawson [Law77] has given an algorithm in $O(n^2)$ to transform any triangulation into a Delaunay triangulation. His algorithm flips the illegal edges of the triangulation into legal ones (an edge of a diagram is said to be *illegal* if its two open neighbor regions $g$ and $h$ are inscribable and such that $g$ is included in the open disk $\omega(h)$ that circumscribes $h$).

We present here a new algorithm to transform any inscribable diagram into the Delaunay diagram. Our algorithm uses a polygonal line that sweeps the set of sites. This polygonal line shrinks when an edge is created and grows when an edge is deleted. This sweep ends when the line is empty.

J. Akiyama, M. Kano, and M. Urabe (Eds.): JCDCG 2000, LNCS 2098, pp. 205-216, 2001.
© Springer-Verlag Berlin Heidelberg 2001

In section 2 we introduce the notion of a nest of illegal edges: every maximal open connected set that is the union of illegal edges and of their neighbor regions is called a *nest of illegal edges*. We prove that every nest of illegal edges is convex. Our sweep algorithm starts on the frontier of such nests of illegal edges.

In section 3, we present the basic notion of a lob.

Our algorithm is composed of two procedures which we present in sections 4 and 5. The first one searches a lob and the other one reduces a lob. Our lob reduction generalizes Lawson's flip.

In section 6 we prove that, if our algorithm is applied to all the nests of illegal edges of an inscribable diagram, then the resulting diagram is $Del(S)$.

In section 7 we prove that the complexity of our algorithm is in $O(n^2)$ and we give an example of the worst case. In the special case where all the sites of $S$ are extremal (i.e., are vertices of the convex hull $conv(S)$), Aggarwal, Guibas, Saxe, and Shor [AGSS89] have given a linear algorithm. In this case, our algorithm remains quadratic in the worst case but has remarkable practical performances.

## 2    The nests of illegal edges

Let $S$ be a set of sites. The set of inscribable diagrams that admit the sites of $S$ as vertices is denoted by $Dins(S)$.

An edge $c$ of a diagram is said to be *legal* if its two open neighbor regions $g$ and $h$ are inscribable and such that the vertices of $g \setminus c$ do not belong to the open disk $\omega(h)$ that circumscribes $h$. If no four sites of $S$ are cocircular, legality is equivalent to nonillegality.

According to Delaunay [Del34], Edelsbrunner [Ede87], and Schmitt and Spehner [SS99] the Delaunay diagram $Del(S)$ is the only diagram of $Dins(S)$ without illegal edges and, if all the sites of $S$ are extremal, the farthest point Delaunay diagram $Del_{-1}(S)$ is the only diagram of $Dins(S)$ without legal edges.

In a diagram $D$ of $Dins(S)$, every vertex $s$ of a nest $H$ of illegal edges is an extremal point of the closure of $H$. This is trivial if no illegal edge is issued from $s$ and is easy to prove by induction on the number of illegal edges issued from $s$ (see figure 1.a). Hence the following theorem:

**Theorem 1.** *Every nest of illegal edges $H$ of an inscribable diagram $D$ is convex and the restriction of $D$ to the set $T$ of the vertices of $H$ is a refinement of $Del_{-1}(T)$.*

## 3    The lobs of a polygonal line

Let $L = (s_1, ..., s_m)$ be a simple (oriented) polygonal line such that all the sites of $L$ are not cocircular. Let $l = (s_i, ..., s_j)$ be a section of consecutive cocircular sites of $L$ such that, for all $k \in [i, j-2]$, $s_{k+2}$ lies on the left of the straight line $s_k s_{k+1}$; let $\overline{\omega}(l)$ be the closed disk that circumscribes $l$ and $leftof(s_i s_{i+1})$ the half–plane lying on the left side of the straight line $s_i s_{i+1}$. If $s_{i-1} \notin \overline{\omega}(l) \cap leftof(s_i s_{i+1})$ $(i > 1)$ and $s_{j+1} \notin \overline{\omega}(l) \cap leftof(s_{j-1} s_j)$ $(j < m)$ then $l$ is called a *lob* (see figure 1.b).

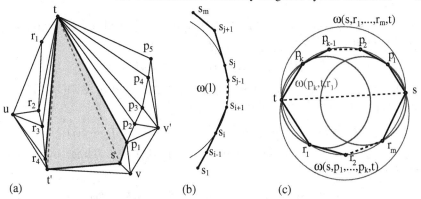

**Fig. 1.** (a) The gray region is a nest of illegal edges. (b) The lob $l = (s_i, ..., s_j)$ of the polygonal line $(s_1, ..., s_m)$. (c) The lobs $(p_k, t, r_1)$ and $(r_m, s, p_1)$ determined by the deletion of the illegal edge $st$.

**Theorem 2.** *A simple and closed polygonal line whose vertices are not cocircular admits at least two lobs.*

*Proof.* The constrained Delaunay diagram (see Chew[Che89] and Lee and Lin [LL86]) of a simple and closed polygonal line $L$, whose vertices are not cocircular, admits at least two regions that have only one edge that does not belong to $L$. Each of these regions defines a lob of $L$. □

## 4 The lob reduction

It is straightforward to prove the following lemma which shows that the deletion of an illegal edge determines a lob (see figure 1.c).

**Lemma 1.** *Let $st$ be the common edge of two inscribable polygons $sp_1...p_k t$ and $tr_1...r_m s$ in a diagram $D$ such that the sites $p_1, ..., p_k$ lie on the right side of $st$. If $st$ is illegal and if the sites $s, p_1, ..., p_k, t, r_1, ..., r_m$ are not cocircular, then the triples $(p_k, t, r_1)$ and $(r_m, s, p_1)$ are two lobs of $L = (s, p_1, ..., p_k, t, r_1, ..., r_m)$.*

Let $l = (s, ..., t)$ be a lob of a polygonal line $L$ of a diagram $D$.

An edge $pq$ of $l$ is said to be *illegal within $l$* if the open region $h$ of $D$ that lies on the right side of the straight line $pq$ is inscribable and included in $\omega(l)$.

The *elementary lob reduction* of $l$ consists in the creation of the edge $st$ when no edge of $l$ is illegal and in the deletion of all illegal edges within $l$ in the opposite case.

In this case, by lemma 1, at least one new lob is determined. More precisely, let $l = (s, ..., s', p, q, r, ..., t)$ be a lob of a polygonal line $L = (s_1, ..., s, ..., t, ..., s_m)$ in a diagram $D$ and suppose that the region $pp_1...p_i q$ of $D$ that lies on the right side of the straight line $pq$ and the region $qq_1...q_j r$ of $D$ that lies on the right side of the straight line $qr$ are inscribable. We now define a lob $l'$ in three cases:

Case 1: $p = s$, $pq$ is illegal (i.e., $p_1 \in \overline{\omega}(l)$) but $qr$ is not illegal (i.e., $q_1 \notin \overline{\omega}(l)$). Then, by lemma 1, $l' = (p_i, q, r)$ is a lob of

$$L' = (s_1, ..., s, ..., p, p_1, ...p_i, q, r, ..., t, ..., s_m).$$

Case 2: $p = s$, $pq$ and $qr$ are illegal (i.e., $p_1 \in \overline{\omega}(l)$ and $q_1 \in \overline{\omega}(l)$). Then, by lemma 1, $l' = (p_i, q, q_1)$ is a lob of

$$L' = (s_1, ..., s, ..., p, p_1, ...p_i, q, q_1, ..., q_j, r, ..., t, ..., s_m).$$

Case 3: $p \notin \{s, s'\}$ and $pq$ is the first illegal edge of $l$. Then, by lemma 1, $l' = (s', p, p_1)$ is a lob of

$$L' = (s_1, ..., s, ..., s', p, p_1, ...p_i, q, r, ..., t, ..., s_m).$$

$l'$ is called the *first lob generated by the elementary lob reduction* of $l$.

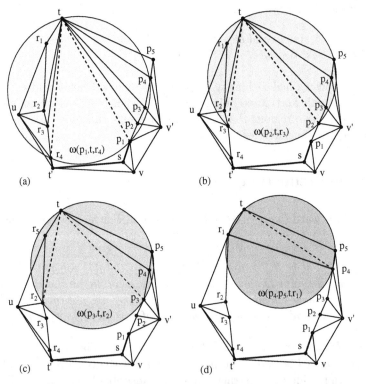

**Fig. 2.** (a) The first elementary lob reduction deletes the edges $tp_1$ and $tr_4$. (b) The second deletes $tp_2$ and $tr_3$. (c) The third deletes $tp_3$ and $tr_2$. (d) The fourth deletes $tp_4$ and the fifth creates the edge $r_1p_3$. The polygonal line $L = (s, p_1, ..., p_4, t, r_1, ..., r_4, t')$ is not convex: $p_1$ and $p_2$ are exceeded concave sites and $r_3$ is a stopping concave site.

Let $l_1, l_2, ..., l_k$ be a sequence of lobs such that, for every $i \in \{1, ..., k-1\}$, the elementary reduction of $l_i$ deletes at least one edge and generates the first lob $l_{i+1}$ and the elementary reduction of $l_k$ creates one edge. Then the sequence of the elementary reductions of $l_1, l_2, ..., l_k$ is called a *lob reduction* (see figure 2).

## 5   The lob searching strategy

Our lob searching strategy is based on the notion of contracting polygonal line.

A *polygonal line* $L = (s_1, ..., s_m)$ is said to be *contracting* if, for every $i \in [1, ..., m - 3], s_{i+3}$ lies on the left side of the polygonal line $(s_1, ..., s_{i+2})$ and belongs to the closed disk $\overline{w}(s_i, s_{i+1}, s_{i+2})$.

A contracting polygonal line $K = (s_1, ..., s_i, p, q, ..., r)$ is simple and admits a unique lob $l$ and this lob is terminal. Moreover, if no edge of $l$ is illegal, then the lob reduction of $l$ transforms $K$ into a new contracting polygonal line.

The deletion of an edge can generate a non-interesting lob.

Let $K = (s_1, ..., s_k, s, t, ..., t')$ be a contracting polygonal line that admits $la = (s, t, ..., t')$ as a terminal lob. If during the lob reduction of $la$ the edge $st$ is deleted and if the region $sp...p't$ that lies on the right side of $st$ is such that the terminal lob $ld = (w, ..., s_k, s)$ of $(s_1, ..., s)$ is also a lob of $K' = (s_1, ..., s_k, s, p, ..., p', t)$ and $p \notin \overline{w}(ld)$, then the lob $ld$ is said to be *exceeded* and $st$ is called the *revealing edge* of $ld$ (see figure 3.a).

In this case $ws$ is not an edge of $Del(S)$ since every circle passing through $w$ and $s$ contains either $s_k$ or $t'$ in its interior, therefore we do not reduce the lob $ld$.

A *site* $q$ of $L$ is said to be *concave* (resp. *convex*) if his neighbor sites $p$ and $r$ are such that $r$ is on the right (resp. left) side of the straight line $pq$ and the bounded region $H$ delimited by $L$ is on the left side of $pq$ and $qr$.

If $w$ and $s$ are consecutive concave sites of $L$, $ld = (w, s)$ is called a *degenerated exceeded lob* and $w(ld)$ denotes the half-plane that lies on the left side of $sp$. Moreover $ld = (w, s)$ is also a (degenerated) contracting section of $L$.

A polygonal line $K$ that is the product of contracting sections $K_1, ..., K_k$ such that the terminal lobs of $K_1, ..., K_{k-1}$ are exceeded lobs of $K$ is called a *piecewise contracting polygonal line*.

After the selection of an initial site and a search direction, our lob searching strategy is to keep the section of $L$ between the initial site and the last site of the current lob piecewise contracting.

## 6   The planarity problem

If $P$ is a convex polygon, the graph constructed by the algorithm is planar because the current polygonal line remains convex.

When the algorithm is applied to a nest of illegal edges of a diagram of $Dins(S)$, the initial polygonal line is convex by theorem 1, but during the algorithm the current polygonal line may become non-convex.

If, after the deletion of an edge $st$ of $L$, $s$ becomes concave, $s$ is called an *exceeded concave site* of $L$, and if $t$ becomes concave, $t$ is called a *stopping concave site* of $L$ (see figure 2.d).

After an elementary reduction of a lob $l = (s, ..., t)$, only $s$ and $t$ may become concave.

The following lemmas are useful to prove that the constructed graph is planar. The proof of the first lemma is straightforward.

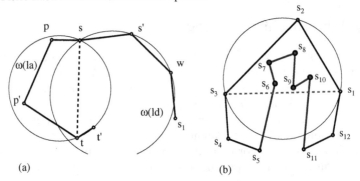

(a)                                                    (b)

**Fig. 3.** (a) After the deletion of $st$, $ld = (w, s', s)$ becomes an exceeded lob. (b) What our algorithm is not supposed to do: the edge $s_1 s_3$ that closes the lob $(s_1, s_2, s_3)$ cuts the current polygonal line $L$ and the concave sites $s_7$, $s_8$ and $s_{10}$ belong to the created region.

**Lemma 2.** *Every edge that does not belong to the current polygonal line and that has been visited during a sequence of lob reductions is legal relatively to the current graph.*

**Lemma 3.** *An edge deleted by a lob reduction cannot be re-created by another lob reduction.*

*Moreover, if the lob reduction is applied to a nest $H$ of illegal edges, each created edge cuts $H$.*

*Proof.* (i) We generalize the proof given by Fortune [For92] for the flip algorithm which maps the sites of the current diagram $D$ onto a paraboloid in dimension 3. Every edge $st$ of $D$ is mapped to the line segment $\lambda(st) = \lambda(s)\lambda(t)$ and every inscribable region $f$ of $D$ to a planar face $\lambda(f)$: $\lambda(D)$ is a polyhedral surface.

A legal (resp. illegal) edge $st$ is then mapped to a convex (resp. concave) edge $\lambda(st)$ of $\lambda(D)$ (see figure 4.a). Let $D'$ be the diagram obtained by the deletion of the edge $st$, then $\lambda(st)$ lies above the surface $\lambda(D')$ and this remains true for every diagram obtained from $D'$ by lob reduction. Hence $st$ cannot be re-created by another lob reduction.

(ii) It follows from (i) that, if $s$ and $t$ are sites of $D$ such that the line segment $st$ crosses only legal edges of $D$, $\lambda(st)$ lies above the surface $\lambda(D)$ and $st$ cannot be created by lob reduction. □

**Lemma 4.** *Let $L$ be a polygonal line such that the section $L'$ of $L$ between the initial site and the last site of the current lob is piecewise contracting, let $ld = (w, ..., s)$ its last exceeded lob and $st$ its revealing edge and suppose that the graph $D$ remains planar until the current lob reduction. If the section $K$ of $L$ between $s$ and the last site of the current lob does not admit any exceeded lob, then $K$ is a contracting line and has the form*

$$K = (s, p_1, ..., p_i, r_1, ..., r_j)$$

*where $p_1 \notin \omega(ld)$, $p_1, ..., p_i$ are on the right side of $st$, and $r_1, ..., r_j$ are on the left side of $st$ and belong to $\omega(ld)$.*

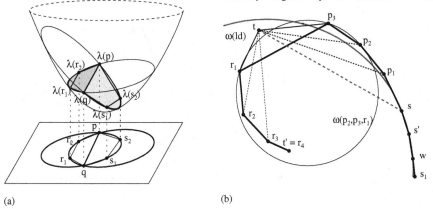

(a)                                                                    (b)

**Fig. 4.** (a) An illegal edge is mapped to a concave edge. (b) $R \subseteq \omega(ld) \cap leftof(st)$.

*Proof.* (i) During the lob reduction which deletes the edge $st$ and reveals the exceeded lob $ld$, other edges having $t$ as endpoint can be deleted and an edge $p_i r_1$ is created. After this reduction, $K$ has the form $(s, p_1, ..., p_i, r_1, ..., r_j)$ where $p_1 \notin \omega(ld)$, $p_1, ..., p_i$ are on the right side of $st$ and $r_1$ lies on the left side of $st$.

Since, by lemma 3, no edge between two sites on the left of $st$ can be created, this remains true after every new lob reduction which preserves the lob $ld$ and does not create any other exceeded lob.

(ii) We prove now that $R$ belongs to $\omega(ld) \cap leftof(st)$ (see figure 4.b).

By lemma 2, for any point $z$ in the triangle $p_{i-1} p_i r_1$, the line segment $zt$ does not cut an illegal edge. Hence, by Delaunay's proof [Del34] (see also [SS99]), $t \notin \overline{\omega}(p_{i-1}, p_i, r_1)$. Since $K$ is contracting, $s \notin \omega(p_{i-1}, p_i, r_1)$ and

$$R = \{r_1, ..., r_j\} \subseteq \overline{\omega}(p_{i-1}, p_i, r_1) \cap leftof(p_i r_1).$$

Let $D$ be the graph before the deletion of the edge $st$. There exists a polygonal line $P$ in $D$ passing through the sites of $R$. Since $D$ is planar, $P$ cannot cut the edge $st$. But this edge cuts twice the circumscribing circle of $p_{i-1} p_i r_1$ and $R \subseteq \overline{\omega}(p_{i-1}, p_i, r_1)$. Thus $R \subseteq leftof(st)$ and

$$R \subseteq \overline{\omega}(p_{i-1}, p_i, r_1) \cap leftof(st, p_i r_1) \subseteq \overline{\omega}(s, p_i, r_1) \cap leftof(st).$$

Since $t \in \omega(ld) \setminus \overline{\omega}(s, p_i, r_1)$ and $p_1 \in \overline{\omega}(s, p_i, r_1) \setminus \omega(ld)$,

$$R \subseteq \overline{\omega}(s, p_i, r_1) \cap leftof(st) \subseteq \omega(ld) \cap leftof(st).$$

$\square$

**Lemma 5.** *Let $K$ be the piecewise contracting section of $L$ between the initial site and the last site of the current lob, let $ld_1, ..., ld_h$ be the exceeded lobs of $K$ and $s_1 t_1, ..., s_h t_h$ their respective revealing edges and suppose that the graph $D$ remains planar until the current lob reduction.*

*Then $K$ has the form*

$$(s_{init}, ..., s_1, ..., s_h, p_1, ..., p_i, r_1, ..., r_j)$$

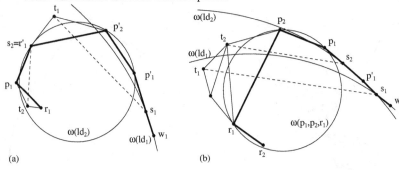

**Fig. 5.** (a) The case with $s_2$ on the left side of $s_1t_1$. (b) The case with $s_2$ on the right side of $s_1t_1$.

where $p_1 \notin \omega(ld_h)$, the points $p_1, ..., p_i$ are on the right side of $s_ht_h$, and the points $r_1, ..., r_j$ are on the left side of $s_1t_1, ..., s_ht_h$ and inside the intersection of the disks $\omega(ld_1), ..., \omega(ld_h)$.

*Proof.* (i) The case h=1 has been proved by lemma 4. We now prove the result for h=2.

Case 1: $s_2$ lies on the left side of $s_1t_1$ (see figure 5.a).

In this case, by the proof of lemma 4 applied to lob $ld_1$ before the deletion of the edge $s_2t_2$, $K' = (s_1, p'_1, ..., p'_u, r'_1, ..., r'_v, s_2, t_2)$ is a contracting line with $\{r'_1, ..., r'_v, s_2\} \subseteq leftof(s_1t_1)$, $t_1 \notin \omega(p'_{u-1}, p'_u, r'_1)$, $s_1 \notin \omega(p'_{u-1}, p'_u, r'_1)$ and

$$\overline{\omega}(p'_{u-1}, p'_u, r'_1) \cap leftof(s_1t_1, p'_ur'_1) \subseteq \omega(ld_1) \cap leftof(s_1t_1) \quad (1).$$

Moreover, since $ld_2$ is a section of $K'$,

$$\omega(ld_2) \cap leftof(p'_u, r'_1, ..., r'_v, s_2) \subseteq \overline{\omega}(p'_{u-1}, p'_u, r'_1) \cap leftof(p'_ur'_1) \quad (2).$$

By the same proof applied to lob $ld_2$, $K$ has the form

$$(s_{init}, ..., s_1, p'_1, ..., p'_u, r'_1, ..., r'_v, s_2, p_1, ..., p_i, r_1, ..., r_j)$$

where $p_1 \notin \omega(ld_2)$, the points $p_1, ..., p_i$ are on the right side of $s_2t_2$, and

$$R = \{r_1, ..., r_j\} \subseteq \omega(ld_2) \cap leftof(s_2t_2) \quad (3).$$

Since the vertices of $L$ are all distinct, $R \cap \{p'_u, r'_1, ..., r'_v, s_2\} = \emptyset$. Moreover, since the current graph is planar, the polygonal line $(r_1, ..., r_j)$ cannot cut the line $(p'_u, r'_1, ..., r'_v, s_2)$. Thus $R \subseteq leftof(p'_u, r'_1, ..., r'_v, s_2)$ and, by (2),

$$R \subseteq \overline{\omega}(p'_{u-1}, p'_u, r'_1) \cap leftof(p'_ur'_1) \quad (4).$$

Let $D$ be the graph before the deletion of the edge $s_1t_1$. There exists a polygonal line $P$ in $D$ passing through the sites of $R$. Since $D$ is planar, $P$ cannot cut the edge $s_1t_1$. Since the edge $s_1t_1$ cuts twice the circle which circumscribes $p'_{u-1}p'_ur'_1$, $R \subseteq leftof(s_1t_1)$ (5).

Hence, by (1), (3), (4), and (5), $R \subseteq \omega(ld_1) \cap \omega(ld_2) \cap leftof(s_1t_1, s_2t_2)$.

Case 2: $s_2$ lies on the right side of $s_1t_1$ (see figure 5.b).

We prove first that $t_2$ also lies on the right side of $s_1t_1$. Otherwise, if $q$ is the neighbor of $s_2$ in the region of $D$ on the right of $s_2t_2$, the polygonal line $(s_1, ..., s_2, q, .., t_2)$ is contracting. This proves that the deletion of $s_2t_2$ cannot reveal an exceeded lob and that two revealing edges do not intersect.

By the proof of lemma 4 applied to lob $ld_2$ , $K$ has the form

$$(s_{init}, ..., s_1, p'_1, ..., s_2, p_1, ..., p_i, r_1, ..., r_j)$$

where $p_1 \notin \omega(ld_2)$ (resp. $p'_1 \notin \omega(ld_1)$), $p_1, ..., p_i$ are on the right side of $s_2t_2$,

$$R = \{r_1, ..., r_j\} \subseteq \overline{\omega}(p_{i-1}, p_i, r_1) \cap leftof(p_ir_1) \subseteq \overline{\omega}(s_2, p_i, r_1) \cap leftof(p_ir_1) \quad (6),$$

$$R \subseteq \overline{\omega}(s_2, p_i, r_1) \cap leftof(s_2t_2) \subseteq \omega(ld_2) \cap leftof(s_2t_2) \quad (7),$$

and $t_2 \notin \omega(p_{i-1}, p_i, r_1)$. Similarly $t_1 \notin \omega(p_{i-1}, p_i, r_1)$.

Since $ld_2$ is the terminal lob of the contracting section $(s_1, p'_1, ..., s_2)$ of $K$, $s_1 \notin \omega(ld_2)$. Thus $s_1 \notin \omega(p_{i-1}, p_i, r_1)$.

Since by lemma 3 no edge between two sites on the left of $s_1t_1$ can be created, the edge $p_ir_1$ which cuts $s_2t_2$ also cuts $s_1t_1$. Thus $r_1 \in leftof(s_1t_1)$. Moreover, by the proof of case 1, we have $R \subseteq leftof(s_1t_1)$ (8).

$s_1 \notin \omega(ld_2)$ and (7) imply that $s_1 \notin \omega(s_2, p_i, r_1)$. Therefore, by (6),

$$R \subseteq \overline{\omega}(s_2, p_i, r_1) \cap leftof(p_ir_1) \subseteq \omega(s_2, s_1, t_1) \cap leftof(p_ir_1) \quad (9).$$

Before the edge $s_1t_1$ has been deleted, $t_1p'_1$ was a legal edge of $D$. Thus $s_2 \notin \overline{\omega}(p'_1, s_1, t_1)$ and $p'_1 \in \overline{\omega}(s_1, s_2, t_1) \setminus \omega(ld_1)$. Since $t_1 \in \omega(ld_1) \setminus \overline{\omega}(s_1, s_2, t_1)$,

$$\omega(s_1, s, t_1) \cap leftof(s_1t_1) \subseteq \omega(ld_1) \cap leftof(s_1t_1) \quad (10).$$

Hence, by (7), (8), (9), and (10), $R \subseteq \omega(ld_1) \cap \omega(ld_2) \cap leftof(s_1t_1, s_2t_2)$.

This result remains true in the special cases when $t_1 = t_2$ and when $ld_1$ (resp. $ld_2$) is degenerated.

(ii) The general case follows by induction on the number $h$ of lobs.   □

**Lemma 6.** *If, during the reduction of a lob* $la = (u', ..., u, v)$*, the last edge* $uv$ *is deleted and* $v$ *becomes a stopping concave site, then as long as* $v$ *remains concave,* $v$ *does not belong to any created region.*

*Proof.* (i) If all the edges deleted during the lob reduction of $la = (u', ..., u, v)$ share $u$ as endpoint, the sites $p_1, ..., p_i$ inserted between $u'$ and $u$ lie on the right side of $u'v$, the sites $r_1, ..., r_j$ inserted between $u$ and $v$ lie on the right side of $uv$ and the created edge $p_ir_1$ is such that $\omega(u, p_i, r_1)$ does not contain any of $u', p_1, ..., p_i, r_1, ..., r_j$ and $v$. Hence $\omega(u, p_i, r_1)$ and the created region do not contain $v$ (see figure 6).

In the opposite case, the created region lies on the right side of $u'u$ and does not contain $v$.

(ii) If the last created edge cuts $uv$, the next edge also cuts $uv$. Indeed, the endpoints of the next lob $l$ also lie on each side of $uv$ since, by lemma 3,

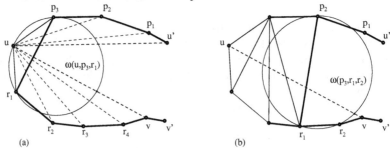

**Fig. 6.** The created edges cut $uv$.

the algorithm cannot build any edge on the right side of $uv$. Moreover the lob reduction algorithm cannot go beyond $v$ before $v$ turns back to a convex site. Hence, the next region created does not contain $v$.                                                      □

**Theorem 3.** *If the lob reduction algorithm is applied to the boundary of a nest of illegal edges of an inscribable diagram, then the resulting graph is planar.*

*Proof.* By lemma 6, every created region does not contain any stopping concave site. Hence, we only have to prove that the created regions contain no exceeded concave site.

Suppose that the graph remains planar before the reduction of the current lob. For every created edge $pr$, the created region $f$ lies on the right side of $pr$. If $s$ is an exceeded concave site, $s$ is the last site of an exceeded lob $ld$ whose revealing edge is $st$. By lemma 5, either $pr$ cuts $st$ and $s$ lies on the left side of $pr$ or $f \subseteq \omega(ld)$, but $s \notin \omega(ld)$. Thus the created region does not contain any exceeded concave site. Therefore, the resulting graph is also planar.                        □

**Theorem 4.** *If the lob reduction algorithm is applied to all the nests of illegal edges of a diagram $D$ of $Dins(S)$ then the resulting diagram is $Del(S)$.*

*Proof.* Indeed, after processing all the nests of illegal edges of $D$, the resulting diagram does not contain any illegal edge and therefore is $Del(S)$.        □

## 7   Complexity of the lob reduction algorithm

**Theorem 5.** *Let $S$ be a set of $n$ sites and $D$ a diagram of $Dins(S)$, then our algorithm which transforms $D$ into the Delaunay diagram of $S$ has $O(n^2)$ worst case running time.*

*Proof.* (i) The construction of the nests of illegal edges is linear in the number of illegal edges, that is $O(n)$.

(ii) Since, by lemma 3, the lob reduction algorithm cannot recreate a deleted edge, it cannot determine twice the same lob and thus, we can attribute to every edge a constant number of operations. Since the complete graph having $S$ as set of vertices admits $O(n^2)$ edges, the complexity of our algorithm is in $O(n^2)$. This bound can be reached as shown in figure 7. A quadratic exemple with only extremal sites can also be given, but it is hard to distinguish the sites on the figure. Other examples can be found in Hurtado, Noy, and Urrutia (see [HNU99]). □

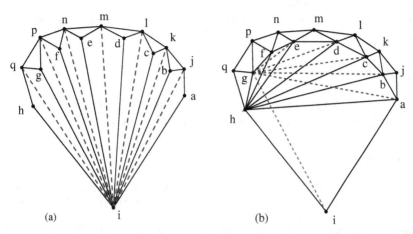

**Fig. 7.** The illegal edges in figure (a) are dashed. Our algorithm transforms the diagram (a) into the Delaunay diagram (b) in $O(n^2)$ time if the nests of illegal edges are processed from right to left. Note that this time is in $O(n)$ when the nests are processed from left to right.

We ran our algorithm on sets of up to 100,000 extremal sites. The number of deleted edges is close to $\frac{2n}{3}$, the number of *incircle* tests is about $7n$ and the computing time is about $2.2 \times 10^{-5}n$ seconds. We have also run an adaptive divide and conquer algorithm [AES96] and a flip algorithm that started with $Del_{-1}(S)$ with the same set of sites. For 50,000 extremal sites, the divide and conquer algorithm ran in 250 s, the flip in 420 s and our algorithm in 1,05 s (with the same optimizations and on the same workstation).

We have also run our algorithm on Delaunay diagrams in which we had flipped about $\frac{n}{2}$ edges. The computing time appeared to be quadratic and took about 100 s for 100,000 sites.

# 8   Conclusion

We have presented an algorithm for constructing the Delaunay diagram of a set $S$ of $n$ sites. This algorithm can be applied either to all the nests of illegal edges of an inscribable diagram of $S$ or to a set $S$ of extremal sites. It has $O(n^2)$ time complexity. If all the sites of $S$ are extremal, it has good practical performances. Moreover, these theoretical and practical performances also hold for the dual construction of the farthest point Delaunay diagram.

Our algorithm can be applied to Fortune's sweep–line algorithm [For87]. We remove both the dictionary that manages the wave front and the priority queue that manages the events that close regions, and we obtain a lazy algorithm that uses the lob reduction algorithm to update the front. This lazy algorithm is in

$O(n^2)$ time (see theorem 5) and therefore is not optimal but it works faster than Fortune's algorithm when applied to sets of 100,000 sites (see [Kau99]). Moreover, it is more robust by using exact computations since its degree is 6 rather than 20 (see [BP97]).

Finally the generalization of our algorithm to the 3–dimensional space seems to be a very interesting problem.

## Acknowledgements

The authors wish to thank the referees and Stéphane Rivière whose advise has greatly helped to enhance the presentation of this paper.

## References

[AES96]   B. Adam, M. Elbaz, and J.-C. Spehner. Construction du diagramme de Delaunay dans le plan en utilisant les mélanges de tris. In *Actes 4ème Journées AFIG*, pages 215–223, 1996.

[AGSS89]  A. Aggarwal, Leonidas J. Guibas, J. Saxe, and P. W. Shor. A linear-time algorithm for computing the Voronoi diagram of a convex polygon. *Discrete Comput. Geom.*, 4(6):591–604, 1989.

[BP97]    J.-D. Boissonnat and F. P. Preparata. Robust plane sweep for intersecting segments. Report TR 3270, INRIA, Sophia Antipolis, September 1997.

[Che89]   L. P. Chew. Constrained Delaunay triangulations. *Algorithmica*, 4:97–108, 1989.

[Del34]   B. Delaunay. Sur la sphère vide. *Bull. Acad. Sci. USSR: Class. Sci. Math. Nat.*, 7:793–800, 1934.

[Ede87]   H. Edelsbrunner. *Algorithms in Combinatorial Geometry*, volume 10 of *EATCS Monographs on Theoretical Computer Science*. Springer-Verlag, Heidelberg, West Germany, 1987.

[For87]   S. J. Fortune. A sweepline algorithm for Voronoi diagrams. *Algorithmica*, 2:153–174, 1987.

[For92]   S. Fortune. Voronoi diagrams and Delaunay triangulations. In D.-Z. Du and F. K. Hwang, editors, *Computing in Euclidean Geometry*, volume 4 of *Lecture Notes Series on Computing*, pages 225–265. World Scientific, Singapore, 2nd edition, 1992.

[HNU99]   F. Hurtado, M. Noy, and J. Urrutia. Flipping edges in triangulations. *Discrete Comput Geom*, 22:333–346, 1999.

[Kau99]   P. Kauffmann. *La construction du diagramme de Delaunay par balayage dans le plan et ses applications*. Ph.D. thesis, Université de Haute–Alsace, Mulhouse, France, 1999.

[Law77]   C. L. Lawson. Software for $C^1$ surface interpolation. In J. R. Rice, editor, *Math. Software III*, pages 161–194. Academic Press, New York, NY, 1977.

[LL86]    D. T. Lee and A. K. Lin. Generalized Delaunay triangulation for planar graphs. *Discrete Comput. Geom.*, 1:201–217, 1986.

[SH75]    M. I. Shamos and D. Hoey. Closest-point problems. In *Proc. 16th Annu. IEEE Sympos. Found. Comput. Sci.*, pages 151–162, 1975.

[SS99]    D. Schmitt and J.-C. Spehner. Angular properties of Delaunay diagrams in any dimension. *Discrete Comput. Geom.*, 5:17–36, 1999.

# Geometric Transformations in Plane Triangulations

Ken-ichi Kawarabayashi[1], Atsuhiro Nakamoto[2],
Yoshiaki Oda[3], and Mamoru Watanabe[4]

[1] Department of Mathematics,
Keio University, Yokohama 223-8522, Japan
[2] Department of Mathematics,
Osaka Kyoiku University, Kashiwara 852-8582, Japan
[3] Department of Mathematics and Computer Science,
Shimane University, Matsue 690-8504, Japan
[4] Department of Computer Science and Mathematics,
Kurashiki University of Sciences and the Arts, Kurashiki 712-8505, Japan

**Abstract.** In this paper, we present several geometric transformations, sometimes called contractions in graph theory, in plane triangulations. Those transformations can be applied for several formalizations of geometric properties (ex. the number of acute triangles) in plane triangulations since they are restricted only for a local region (some adjacent triangles). We refer to such an application slightly.

## 1 Introduction

We consider only *simple plane* graphs, that is, graphs with no loops and no multiple edges embedded in the plane so that any two edges do not cross except their ends. A plane *triangulation* is a plane graph with each face triangular. For a plane graph $G$, we denote its vertex set, edge set and face set by $V(G)$, $E(G)$ and $F(G)$, respectively. We denote the minimum degree of a graph $G$ by $\delta(G)$. A $k$-*cycle* means a cycle of length $k$. We say that $S \subset V(G)$ is a $k$-*cut* if $|S| = k$ and the graph $G - S$ is disconnected. Note that the $k$ vertices in a minimal $k$-cut in a plane triangulation $G$ lie on a common $k$-cycle $C$. In this case, we say that $C$ is a *separating* $k$-cycle and each edge $e \in E(C)$ is *contained* in a $k$-cut. The interior of $C$ is denoted by Int $C$ and the set of the vertices in Int $C$ is denoted by $V(\text{Int } C)$.

Let $G$ be a plane triangulation and let $e$ be an edge of $G$. *Contracting $e$* (or *contraction of $e$*) in $G$ is to remove $e$, identify the two ends of $e$ and replace two pairs of multiple edges by two single edges respectively, as shown in Fig. 1. It is well-known that every plane triangulation can be transformed into a tetrahedron by a sequence of contractions of edges [6].

For plane triangulations with minimum degree $\geq 4$, the following theorem is known. The *removal of an octahedron* is an operation shown in Fig. 2.

J. Akiyama et al. (Eds.): JCDCG 2000, LNCS 2098, pp. 217–221, 2001.
© Springer-Verlag Berlin Heidelberg 2001

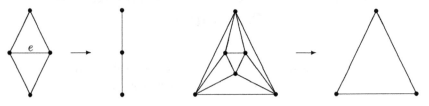

**Fig. 1.** Contraction of an edge $e$          **Fig. 2.** Removal of an octahedron

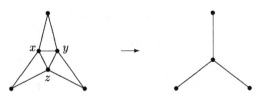

**Fig. 3.** Contraction of a triangle $xyz$

**Theorem 1. (Nakamoto and Negami [5])** *Every plane triangulation can be transformed into an octahedron by a sequence of contractions of edges and removals of an octahedron, preserving the minimum degree $\geq 4$.*

Moreover, it is known that every 4-connected plane triangulation can be transformed into an octahedron by a sequence of contraction of edges, preserving the 4-connectedness [2]. (A related result can be found in [4].)

Let $G$ be a plane triangulation and let $xyz$ be a face of $G$. *Contraction* of a triangle $xyz$ (or *contracting* $xyz$) is to remove three edges $xy, yz$ and $xz$, to identify three vertices $x, y$ and $z$ into a single vertex and to replace three pairs of multiple edges by three single edges respectively, as shown in Fig. 3. In this paper, we deform triangulations by contractions of triangles.

The following two are our main results:

**Proposition 1.** *Every plane triangulation can be transformed into a tetrahedron by a sequence of a contraction of a triangle and a removal of a vertex of degree 3.*

**Theorem 2.** *Every 4-connected plane triangulation can be transformed into an octahedron by a sequence of contraction of a triangle and a contraction of an edge incident to a vertex of degree 4, preserving the 4-connectedness.*

Remark that in Proposition 1 and Theorem 2, only contracting triangles are not enough to transform all triangulations into a tetrahedron and an octahedron, preserving the connectivity. Consider a plane triangulation $G$ obtained from any plane triangulation $T$ by adding a vertex of degree 3 into each face. Every triangle $f$ in $G$ contains an edge $e$ of $T$. Thus, the contraction of $f$ contracts $e$, but $e$ is contained in at least two separating 3-cycles in $G$. Thus, every contraction of a triangle yields multiple edges. And consider a 4-connected plane triangulation $G'$

**Fig. 4.** Subdividing a face by a path of length $t \geq 3$

obtained from any plane *quadrangulation* $Q$ (i.e., a plane graph with each face quadrilateral) by putting a path $v_1 v_2 \cdots v_t$ of length $t \geq 3$ into each face $xyzw$ of $Q$ and adding edges $xv_1, zv_t$ and $v_i y, v_i w$ for $i = 1, \ldots, t$, as shown in Fig. 4. In $G'$, every triangle has an edge included in at least two separating 4-cycles. Thus, contracting any triangle transforms $G'$ into a graph with a 3-cut.

It is easy to see that each transformation is restricted only for a local region (a few vertices, edges and triangles). So, it preserves not only combinatorial structures but also geometric properties for the other regions. For example, by using Theorem 2 the authors proved that every 4-connected plane triangulation with $m$ finite faces other than an octahedron has a straight-line embedding in the plane such that $(m+3)/2$ faces are acute triangles [3] (it has been shown in [1] that every plane triangulation with $m$ finite faces can be drawn in the plane so that $\lceil m/3 \rceil$ of their faces are acute), and the proof of Theorem 2 has also included in [3]. Proposition 1 and Theorem 2 themselves seem to be interesting and strongly expected to be used to give several formalizations of geometric properties in plane triangulations.

## 2  Proofs

Let $G$ be a $k$-connected plane triangulation and let $e$ and $xyz$ be an edge and a face of $G$, respectively. (It is well-known that $3 \leq k \leq 5$ since any triangulation is 3-connected and any plane graph has a vertex of degree at most 5.) We say that an edge $e$ (resp., a triangle $xyz$) is $k$-*contractible* if the graph, denoted by $G/e$ (resp. $G/xyz$), obtained from $G$ by contracting $e$ (resp. $xyz$) is still a $k$-connected plane triangulation.

**Lemma 1.** *Let $G$ be a 4-connected plane triangulation and let $v_0 v_1 v_2$ be a triangle of $G$. Then $v_0 v_1 v_2$ is not 4-contractible if and only if for some $i$, $v_i v_{i+1}$ is contained in a 4-cycle $C$ separating a component $T$ with $V(T) \neq \{v_{i+2}\}$ from $G$, where the subscripts are taken modulo 3.*

*Proof.* The sufficiency is obvious and hence we show the necessity. Since $v_0 v_1 v_2$ is not 4-contractible, $G/v_0 v_1 v_2$ has a separating 3-cycle $C$. Clearly, the vertex $[v_0 v_1 v_2] \in V(G/v_0 v_1 v_2)$ arising from contracting $v_0 v_1 v_2$ in $G$ lies on $C$. Thus, we can put $C = [v_0 v_1 v_2]ab$ for some vertices $a, b \in V(G/v_0 v_1 v_2) \setminus \{[v_0 v_1 v_2]\}$. Obviously, $C$ corresponds to some separating 4-cycle $C'$ in $G$ through $a, b \in V(G) - \{v_0, v_1, v_2\}$. If we suppose that $C' = v_0 v_1 ab$, then neither of components of $G - C'$ is one vertex $v_2$ since $v_2 \in V(G)$ corresponds to the vertex $[v_0 v_1 v_2] \in V(G/v_0 v_1 v_2)$ on $C$. Thus, the lemma follows.

Mimicking the proof of Lemma 1, we can show the following lemma by a suitable modification.

**Lemma 2.** *Let $G$ be a plane triangulation and let $v_0 v_1 v_2$ be a triangle of $G$. Then $v_0 v_1 v_2$ is not 3-contractible if and only if for some $i$, $v_i v_{i+1}$ is contained in a 3-cycle $C$ separating a component $T$ with $V(T) \neq \{v_{i+2}\}$ from $G$, where the subscripts are taken modulo 3.*

Now we prove Proposition 1 and Theorem 2:

*Proof of Proposition 1.*    Let $G$ be a plane triangulation other than a tetrahedron. If $G$ has a vertex $v$ of degree 3, then remove $v$. So, we suppose that $\delta(G) \geq 4$. If $G$ is 4-connected, then every triangle bounding a face is 3-contractible, by Lemma 2. Thus, assuming $G$ has a separating 3-cycle $C = xyz$, we shall show that there exists a 3-contractible triangle in the interior of $C$. We may also assume that $C$ is *minimal*, that is, there is no other separating 3-cycle in $(\text{Int } C) \cup C$. Consider a face $xyv$ in $(\text{Int } C) \cup C$, where $v \in V(\text{Int } C)$. Let $vxp$ be a face in $(\text{Int } C) \cup C$ sharing the edge $vx$. If $p = z$, then by the assumption of $C$, $p$ has degree 3, which is contrary to the minimum degree of $G$. Thus, we have $p \neq z$. Here we claim that the triangle $vxp$ is 3-contractible. If not, then by Lemma 2, some edge of $vxp$ would be contained in a separating 3-cycle. However, since the triangles $vxp$ and $xyz$ share just one vertex, such a 3-cycle is included in $(\text{Int} C) \cup C$, which is contrary to the minimality of $C$.

*Proof of Theorem 2.*    It suffices to prove that if $G$ is a 4-connected plane triangulation other than an octahedron, then $G$ has a 4-contractible edge incident to a vertex of degree 4, or a 4-contractible triangle. Since $G$ is 4-connected, $G$ has no vertex of degree less than 4.

Suppose that $G$ has a vertex $v$ of degree 4. Let $v_1, v_2, v_3$ and $v_4$ be the four neighbors of $v$ in this cyclic order. If $vv_1$ is not 4-contractible (i.e., $G/vv_1$ has a 3-cut), then $G$ has a 4-cycle $vv_1 x v_3$ for some $x \in V(G) \setminus \{v, v_1, \ldots, v_4\}$. Without loss of generality, we may suppose that $v_2$ is in the interior of the cycle $vv_1 x v_3$. Similarly, if $vv_2$ is not 4-contractible, then we can find a 4-cycle $vv_2 y v_4$ for some $y \in V(G) \setminus \{v, v_1, \ldots, v_4\}$. By Jordan Curve Theorem, any two closed curves in the plane cannot cross at only one point, and hence we have $x = y$. In this case, the graph induced by $\{v, v_1, v_2, v_3, v_4, x(= y)\}$ is isomorphic to an octahedron. Since $G$ is 4-connected, $G$ must be an octahedron, which contradicts to the assumption. Thus, at least one of the edges $vv_1$ or $vv_2$ is 4-contractible.

Now we consider the case when $G$ has no vertex of degree 4. By Lemma 1, if $G$ is 5-connected, then every triangle is 4-contractible, and hence we suppose that $G$ has a separating 4-cycle $C = xyzw$. Choose $C$ to be *minimal*, that is, there is no other separating 4-cycle in $(\text{Int } C) \cup C$. Let $vxy$ be a triangular face in $(\text{Int } C) \cup C$, where $v \notin V(C)$ since $C$ is a 4-cut. Let $pvx$ be a face in $(\text{Int} C) \cup C$ sharing $vx$ with the face $vxy$. Note that $p \notin V(C)$ since $C$ is minimal or since $G$ is

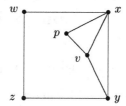

**Fig. 5.** Structure of $F$

4-connected. (See Fig. 5.) Now we claim that the triangle $pvx$ is 4-contractible. To do so, we have only to check, by Lemma 1, that each of $pv, vx, px$ is not contained in a separating 4-cycle.

For getting a contradiction, we first suppose that the edge $pv$ lies on a separating 4-cycle $C'$. We may suppose that $C'$ is not contained in $(\text{Int}C) \cup C$. (For otherwise, $C$ would not be minimal, a contradiction.) Observe that neither $p$ nor $v$ are adjacent with $z$. (For otherwise, i.e., if $p$ is adjacent with $z$, then $xpzy$ would be a separating 4-cycle, contrary to $C$ being minimal.) Thus, we have $C' = pvyw$. However, if so, then $xyw$ would form a 3-cycle of $G$. This implies that $xyw$ is a separating 3-cycle or the degree of $x$ is exactly 4, contrary to $G$ being 4-connected or $\delta(G) > 4$.

For the edges $xv$ and $xp$, we have only to consider the case when $xv$ lies on some separating 4-cycle $C'' = xvab$ for some $a, b \in V(G) - \{x, v\}$. We may assume that $C''$ is not contained in $F$, by the minimality of $C$. Since $v$ is not adjacent to $z$ by the same argument as above, the vertex $a$ on $C''$ is in Int $C$. Since $C''$ is separating, we have $ay \notin E(C'')$. We also have $aw \notin E(C'')$ since $C''$ is minimal. The remaining possibility is that $az \in E(C'')$ and $C'' = xvaz$. Since a quadrilateral $vazy$ cannot be a separating 4-cycle by the minimality of $C$, it has a diagonal either $vz$ or $ay$. If $vz \in E(G)$, then $G$ would have a 3-cut $\{x, v, z\}$, and if $ay \in E(G)$, then $\{x, y, z\}$ would be a 3-cut, or the degree of $y$ would be exactly 4. In both cases, we have a contradiction. Thus, the face $pvx$ is 4-contractible.

# References

1. A. Kaneko, H. Maehara and M. Watanabe, On the number of acute triangles in a straight-line embedding of a maximal plane graph, *J. Combin. Theory, Ser. B* **75** (1999), 110-115.
2. K. Kawarabayashi and A. Nakamoto, Private communication.
3. K. Kawarabayashi, A. Nakamoto, Y. Oda and M. Watanabe, Acute triangles in 4-connected maximal plane graphs, preprint.
4. J. Maharry, An excluded minor theorem for the octahedron, *J. Graph Theory* **31** (1999), 95–100.
5. A. Nakamoto and S. Negami, Generating triangulations on closed surface with minimum degree at least 4, to appear in *Discrete Math.*
6. E. Steinitz and H. Rademacher, *Vorlesungen üder die Theorie der Polyeder*, Springer, Berlin, 1934.

# Separation Sensitive Kinetic Separation Structures for Convex Polygons

David Kirkpatrick and Bettina Speckmann

Department of Computer Science
University of British Columbia

**Abstract.** We extend the kinetic data structure for collision detection between moving simple polygons introduced in [14] to incorporate a hierarchical representation of convex chains. This permits us to define and maintain an adaptive hierarchical outer approximation for convex polygons. This representation can be exploited to give separation sensitive complexity bounds for kinetic collision detection comparable to those of Erickson et al. [11] who deal with pairs of convex polygons. More importantly, it forms the basis of a more general representation, developed in a companion paper, that applies to collections of general (not necessarily convex) polygonal objects.

## 1 Introduction

Collision detection is a basic and unavoidable computational problem arising in all areas of geometric modeling involving objects in motion. Even if we restrict attention to objects in the plane the complexity of collision detection is far from being completely understood.

Consider, for example, the problem of detecting collisions between moving polygonal objects in the plane. Most approaches to collision detection work in two phases. First, a "broad phase" (filtering) algorithm and data structure are used to determine pairs of objects that might possibly collide. A different "narrow phase" (refinement) algorithm and data structure then tests each pair. In general, such approaches force the objects into a representation best suited for one of the two phases or involves some kind of hybrid representation not ideally suited for either.

Recently a number of kinetic data structures (KDS) [4, 13] have been proposed for two-dimensional collision detection. These attempt to avoid arbitrary distinctions between broad and narrow phases and, instead, adapt to changes in the nature and degree of separation between objects. Kinetic data structures exploit the coherence of real motion and maintain, over time, a collection of elementary geometric tests (*certificates*) that together certify that the objects are disjoint. Objects are assumed to have fixed (but changeable) motion trajectories (*flight plans*) and certificates are maintained in a priority queue based on the time of expiration calculated from the current flight plans. When a certificate expires – its test no longer holds – or when a flight plan is changed, the data structure must be updated.

J. Akiyama, M. Kano, and M. Urabe (Eds.): JCDCG 2000, LNCS 2098, pp. 222-236, 2001.

Three natural parameters serve to describe the complexity of a given configuration of objects: $k$ – the number of objects, $N$ – their total number of vertices, and $\kappa$ – the size of the *minimum link subdivision* [18] separating the objects.

Kinetic data structures and their associated maintenance algorithms can be evaluated and compared with respect to four desired characteristics. A good KDS is *compact* if it uses little space in addition to the input, *responsive* if the data structure invariants can be restored quickly after the failure of a certificate, *local* if it can be updated easily if the flight plan for an object changes, and *efficient* if the worst-case number of events handled by the data structure for a given motion is small compared to some worst-case number of "external events" that must be handled for that motion.

Of particular relevance to the work of this paper are the paper of Basch et al. [3], where pairs of not necessarily convex polygons are maintained using balanced geodesic triangulations, and its extension to collections of polygons by Agarwal et al. [2]. The number of certificates maintained in these schemes is $O(\kappa \log N)$. Furthermore, they have the additional feature that they are canonical (the structures depend only on the current state of the polygons and not on their motion history) which facilitates a non-trivial analysis of their efficiency.

In contrast, the kinetic separation structure (KSS) introduced in [14] maintains a certificate set of size $O(\kappa)$, but is non-canonical (i.e. history dependent) and hence difficult to analyse in terms of efficiency.

Unfortunately, both structures as described are unable to exploit the *degree* of separation of neighboring objects, which is a significant factor in their potential kinetic inefficiency. This shows up, for example, in the fact that the certificate set certifying disjointness of two widely separated translating convex objects (with a total of $N$ vertices) may need to be updated $\Theta(N)$ times, despite the fact that a constant number of certificates would suffice for the entire motion.

One approach to incorporating metric information into kinetic separation structures involves the maintenance of Voronoi diagrams. For example, Guibas et al. [12] study the maintenance of the compact Voronoi diagram of McAllister et al. [16] for a set of disjoint convex polygons moving in the plane. While this structure provides a succinct proof of disjointness, it does so at the expense of a potentially large number of certificate updates even when the objects are widely separated. Furthermore, it has no obvious generalization to collections of non-convex objects.

Another natural approach to exploiting the degree of separation of objects is to approximate each individual object by some kind of coarse outer approximation whose coarseness depends on the separation. The disjointness of the approximations implies the disjointness of the objects but the former may be less expensive to maintain. For isolated convex objects (viewed as the intersection of half spaces determined by their bounding edges) it is possible to construct a hierarchy of successively coarser approximations based on the simple idea of progressive relaxation of half space constraints; a familiar example of this in computational geometry is the Dobkin-Kirkpatrick polyhedral approximation hierarchy [10]. Erickson et al. [11] developed a kinetic collision detection structure for pairs of *convex* objects based on distance-sensitive variants of such hierarchies, whose performance is thereby made sensitive to the degree of separation between the objects relative to their diameter.

For convex objects in the context of other convex objects or, more generally, for collections of disjoint but possibly interleaved non-convex objects it is not immediately clear what constitutes a useful notion of hierarchical representation or how such a notion could be exploited to provide separation sensitive complexity bounds for kinetic collision detection.

Our objective, in this and a companion paper [15], is to demonstrate that the KSS structure lends itself in a natural way to modifications that provide a measure of sensitivity to the degree (as well as the nature) of the separation of a collection of moving polygons. In this paper we develop the foundation of such a structure by focusing on collections of convex polygons. We exploit hierarchical approximations by replacing convex chains within pseudo-triangles by hierarchical approximation structures. As a consequence, we achieve a reduction in the number of certificate update events as well as more efficient implementations of the pseudo-triangulation update primitives that form the core of the KSS.

Taken together this collection of chain approximations provide an approximation structure for collections of moving convex polygons that is adaptive in the sense that individual polygons are represented at a level of detail sufficient to separate them from their (dynamically changing) set of neighbours. The total representation size (over the entire collection of polygons) is proportional to the number of objects independent of the complexity of the individual objects.

As the objects move the structure shifts its level of detail so as to provide a closer approximation in areas where the polygons are close to each other while coarsening in areas where the polygons are well separated. The approximation is maintained in such a way that it conforms to the partition of the free space into pseudo-triangles. This automatically provides us with the means to detect the parts of the current boundary of the polygons where we can extend the approximation or where we have to increase its level of detail.

A complete analysis of the efficiency of our hierarchical KSS (like that of the unmodified KSS structure) remains a challenge. Nevertheless, it is possible to provide some partial results that demonstrate the potential of this structure. In particular, its performance is comparable with that of the separation sensitive structure of Erickson et al. [11] that was designed for maintaining the separation of *two* convex objects. Specifically, for any constant number of disjoint convex polygons in the plane, let $D$ be their maximum diameter, let $N$ be their total number of vertices, and let $s$ be the minimum distance between all pairs of polygons during their motion. Our structure processes $O(\min\{\log(D/s), \log(N)\})$ events when each of the polygons moves in a linear trajectory. All events can be processed at a cost of $O(\min\{\log(D/s), \log(N)\})$ time per event.

In the next section we recall the important features of the KSS kinetic separation structure. Section 3 sets out the the essential properties of the convex chain approximation hierarchies needed in our separation sensitive augmentation of the KSS, and illustrates several potential representations. Section 4 addresses the issue of maintaining our augmented KSS, with an emphasis on the dynamic maintenance of the chain approximations that accompany changes in the pseudo-triangulation of the free space and the new kinetic events that accompany the introduction of approximate polygons. Section 5 summarizes the various kinetic attributes of our separation sensitive structure. Finally, section 6 sets out some directions for future work. The Ph.D. dissertation of the second author, currently in preparation, will contain further details on results in this paper and related work.

## 2    The Kinetic Separation Structure

In this section we revisit the Kinetic Separation Structure (KSS) as it was presented in [14], emphasizing those features that impact its adaptation as a separation sensitive structure. As in [14], we describe how the KSS is built starting from an arbitrary triangulation $\mathcal{T}$ of the free space $\mathcal{E}$ separating a given set of $k$ disjoint simple polygons in the plane together with a point at infinity. This serves to introduces the KSS features in a disciplined fashion. It also makes clear the extent to which the structure is canonical and suggests which features are necessary only for the efficiency (as opposed to the correctness) of the structure.

A KSS can be constructed by first identifying $2k-2$ *partition triangles* in $\mathcal{T}$. Partition triangles are named for the fact that they partition the space between objects (the complement of the polygons and the point at infinity) into disjoint regions called *corridors* (see Fig. 1).

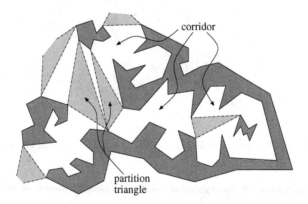

**Fig. 1.** Decomposition of the free space between polygons.

**Lemma 1.** *Every corridor is bounded by exactly two edges of partition triangles (*end-edges*) and by exactly two (possibly degenerate) chains of edges of polygons.*

The condition that states that a partition triangle has a positive area is called a *triangle certificate*. Similarly, a *corridor certificate* asserts that none of the polygonal chains or end-edges of a corridor overlap. It is shown that these certificates suffice to detect all changes in the topological structure of the configuration of polygons.

**Theorem 1.** *The topological structure of a given configuration of polygons is invariant under motions of the polygons that do not violate any triangle or corridor certificate.*

To establish and maintain corridor certificates a data structure is built inside each corridor that subdivides the empty space between its two defining polygonal chains and connects its two end-edges through a sequence of *pseudo-triangles* (three convex vertices, *corners*, joined by three concave chains).

Consider a triangulated corridor with two bounding polygonal chains labeled $\gamma$ and $\delta$. Two different kinds of triangulation edges can be distinguished: *bridges* connecting $\gamma$ with $\delta$ and *chords* connecting $\gamma$ with $\gamma$ or $\delta$ with $\delta$.

The triangles adjacent to bridges and end-edges form a *pathway* as illustrated in Fig. 2. If the pathway is removed from a corridor what remains is a (possibly empty) set of empty pockets, each bounded by a chord (called the *extreme chord* for that pocket). In the dual graph, the edges dual to chords induce a tree (called the *chord tree*) in each pocket (see Fig. 2).

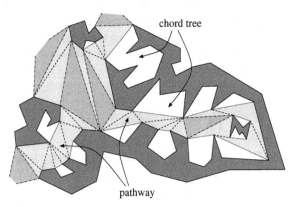

**Fig. 2.** Decomposition of the corridors into pathways and chord trees.

A pathway is said to be *degenerate* if any of its end-edges or bridges overlap with a chord or polygon edge. Thus to maintain a corridor certificate it suffices to maintain the corresponding non-degenerate pathway. In a similar way the non-degeneracy of a pathway is certified by the non-degeneracy of each of its constituent (pseudo-)triangles.

In order to allow for efficient flight plan updates the set of certificates which at any given time certify the disjointness of objects (the *active set*) should be kept as small as possible. To reduce the size of the active set associated with each corridor the triangles in its associated pathway are merged into a set of pseudo-triangles. The size of the resulting set is linearly related to the size of a minimum link separator for the two polygonal chains defining the corridor.

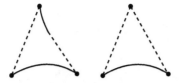

**Fig. 3.** The two types of object triangles (bridges/end-edges dashed).

This pseudo-triangulation is constructed and maintained by distinguishing two types of (pseudo-)triangles in a given pathway:

- *corner triangles* that have end-edges/bridges on only one side, chords/polygon edges of the same object on the other two; and
- *object triangles* that contain end-edges/bridges on two sides and chords/polygon edges of the same object on the third.

Note that every object triangle in a pathway has at least one side that consists only of a bridge or end-edge (see Fig. 3).

The pseudo-triangles (initially triangles) in a pathway are merged or swapped in order to establish the following invariant:

**Invariant 1 (Pathway invariant).** *Any three consecutive pseudo-triangles in a pathway must include a corner triangle and any two adjacent object triangles must belong to different objects.*

See Fig. 4 for a fully merged version of the pathways of our example that conforms to the pathway invariant.

The non-degeneracy of a pseudo-triangle can be certified by the non-degeneracy of angles formed at each of its three corners. Thus, the pathway invariant guarantees that the active set for a given corridor has, up to a small additive constant, size proportional to the number of corners in the associated pathway. This is related, in turn, to the size the the minimum link separator for the corridor through the following:

**Lemma 2.** *A minimum link separator for the two bounding chains of a corridor contains, up to a small additive constant, between one and six times as many segments as there are corner triangles inside the associated pathway.*

Although the number of certificates in the active set is bounded by something close to what could be argued is essential for the certification of disjointness, it is not hard to see that the number or certificate updates required for even very simple motions can vastly exceed what would reasonably be described as necessary. Consider the case of two convex objects with a total of $N$ vertices translating along a straight line (see Fig. 5). As the objects move, each chord and end-edge has to attach to each vertex on each of the polygons. This corresponds to $\Theta(N)$ certificate failures. It seems obvious that a constant number of certificate updates should suffice to certify disjointness in this setting, since the nature of the separation between the objects (i.e. their minimum-link separator) does not

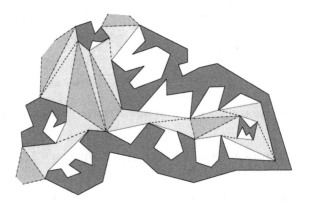

**Fig. 4.** Minimal pseudo-triangulation for the pathways.

change during the motion and their separation is at all times large with respect to their size.

In the event, of concern in this paper, that all objects are convex, pockets and corner triangles do not exist. It follows from Invariant 1 that every pathway consists of at most two pseudo-triangles, the sides of which contain contiguous fragments of the boundary of at most two objects. In the next section we discuss the construction of dynamic hierarchical representations of these fragments which, in turn, makes possible the formulation separation sensitive bounds on both the responsiveness and efficiency of the structure.

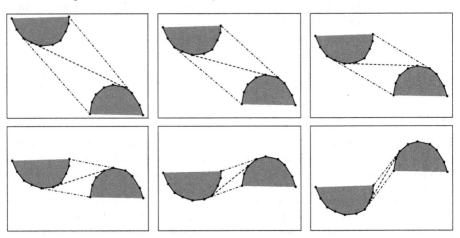

**Fig. 5.** Two convex objects moving along straight lines.

## 3   Hierarchical representations of convex chains and relative convex hulls

One of the fundamental reasons for focusing attention on convex objects (either directly or through decomposition) in geometric modeling is the fact that they lend themselves to hierarchical representations. Hierarchical representations of static convex objects permit efficient response to basic intersection (with respect to points or lines) and separation queries [5, 8, 9, 7] and collision detection queries for elementary motions [10]. It is also possible to build hierarchical representations for non-convex objects in terms of their convex pieces to provide responses to separation queries whose cost is a function of the complexity (form) of the separation (measured in terms of the size of the minimal separating chain) [17].

In the context of kinetic collision detection, Erickson et. al [11] described the use of hierarchical representations of individual convex objects to provide separation sensitive maintenance of disjointness certificates for pairs of such objects under a variety of motion models. Critical for this application is additional constraint on outer-hierarchical representations that they be distance sensitive in the sense that the distance of successive levels of the approximation hierarchy from the underlying polygon grows in a controlled fashion that is bounded by some (exponential) function of the level index. This ensures, for example, that two convex objects of diameter at most $D$ and with separation $s$ admit disjoint approximations within their respective hierarchies totalling at most $O(\log(D/s))$ edges.

The hierarchical representations (so-called boomerang hierarchies) proposed by Erickson et. al depend on both the convexity and context independence of the individual objects. Our goal is to bring the benefits of hierarchical representations to a richer context in which we have collections of convex objects in motion. In this setting the hierarchical representations of individual objects must be context dependent and malleable.

We achieve context dependence by building our hierarchical representations in conjunction with the kinetic separation structure (KSS) described in the preceding section. The resulting structure maintains an outer approximation of each object in the collection. Furthermore, at any time the total size of the combined approximations is to within a (small) constant multiple of the total number objects (independent of their complexity).

As was the case with Erickson et. al, we can make use of any hierarchical structure for convex polygonal chains that satisfies certain basic properties. First, we need distance sensitivity as defined above. It will suffice for our present purposes to have a hierarchical representation of chains that is realized as a sequence of $h = \Theta(\log n)$ layers with a total of $\Theta(n)$ edges, where $n$ is the size of the underlying chain (the *base chain*). The first layer is the chain itself and the $(i + 1)$-st layer contains at most some fixed fraction of the number of segments of its predecessor. Furthermore, every segment on the $(h - i)$-th layer contains points of distance at most $O(D/2^i)$ from the base chain, where $D$ denotes the diameter of the base chain. The second property that we need is malleability; our approximation for any one polygon is made up of (possibly many) pieces which may need to be split and appended as the object moves in relation to its neighbours. For our analysis we need a structure that permits a hierarchical representation of a chain to be decomposed into its $O(1)$ top-level subchains in $O(1)$ time. Furthermore, it must support efficient merging of the representations of adjacent co-convex chains (to form a representation of the chain formed by appending the two at their shared endpoint).

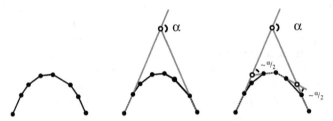

**Fig. 6.** Constructing the hierarchy.

Any of the boomerang hierarchies proposed by Erickson et. al could be used (or be easily adapted) for our purposes. An even simpler hierarchy is constructed by associating with each vertex in the base chain (except the last) the external angle formed by the extension of its incoming edge and its out-going edge (see Fig. 6). Vertices with angle $\alpha$ greater than $2\pi/n$ are replicated $\lceil \alpha n/(2\pi) \rceil$ times (by the insertion of zero length edges at appropriate angles) so that every (base chain) vertex has angle at most $2\pi/n$.

Since the objects being represented are all convex the relative convex hull of individual objects never changes. This permits an implicit hierarchical representation where the entire boundary of each polygon (after suitable replication of vertices as above) is represented as a simple array. In this case, each base chain is represented by a constant number of pointers (with updates, including merges, in $O(1)$ time) and its hierarchical representation is implemented implicitly using binary search.

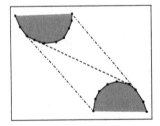 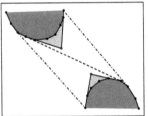 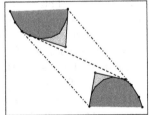

**Fig. 7.** Creating approximations for the convex chains of the objects in Fig. 5.

We initialize our separation sensitive KSS structure by replacing convex chains in each pseudo-triangle by a hierarchical representation (see Fig. 7). However, this segmentation of the hierarchical representation at pseudo-triangle boundaries is not maintained; doing so would require having every pseudo-triangle corner be a real (as opposed to approximate) polygon vertex, which

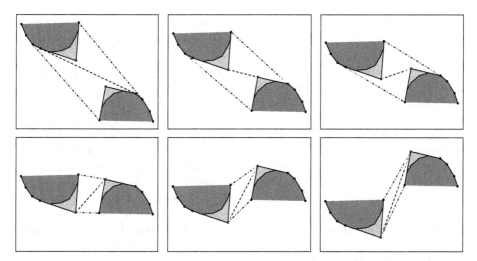

**Fig. 8.** Two hierarchically represented objects moving along straight lines.

would negate much of the potential benefit of the hierarchical representation. Instead, pseudo-triangles are free to attach at arbitrary approximation vertices subject only to the constraint that the number of approximation chains contained within a single pseudo-triangle is at most some fixed constant. Thus approximation chains are sometimes split, shifted between adjacent pseudo-triangles, and merged, all in $O(1)$ time by suitable adjustment of the pointers into the array structure (see Fig. 9 for a detailed example).

Fig. 8 repeats the motion of the objects of Fig. 5 for two hierarchically represented objects. Note that in this case only a constant number of pseudo-triangle changes occur.

## 4   Maintenance

We now assume that we are given $k$ convex polygons moving with continuous motion and that we are able to compute the failure time of any certificate associated with those polygons in constant time.

Since we do not distinguish between polygon and approximation vertices as the attachment points of end-edges and bridges all of the KSS update operations as described in [14] can be executed as before. When any certificate fails, we are able to update the data structure and restore the invariants by computation on a constant number of pseudo-triangles. Thanks to the approximation structures relocation of bridges and end-edges and the restoration of invariants can be performed in constant time. As noted earlier, the restructuring of the hierarchies of the pseudo-triangles whenever a complete base chain moves from one triangle into the next can be done in $O(1)$ time by exploiting the implicit nature of our hierarchical representation.

The hierarchy however introduces two new events: (i) a polygon edge or chord collides with an approximation vertex, and (ii) a polygon vertex collides with an edge that ends on an approximation vertex.

Events of type (i) can be handled in $O(1)$ time by "popping" the approximation vertex and restoring the invariants using the convex chain that was previously hidden. See Fig. 9 for an illustration and note how the hierarchy shifts after the collision.

Note that any number of partition triangles can be attached to an approximation vertex whenever it is popped. Since no partition triangle starts out being attached to an approximation vertex the relocation of the partition triangles to a polygon vertex contributes only $O(1)$ amortized cost to this restructuring event.

Events of type (ii) might require multiple approximation vertices (from within the same hierarchy) to be popped in order to determine if a "real" collision occurred. How many layers of the hierarchy have to be traversed is directly proportional to the separation of the objects with respect to their diameters. Specifically, if $D$ is the maximum diameter of the objects and $s$ is their actual object separation at the moment their approximations collide, then $O(\min\{\log(D/s), \log N\})$ layers of the hierarchy will be traversed. Restructuring the hierarchies will again take $O(1)$ time.

## 5    Kinetic Data Structure Properties

In this section we analyse the properties of our hierarchical KSS structure and highlight the differences with the original structure. Recall that we have $k$ convex polygonal objects consisting of a total of $N$ vertices. Note that the minimum link subdivision for such a collection of objects has size $\Theta(k)$. Our data structure has the following desirable properties for a KDS.

### Compactness

There are $\Theta(k)$ partition triangles each of which will contribute a certificate to the active set. Each polygon $P$ contributes $\Theta(\kappa_P)$ certificates, where $\kappa_P$ is the size of a minimum link cycle separating polygon $P$ from the other objects. Therefore the active set contains $\Theta(k)$ certificates at any time.

### Locality

Observe that each polygon is surrounded by a cycle of pathways connected by partition triangles. Due to Lemma 2 we can therefore state that the number of certificates each polygon $P$ appears in is $\Theta(k)$ in the worst case.

### Responsiveness

Our data structure uses two kinds of certificates: triangle certificates and corridor (i.e. corner) certificates. Without the hierarchical representation the update of both kinds of certificates can be done in $O(\log n)$ time, where $n$ is the number of vertices of the (at most five) pseudo-triangles involved. With the hierarchical representation, the corresponding relocation of bridges and end-edges, as well as the restoration of invariants and (implicit) restructuring of the hierarchy, can be accomplished in $O(1)$ time.

In all cases the update involves the destruction and creation of a constant number of edges and associated corner certificates. Each corner and triangle certificate that is disturbed during this process needs to be descheduled or rescheduled in the event queue which takes additional time logarithmic in the size of the active set.

## Efficiency

A key performance measure for kinetic data structures is the number of events processed in the worst case. As already pointed out by Agarwal et al. in [1] the notion of efficiency in the case of collision detecting data structures is not clear, since there is no canonical discrete attribute against which to compare the performance of the KDS, i.e. it is not canonically defined what constitutes an external event.

Nevertheless, as was the case with our original structure, it is possible to give an easy upper bound for the number of certificate failures if the polygons translate along algebraic trajectories of bounded degree.

**Theorem 2.** *If $k$ simple polygons with a total of $N$ vertices translate along algebraic trajectories of degree $d$ then the number of events processed by the KDS is $O(dN^3)$.*

By introducing hierarchical representations of convex chains the efficiency of our structure becomes separation sensitive. In the case of a constant number of disjoint polygons in the plane the efficiency is comparable to the bounds obtained by Erickson et al. in the even more resticted case when the number of convex polygons is restricted to two. Let $D$ be the maximum diameter of all the polygons, let $N$ be their total number of vertices, and let $s$ be the minimum distance between all pairs of polygons during their motion.

**Theorem 3.** *The number of events processed is $O(\min\{\log(D/s), \log(N)\})$ when each of some constant sized set of polygons moves in a linear trajectory. All events can be processed at a cost of $O(\min\{\log(D/s), \log(N)\})$ time per event.*

*Proof.* (sketch) Partition triangles and bridges will only attach once to every vertex on a given pathway. Therefore it is sufficient to bound the number of vertices appearing on the pathway (due to a shift in the hierarchy) during the motion. The result follows from the fact that there are at most $D/s$ vertices in the hierarchy at distance at least $s$ from a given polygon. □

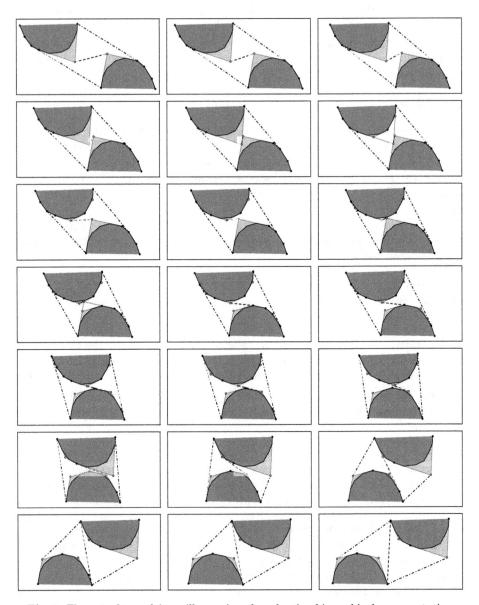

**Fig. 9.** Flypast of two objects illustrating the adaptive hierarchical representation

# 6  Future Work

In a companion paper [15] we explore the generalization of the results of this paper to collections of non-convex objects. In addition, we hope to obtain tighter efficiency bounds for various kinds of motions by exploiting the properties of our hierarchical KSS structure. The simplicity of our structure seems promising for extension to two and a half or three dimensions, but the analysis will undoubtedly be even more of a challenge.

# References

1. P. K. Agarwal, J. Basch, M. de Berg, L. J. Guibas, and J. Hershberger. Lower bounds for kinetic planar subdivisions. In *Proc. 15th ACM Sympos. Comp. Geom.*, pages 247–254, 1999.

2. P. K. Agarwal, J. Basch, L. J. Guibas, J. Hershberger, and L. Zhang. Deformable free space tilings for kinetic collision detection. In *Proc. 5th Workshop Algorithmic Found. Robotics*, 2000.

3. J. Basch, J. Erickson, L. J. Guibas, J. Hershberger, and L. Zhang. Kinetic collision detection for two simple polygons. In *Proc. 10th ACM-SIAM Sympos. Discrete Algorithms*, pages 102–111, 1999.

4. J. Basch, L. Guibas, and J. Hershberger. Data structures for mobile data. In *Proc. 8th ACM-SIAM Sympos. Discrete Algorithms*, pages 747–756, 1997.

5. B. Chazelle and D. P. Dobkin. Intersection of convex objects in two and three dimensions. *J. ACM*, 34(1):1–27, Jan. 1987.

6. T. H. Cormen, C. E. Leiserson, and R. L. Rivest. *Introduction to Algorithms*. MIT Press, Cambridge, MA, 1990.

7. D. Dobkin, J. Hershberger, D. Kirkpatrick, and S. Suri. Implicitly searching convolutions and computing depth of collision. In *Proc. 1st Annu. SIGAL Internat. Sympos. Algorithms*, volume 450 of *Lecture Notes Comput. Sci.*, pages 165–180. Springer-Verlag, 1990.

8. D. P. Dobkin and D. G. Kirkpatrick. Fast detection of polyhedral intersection. *Theoret. Comput. Sci.*, 27(3):241–253, Dec. 1983.

9. D. P. Dobkin and D. G. Kirkpatrick. A linear algorithm for determining the separation of convex polyhedra. *J. Algorithms*, 6:381–392, 1985.

10. D. P. Dobkin and D. G. Kirkpatrick. Determining the separation of preprocessed polyhedra—A unified approach. In M. S. Paterson, editor, *Automata, Languages and Programming, 17th International Colloquium*, volume 443 of *Lecture Notes in Computer Science*, pages 400–413, Warwick University, England, 16–20 July 1990. Springer-Verlag.

11. J. Erickson, L. Guibas, J. Stolfi, and L. Zhang. Separation-sensitive collision detection for convex objects. In *Proc. 10th ACM-SIAM Sympos. Discrete Algorithms*, pages 327–336, 1999.

12. L. Guibas, J. Snoeyink, and L. Zhang. Compact voronoi diagrams for moving convex polygons. In *To appear Proc. 7th Scandinavian Workshop on Algorithm Theory*, 2000.

13. L. J. Guibas. Kinetic data structures: A state of the art report. In *Proc. 3rd Workshop Algorithmic Found. Robotics*, pages 191–209, 1998.

14. D. Kirkpatrick, J. Snoeyink, and B. Speckmann. Kinetic collision detection for simple polygons. In *Proc. 16th ACM Sympos. on Comp. Geom.*, pages 322–330, 2000.

15. D. Kirkpatrick and B. Speckmann. Separation sensitive kinetic collision detection for simple polygons. In preparation.

16. M. McAllister, D. Kirkpatrick, and J. Snoeyink. A compact piecewise-linear voronoi diagram for convex sites in the plane. *Discrete Comp. Geom.*, 15:73–105, 1996.

17. D. M. Mount. Intersection detection and separators for simple polygons. In *Proc. 8th Annu. ACM Sympos. Comput. Geom.*, pages 303–311, 1992.

18. S. Suri. *Minimum link paths in polygons and related problems.* PhD thesis, Dept. Comp. Sci., Johns Hopkins Univ., 1987.

19. G. T. Toussaint. Shortest path solves translation separability of polygons. Technical Report SOCS-85.27, School Comput. Sci., McGill Univ., 1985.

# On Acute Triangulations of Quadrilaterals

Hiroshi Maehara

Ryukyu University, Nishihara, Okinawa, Japan

**Abstract.** An acute triangulation of a polygon $\Gamma$ is a triangulation of $\Gamma$ into acute triangles. Let $f(\Gamma)$ denote the minimum number of triangles necessary for an acute triangulation of $\Gamma$, and let $f(n)$ denote the the maximum value of $f(\Gamma)$ for all $n$-gons $\Gamma$. We prove $f(4) = 10$.

## 1 Introduction

By a *triangulation* of a polygon, we mean a subdivision of the polygon into a finite number of non-overlapping triangles in such a way that any two distinct triangles are either disjoint, have a single vertex in common, or have one entire edge in common. Baker, Grosse and Rafferty [1] proved that every polygon admits a triangulation into *non-obtuse* triangles, and Bern, Mitchell and Ruppert [2] gave an algorithm for triangulating $n$-gons into $O(n)$ non-obtuse triangles. An *acute* triangulation of a polygon is a triangulation whose triangles are all acute triangles. Then, at least how many acute triangles are necessary for an acute triangulation of an obtuse triangle? This problem was proposed by Martin Gardner in 1960 (see, [4] pp. 39–42), and Wallace Manheimer [6] gave a solution that the number is 7. Cassidy and Lord [3] showed that a square can be triangulated into eight acute triangles, eight is the minimum number of acute triangles for a square, and the triangulation into eight acute triangles is unique in a sense. Does every polygon admit an acute triangulation? This is answered affirmatively in Maehara [5].

Now, for a polygon $\Gamma$, let $f(\Gamma)$ denote the minimum number of acute triangles necessary for an acute triangulation of $\Gamma$, and let $f(n)$ denote the maximum value of $f(\Gamma_n)$ for all $n$-gons $\Gamma_n$. Thus $f(3) = 7$ by [6]. In this paper, we prove that $f(4) = 10$.

## 2 Preliminaries

A vertex $P$ of a polygon $\Gamma$ is called

an *acute corner* if $\angle P < \pi/2$,

an *obtuse corner* if $\pi/2 < \angle P < \pi$, and

a *concave corner* if $\pi < \angle P < 2\pi$,

where $\angle P$ denotes the interior angle of the polygon. Let $\mathcal{T}$ be an acute triangulation of a polygon $\Gamma$. The number of triangles in $\mathcal{T}$ is called the *size* of $\mathcal{T}$. Mostly, we regard $\mathcal{T}$ as a plane graph, that is, a planar graph embedded in the plane. A vertex $P$ of $\mathcal{T}$ is called

J. Akiyama, M. Kano, and M. Urabe (Eds.): JCDCG 2000, LNCS 2098, pp. 237-243, 2001.

a *corner vertex* if $P$ is a vertex of $\Gamma$,
a *side vertex* if $P$ lies within a side of $\Gamma$, and
an *interior vertex* if $P$ lies inside $\Gamma$.

In a graph, the number of those vertices that have degree $i$ is denoted by $\nu_i$. The following lemma will be clear.

**Lemma 1.** *Let $T$ be an acute triangulation of an $n$-gon $\Gamma_n$. Then*

(1)  *the minimum degree of $T$ is at least 2,*

(2)  $\deg P \geq \begin{cases} 5 & \text{if } P \text{ is an interior vertex} \\ 4 & \text{if } P \text{ is a concave corner or a side vertex} \\ 3 & \text{if } P \text{ is an obtuse corner} \end{cases}$

(3)  $\nu_2 + \nu_3 \leq n - (\# \text{ of concave corners})$, *and* $\nu_2 \leq (\# \text{ of acute corners})$.

□

**Lemma 2.** *Let $T$ be an acute triangulation of a polygon, and suppose that* (1) *$T$ has a single interior vertex, and* (2) *$\nu_2 + \nu_3 \leq 3$, $\nu_2 \leq 2$. Then $T$ is a plane graph isomorphic to* .

*Proof.* Let $P$ the interior vertex, and let $k = \deg P$. Let $G$ be the subgraph of $T$ induced by $P$ and its neighbors. Then $G$ is a $k$-wheel, and $T$ is a plane graph obtained from $G$ by attaching some triangles from the outside. Note that in $G$, $\nu_2 + \nu_3 = \nu_3 = k \geq 5$, while in $T$, $\nu_2 \leq 2$, $\nu_2 + \nu_3 \leq 3$. If we attach a triangle from the outside of $G$ at an edge of $G$, then the number $\nu_2 + \nu_3$ decreases by one. To decrease $\nu_2 + \nu_3$ further, we have to attach a triangle at an edge whose both ends have degree 3. Thus, to get a plane graph with only one interior vertex and satisfying $\nu_2 \leq 2$, $\nu_2 + \nu_3 \leq 3$, the degree $k$ of $P$ must be 5, and just two triangles must be attached at nonadjacent edges of $G$. This proves the lemma.
□

For an $\varepsilon > 0$, the *$\varepsilon$-neighborhood* of a point $P$ is a set of points within distance $\varepsilon$ from $P$. The following lemma will be clear.

**Lemma 3.** *If $ABC$ is an acute triangle, then there is an $\varepsilon > 0$ such that for any point $A_1$ in the $\varepsilon$-neighborhood of $A$, the triangle $A_1BC$ is an acute triangle.*
□

## 3   Triangles

Theorem 1 below is one half of the solution of $f(3) = 7$ by W. Manheimer [6], and Theorem 2 covers the remaining half of his proof. Though Theorem 2 is not simpler than his argument, it can be applied to the quadrilateral case.

**Theorem 1** *For any non-acute triangle $\Gamma_3$, $f(\Gamma_3) \geq 7$.*

*Proof.* Let $m$ be the minimum value of $f(\Gamma_3)$ for all non-acute triangles $\Gamma_3$, and let $ABC$ be a non-acute triangle such that $f(ABC) = m$. Let $\mathcal{T}$ be an acute triangulation of $ABC$ of size $m$. Then $\mathcal{T}$ has at least one interior vertex. For otherwise, an edge emanating from the non-acute corner of $ABC$ divides it into two triangles, at least one of which is a non-acute triangle. Since $f(\Gamma_3) \geq m$ for any non-acute triangle $\Gamma_3$, we have $f(ABC) \geq m + 1$, a contradiction.

Around any interior vertex of $\mathcal{T}$, there are at least 5 triangles by Lemma 1(2). Hence, if $\mathcal{T}$ has two or more interior vertices, then the size of $\mathcal{T}$ is at least 8. If $\mathcal{T}$ contains only one interior vertex, then the size of $\mathcal{T}$ is 7 by Lemma 2. □

**Theorem 2** *Let $ABC$ be a triangle with non-acute corner $A$. Then, for any point $P$ on the side $AC$, there is an acute triangulation $\mathcal{T}$ of $ABC$ with size 7 such that $P$ is the only side vertex lying on $AC$, and $\deg C = 2$ in $\mathcal{T}$.*

*Proof.* Let $F$ be the foot of perpendicular from $A$ to the side $BC$.

(1) First, suppose that $\angle FPC$ is acute. Let $R$ be the point on the side $AB$ such that $RP$ is parallel to $BC$, $Q$ be the intersection of $AF$ and $RP$, $S$ be the foot of perpendicular from $R$ to the side $BC$, see Figure 1(a). Then, the line-segments $QA, QP, QF, QS, QR, PF$ and $RS$ divide the triangle $ABC$ into 7 *non-obtuse* triangles. Now, slide $S$ toward $F$ slightly, and slide $F$ toward $C$ slightly so that the 5 triangles $RBS, RSQ, QSF, QFP, PFC$ become acute triangles. Then, by Lemma 3, we can slide $Q$ slightly in the direction $\overrightarrow{AQ}$ so that all triangles become acute triangles.

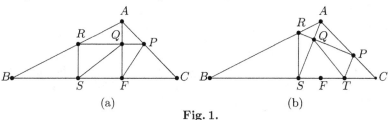

**Fig. 1.**

(2) Next, suppose that $\angle FPC$ is not acute, see Figure 1(b). For a (variable) point $R$ on the side $AB$, let $S = S(R)$ be the foot of the perpendicular from $R$ to the side $BC$, $Q = Q(R)$ be the intersection of $AS$ and $PR$. If $R$ goes near $A$, then $\angle PQA$ becomes obtuse, whereas, if $R$ goes near $B$, then $\angle PQA$ becomes acute. Hence there is a position of $R$ between $A$ and $B$ for which $\angle PQA = \pi/2$. Fix $R$ and $S = S(R), Q = Q(R)$ at such positions, and let $T$ be a point on $BC$ such that $AS$ and $PT$ are parallel. Then, since $\angle TPC = \angle QAP$ is acute, $T$ lies between $F$ and $C$. Hence $\angle QTS$ is acute. Thus, the line-segments $QA, QP, QT, QS, QR, PT$ and $RS$ divide $ABC$ into seven *non-obtuse* triangles. Let $M$ be the midpoint of $AP$. Now, since $QTS$ is an acute triangle, we can slide $Q$ slightly in the direction perpendicular to $QM$ and away from $A$ so that $QTS$ remains to be an acute triangle. Then $AQP, QPT, AQR$ become acute triangles, and $\angle AQR + \angle RQS$ becomes less than $\pi$. Finally, slide $R$ toward $B$ so that both angles $\angle AQR$, and $\angle SQR$ become acute. Then we have a desired acute triangulation. □

**Corollary 1 ([6])** *For any non-acute triangle ABC, $f(ABC) = 7$ holds, and $f(3) = 7$.* □

We will need some additional lemmas about triangles for dealing with quadrilaterals in the next section.

**Lemma 4.** *Let ABC be an acute triangle. Then, for any point P on the side AC, there are two points Q, R (Q on AB and R on BC) such that the line-segments PQ, QR, RP divide ABC into 4 acute triangles.*

*Proof.* Let $M$ be the midpoint of $AC$. If $P = M$, then we can take as $Q, R$, the midpoints of $AB$ and $BC$, respectively. So, we may assume that $P$ lies between $A$ and $M$. Let $Q$ be the point on $AB$ such that $QP$ is parallel to $BC$, see Figure 2. Let $P_1, Q_1$ be the points on $BC$ such that $PP_1 \parallel AB$ and $QQ_1 \parallel AC$. Then, since $P$ lies between $A$ and $M$, the line-segments $PP_1$ and $QQ_1$ intersect. Let $X$ be the intersection point, and let $R$ be the foot of perpendicular from $X$ to $BC$. Then, it will be easy to see that the triangles $APQ, BQR, CRP, PQR$ are all acute triangles. □

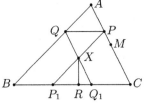

**Fig. 2.**

**Corollary 2** *Let ABC be a triangle, and let P be a point on the side AC. If $\angle B$ is acute, then there is an acute triangulation $\mathcal{T}$ of ABC with size at most 7 such that P is the only side vertex lying on AC.* □

## 4    Quadrilaterals

**Theorem 3** *For any quadrilateral ABCD, $f(ABCD) \leq 10$ holds.*

*Proof.* For any rectangle, there is an acute triangulation of size 8 [3] such as

. Hence we assume that $ABCD$ is not a rectangle. Then it has at least one acute corner, say, $B$. (If $ABCD$ is a concave quadrilateral, we assume that $B$ is an acute corner neighboring to the concave corner.) Then the diagonal $AC$ divides $ABCD$ into two triangles.

(1) First suppose that $\angle D \geq \pi/2$.

Let $P$ be the foot of perpendicular from $D$ to $AC$. Since $\angle B$ is acute, Corollary 2 implies that there is an acute triangulation of $ABC$ with size at most

7 such that $P$ is the only vertex lying on $AC$. Thus $ABCD$ can be triangulated into at most 9 non-obtuse triangles among which the triangles other than $ADP, CDP$ are all acute triangles. Now, slide the point $P$ slightly in the direction $\overrightarrow{DP}$. If the movement of $P$ is sufficiently small, then by Lemma 3, we get an acute triangulation of $ABCD$ into at most 9 triangles. Notice that in this acute triangulation, no vertex appears on the side $CD$, and on the side $DA$.

(2) Suppose that $\angle D$ is acute. We may assume that $CD$ is not shorter than $DA$. Choose a point $D_1$ on $CD$ so that $AD_1D$ is an acute triangle. Then in the quadrilateral $ABCD_1$, $D_1$ is an obtuse corner. Hence, by the argument in (1), there is an acute triangulation of $ABCD_1$ whose size is at most 9, and no vertex appears on $CD_1$. Since $CD_1D$ is an acute triangle, we have then an acute triangulation of $ABCD$ into at most 10 triangles. □

**Theorem 4** *There is a concave quadrilateral $ABCD$ such that $f(ABCD) \geq 10$.*

*Proof.* Let $ABCD$ be a concave quadrilateral such that $AB = AD$, $BC = CD$ and

$$\angle A = 2\tan^{-1}3 \,(\approx 143.1°), \quad \angle C = 2\pi - 2\tan^{-1}5 \,(\approx 202.6°).$$

Then, $C$ is the concave corner. We are going to show that $f(ABCD) \geq 10$.

Suppose that there is an acute triangulation $\mathcal{T}$ of $ABCD$ whose size is less than 10. Since $\angle A$ is obtuse, there must be an edge emanating from $A$ into the quadrilateral $ABCD$. The line segment $AC$ is not an edge of $\mathcal{T}$, for otherwise, the size of $\mathcal{T}$ is at least $f(ABC) + f(ACD) = 7 + 7 > 10$. Since $C$ is the obtuse corner of both triangles $ACB$ and $ACD$, there must be an edge from $C$ into the triangle $ABC$ and also an edge from $C$ into the triangle $ACD$. Thus, if $\mathcal{T}$ has no interior vertex, then an edge emanating from $A$ and an edge emanating from $C$ would cross each other. Therefore $\mathcal{T}$ has at least one interior vertex. Since the size of $\mathcal{T}$ is at most 9 by assumption, the number of interior vertices of $\mathcal{T}$ is at most 2.

(1) First suppose that $\mathcal{T}$ contains a single interior vertex $P$. Then, by Lemma 2, $\mathcal{T}$ is isomorphic to the graph illustrated in Figure 3. Since $\deg A \geq 3, \deg C \geq 4$

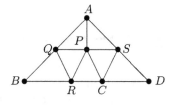

**Fig. 3.**

in $\mathcal{T}$, we may assume that $A, B, C, D$ are assigned as in Figure 3. Label the remaining vertices with $P, Q, R, S$ so that $AQBRCDS$ is the outer cycle of the graph $\mathcal{T}$.

Now, let us consider the *geometric shape* of the quadrilateral $ABCD$ in Euclidean plane. We may suppose that $C = (0,0)$, $A = (0,10)$, and $D$ has a positive abscissa. Let $X$ be the foot of perpendicular from $C$ to the edge $AD$, see Figure 4. Then $X = (3,9)$. Since $\angle CSD$ is acute, $S$ lies between $A$ and $X$. Let $Y, Z$ be the intersection points of the circle with diameter $CX$ and the line passing through $A$, parallel to $CX$. Then $\angle CYX = \angle CZX = \pi/2$. Since the equations of the line and the circle are $y = 3x + 10$, $(x - 3/2)^2 + (y - 9/2)^2 = 90/4$, respectively, we have

$$Y = \left(\frac{-3 + \sqrt{5}}{2}, \frac{11 + 3\sqrt{5}}{2}\right), \quad Z = \left(\frac{-3 - \sqrt{5}}{2}, \frac{11 - 3\sqrt{5}}{2}\right).$$

Let $U$ be the intersection of the line $AZ$ and the side $BC$, $V$ be the intersection of the line $CZ$ and the side $AB$, and $W$ be the intersection of the line $XZ$ and the side $BC$, then

$$U = \left(\frac{25}{7}, \frac{-5}{7}\right), \quad V = \left(\frac{-555 - 225\sqrt{5}}{122}, \frac{1035 - 75\sqrt{5}}{122}\right),$$

$$W = \left(\frac{975 - 525\sqrt{5}}{38}, \frac{195 - 105\sqrt{5}}{38}\right).$$

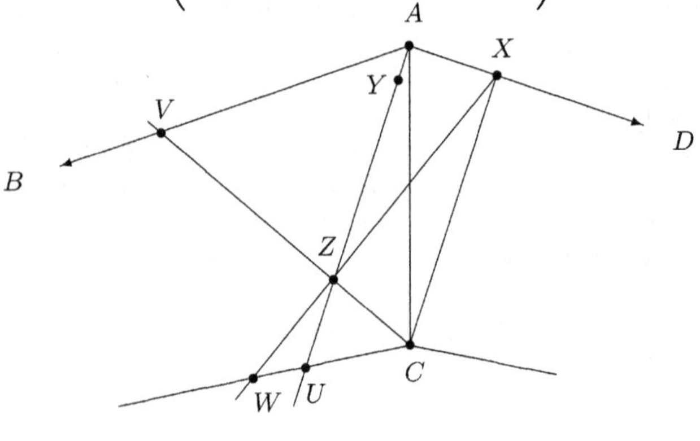

Fig. 4.

Since $\angle SAP = \angle XAP$ is acute, $P$ must lie in the trapezoid $AXCU$. Since $\angle XPC < \angle SPC$ and $\angle SPC$ is acute, $P$ must lie outside the circle with diameter $CX$. Thus, $P$ must lie either in the triangle $AYX$, or in the triangle $CZU$. Now, since $\angle PCR = \angle PCU$ and since $\overrightarrow{CY} \cdot \overrightarrow{CU} = (20 - 40\sqrt{5})/14 < 0$, $P$ cannot lie in the triangle $AYX$. Hence $P$ lies in the triangle $CZU$. Since $Q$ must lie in the same side as $R$ with respect to the line $CP$ (for otherwise, one of $\angle RPC, \angle RPQ$ is not acute), $Q$ must lie between $B$ and $V$. Similarly, since $R$ must lie in the same side as $C$ with respect to the line $SP$, $R$ must lie between $W$ and $C$. Then, since $\overrightarrow{VB} \cdot \overrightarrow{VW} = \lambda(-3,-1) \cdot \overrightarrow{VW} \approx -2.16\,\lambda$ for some $\lambda > 0$, $\angle BVW$ is obtuse, and hence $\angle BQR$ cannot be acute, a contradiction.

(2) Suppose that $\mathcal{T}$ has two interior vertices $P_1, P_2$. Then they are adjacent, for otherwise, the size of $\mathcal{T}$ is $\geq 10$. Since $\nu_2 + \nu_3 \leq 3$, $\nu_2 \leq 2$ in $\mathcal{T}$, it is not be difficult to see that $\mathcal{T}$ is isomorphic to the graph illustrated in Figure 5. Since only

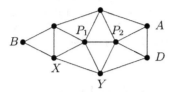

**Fig. 5.**

$B$ or $D$ can have degree 2, we may assume that the vertices $A, B, D, P_1, P_2$ are assigned as in Figure 5. Then $C$ is either $X$ or $Y$ in Figure 5. Now, (considering the geometric shape of the quadrilateral $ABCD$) since $\angle AP_2D < \pi/2$, $P_2$ must lie in the triangle $ABC$. In this case, however, it follows that $\angle P_2YD > \pi/2$, a contradiction.

This completes the proof. $\square$

From Theorems 3, 4, we have the following.

**Corollary 3** $f(4) = 10$. $\square$

*Remark.* It is not difficult to prove that for every *convex* quadrilateral $\Gamma_4$, $f(\Gamma_4) \leq 9$ holds. But I could not find any convex quadrilateral that requires 9 acute triangles.

**Problem.** Does $f(\Gamma_4) \leq 8$ hold for every *convex* quadrilateral $\Gamma_4$?

# References

1. B. S. Baker, E. Grosse and C. S. Rafferty, Nonobtuse triangulations of polygons, *Discrete Comput. Geom.* 3 (1988) 147–168.
2. M. Bern, S. Mitchell and J. Ruppert, Linear-size nonobtuse triangulation of polygons, *Discrete Comput. Geom.* 14 (1995) 411–428.
3. Charles Cassidy and Graham Lord, A square acutely triangulated, *J. Rec. Math.*, 13(1980/81) 263–268.
4. Martin Gardner, *New Mathematical Diversions*, Mathematical Association of America, Washington, D.C., 1995.
5. H. Maehara, Acute triangulations of polygons, submitted.
6. Wallace Manheimer, Solution to Problem E1406: Dissecting an obtuse triangle into acute triangles, *American Mathematical Monthly*, 67 (1960) 923.

# Intersecting Red and Blue Line Segments in Optimal Time and Precision

Andrea Mantler and Jack Snoeyink

UNC-CH Computer Science
CB 3175 Sitterson Hall
Chapel Hill, NC 27599–3175 USA
{mantler,snoeyink}@cs.unc.edu

**Abstract.** A common geometric problem in computer graphics and geographic information systems is to compute the arrangement of a set of $n$ segments that can be colored red and blue so that there are no red/red or blue/blue crossings. We give a sweep algorithm that uses the minimum arithmetic precision and runs in optimal $O(n \log n + k)$ time and $O(n)$ space to output an arrangement with $k$ vertices, or $O(n \log n)$ time to determine $k$. Our initial implementation in Java can be found at http:\\www.cs.unc.edu\~snoeyink\demos\rbseg.

## 1 Introduction

Important cases of the problems of polygon clipping in computer graphics [4], boolean operations in 2D computer-aided design (CAD/CAM) [19, 23], and map overlay in geographic information systems (GIS) [7] can all be abstracted as the problem of building an arrangement of $n$ red and blue line segments in which there are no red/red or blue/blue crossings.

This problem is often attacked by first finding all $k$ intersections of line segments, then building the arrangement, which require sorting intersections along each line if that was not done in the first step. Several algorithms have been developed that approach or achieve the optimal, output-sensitive running time of $O(n \log n + k)$ in the general (uncolored) case for finding intersections [1, 2, 5, 9] and for computing arrangements [3, 6, 7, 10] and in the red/blue case for finding intersections [8, 11, 12, 21].

It is surprisingly difficult to guarantee correct implementations of these algorithms. One reason is a large number of degenerate cases, in which an endpoint of one line segment lies inside another or the intersection of two line segments is a segment rather than a point. A second reason is that classical segment intersection algorithms use primitives with relatively high algebraic degree [6]. In general, if line segments are specified by the coordinates of their endpoints, then computing an arrangement requires four times the input precision, and computing a trapezoidation of the arrangement, as in the Bentley-Ottman sweep [3], requires five times the input precision. Floating point implementations of these algorithms will encounter roundoff error, occasionally resulting in inaccurate comparisons and incorrect results.

J. Akiyama et al. (Eds.): JCDCG 2000, LNCS 2098, pp. 244-251, 2001.
© Springer-Verlag Berlin Heidelberg 2001

In developing a new algorithm, therefore, our primary aims were to minimize arithmetic precision and the effort to handle degeneracies. Note that double precision is the lower limit finding intersections by any means, since testing a pair of segments requires the evaluation of an irreducible quadratic polynomial [6]. In the red/blue case, Chan's trapezoid sweep [8] can compute intersections with only three times input precision. Boissonnat and Snoeyink [5] sketch one way to reduce the precision requirements of Chan's algorithm. In this paper, we describe a more symmetric variant of that algorithm that produces the arrangment, not just the intersections. It handles degeneracies by breaking segments exactly.

Most segment intersection algorithms, whether for the general case or the red/blue case, aim for an output-sensitive running time of $O(n \log n + k)$ to compute $k$ intersections of $n$ segments. The hereditary segment tree techniques [11, 21] can count intersection in $O(n \log n)$ time but do not compute an arrangement. Mairson and Stolfi [18] count and compute a bundled representation of an arrangment, but run in suboptimal $O(n(\log n + \sqrt{k}))$ time. Our algorithm can count intersections and compute a bundled form of the arrangement in optimal $O(n \log n)$ time.

## 2    Preliminaries

Our algorithm uses a plane sweep, which turns the static 2D problem of line segment intersection into a dynamic 1D problem of maintaining the list of segments that intersect a vertical sweepline as the sweepline moves from left to right across the plane.

We first recall the invariants maintained by Bentley and Ottmann's [3] sweep algorithm, then define witness points which play an key role in the invariants for our algorithm.

### 2.1    The Bentley-Ottmann sweep

Bentley-Ottmann's classic sweep algorthm [3] maintains two invariants as it moves a sweepline from left to right across the plane: that the $y$ coordinate order of the segments intersecting the sweepline is known, and that all intersections to the left of the sweepline have been reported. Intersections between segments reveal themselves as changes in order, so these invariants must be updated when the sweepline encounters a segment endpoint or intersection point.

Their algorithm maintains two data structures: $L$, a list ordered by $y$ coordinate of the segments that intersect the sweepline, and $Q$, a priority queue ordered by $x$ coordinate of all segment endpoints and the intersection points right of the sweepline that are defined by adjacent segments in $L$. When the sweepline encounters the left endpoint of a line segment, it can search $L$ to determine where to add that segment, then update $Q$ to have only the intersections between adjacent segments in $L$. For a right endpoint, a segment is removed from $L$ and $Q$ is updated. For an intersection point, the segments involved are swapped in $L$, and again $Q$ is updated. If $L$ and $Q$ support logarithmic time search, insertion, and deletion, then the algorithm runs in $O((n + k) \log n)$ time.

To maintain the list $L$, it is sufficient to be able to test whether an endpoint lies above or below a line, which is a degree two polynomial when the segments are specified by their endpoint coordinates. For $Q$, however, one must compare $x$ coordinates of endpoints and intersection points, which is a degree three test, and $x$ coordinates of interseciton points, which is a degree five test.

## 2.2   Witnesses to intersection

By limiting our computation to double precision, we lose the ability to compare $x$ coordinates of intersections to either intersections or endpoints. We can only compare $x$ coordinates of endpoints, and test whether an endpoint is above or below a given segment.

In fact, we can deform the input line segments, as long as they remain monotone and do not pass over any endpoints. The tests we use cannot distinguish a difference [5, 6]. Figure 1 illustrates that deformation can move the intersection point of two segments, $a \cap b$, so our algorithms cannot rely on ordering intersections by $x$ coordinates. Figure 1 also illustrates, however, that if we deform segments $a$ and $b$ to push their intersection point as far to the right as possible, this intersection point must stop before it reaches the leftmost point in a wedge be-

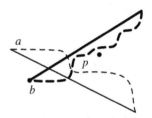

**Fig. 1.** Point $p$ witnesses $a \cap b$

tween the two segments. We call this point ($p$ in the figure) the *witness* for the intersection of $a$ and $b$, because it is the first point that certifies that these segments have swapped from their initial order. Note that the right endpoints of segments will serve as witnesses if there are no earlier points in the wedge, so every intersection between a pair of segments has exactly one witness point.

Under different terminology, witnesses also play an important role in Boissonnat and Preparata's lazy sweep [6]. In their paper, a pair of segments is called *prime* for $p$, when $p$ is the witness of their intersection.

## 2.3   From the general to the red/blue case

In the general case, we cannot hope to build an arrangement in double precision. The main reason is that quadruple precision is needed to determine the orientation of a triangle formed by three segments. Also, for curve segment intersection, there is a family of lower bound examples [5] with $n$ segments and $k$ total intersections that require $\Theta(n\sqrt{k})$ operations to count all intersections, where the hidden constant is independent of $n$ and $k$.

The special case of red/blue segment intersection, in which the segments can be colored red and blue with no red/red or blue/blue crossing pairs, we do have hope. Without red/red crossings, for example, any deformation of segments may change the locations where a single blue segment intersects the reds, but will not change the intersection order. We exploit this in the next section.

# 3    Optimal red/blue segment intersection

We turn now to the special case of red/blue segment intersection, where we can compute the arrangement using only double precision tests. We describe the invariants, events, data structure, and processing for an algorithm to count the intersections or report the arrangment explicitly. The analysis of running time is simple. We also comment on computing a compact representation of the arrangement, and on the handling of degeneracies.

## 3.1    Invariant and events

Our algorithm performs a sweep, maintaining the invariants that all intersections whose witnesses are left of the sweepline have been reported, and that the order of segments along the sweepline is consistent with pushing all intersections as far right as possible.

The events during this sweep occur only at the endpoints of segments. This gives the potential for a practical advantage over a Bentley-Ottmann sweep: we can pre-sort the events and avoid a dynamic priority queue. On the other hand, our structure containing the segments ordered along the sweepline becomes correspondingly more complex, since it must use an order consistent with a deformation, and not an order given by actual segment positions.

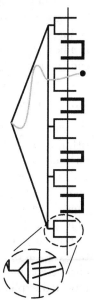

## 3.2    Data structure and processing

The data structure that stores the ordered list of segments crossing the sweepline consists of three pieces, which are illustrated schematically in Figure 2.

First, we group segments into red and blue *bundles*, which are the maximum consecutive sequences of segments of the same color. Notice that since there are no crossings between segments of the same color, segments within a bundle will remain ordered even if they are deformed. Each bundle stores the segments in a small balanced tree structure, and keeps pointers to the top-most and bottom-most. Each bundle must support insert, delete, split, and merge operations.

**Fig. 2.** Bundles

Second, we place all bundles into a doubly-linked list in order. Bundles will alternate colors.

Third, we organize red bundles into a *bundle tree*, a balanced search tree supporting split and merge operations.

We now describe the event processing when the sweepline reaches endpoint $p$. We include the handling of degenerate cases, such as segments sharing or containing endpoints.

First, we locate $p$ among the red bundles using the bundle tree. We will find either that $p$ is between two red bundles, inside one red bundle, or inside at most two red bundles with any intermediate red bundles consisting of segments that all end at $p$. If we search and split the at most two bundles that contain point $p$, then each red bundle is either 'above,' 'ending,' or 'below' at $p$.

Next, we locate $p$ among the blue bundles using the linked list, again splitting at most two blue bundles so that each blue bundle is either 'above,' 'ending,' or 'below' at $p$. Although this is a linear search, the bundles involved will be merged in the next step, so we can charge the search cost to the merging.

Once we know all bundle positions relative to $p$, we can look for adjacent bundles, one red and one blue, that are in the wrong positions. These bundles have intersections witnessed by $p$. For example, if a red 'above' bundle $R$ is below a blue 'ending' or 'below' bundle $B$, then every pair in $R \times B$ has an intersection that is witnessed by $p$. We can report these intersections, then move $R$ up and past $B$ by merging $R$ with the red bundle immediately above, merging $B$ with the blue bundle below, and repairing the bundle tree and linked list to reflect the actions to bundles $R$ and $B$.

Finally, after all bundles are in the correct relative positions, we remove those ending at point $p$, and, if needed, form new bundles for segments starting at $p$. If $p$ lies exactly on a segment, then we break that segment into two, so that one ends and one starts at $p$. This completes the processing at $p$.

### 3.3   Analysis

We claim that, if the data structures are implemented correctly, then the algorithm correctly maintains the invariants, and therefore correctly computes all intersections.

The analysis of running time is not difficult. Each endpoint causes a constant number of tree searches and splits, each of which can be carried out in $O(\log n)$ time. Because each merge joins two trees that were created by a segment insertion or a split, the total number of merges is also $O(n)$. List searches can be charged to bundle merges, as can the counting of intersections. Therefore, all operations take a total of $O(n \log n)$ time. The reporting of intersections can be charged to output complexity, giving a total of $O(n \log n + k)$ time if the arrangement is reported explicitly.

In our implementation, we use splay trees [22] for the bundles and bundle trees. It is not hard to extend an amortized analysis for splay trees [15] to the merges and splits used by our algorithm.

### 3.4   Compact representation of the arrangement

By using an idea from persistent data structures [13] we can record intersections in bundles using $O(n \log n)$ time and space, even if there are quadratically many intersections. Each time we are to split or merge a bundle, we copy the root-to-leaf path that is to change, and make our modifications to the copy. This preserves access to the old bundle, and the intersections can be recorded by recording at each endpoint $p$ which bundles intersections were witnessed by $p$.

Mount developed a data structure for storing a planar subdivision of size $n$ on a polyhedron of size $n$ in $O(n \log n)$ space, while still supporting point location [20]. He decomposes the arrangement of their overlay into grids where bundles of subdivision and polyhedron edges all cross. He notes that it would be interesting to compute this bundle structure directly to avoid a quadratic worst case in applications such as computing the Voronoi diagram on the surface of a polyhedron. Although our computation does give a bundle structure directly, we cannot use it for point location as Mount does. Our structure represents a deformation of the arrangement, and if a new query point is considered, then the deformation may need to change to accomodate it.

### 3.5   Degeneracies

We close with one observation on how we have handled degenerate cases. In an intersection algorithm, new points are calculated along segments. The implementer is faced with a choice of whether to break line segments at these new intersection points, or to hold the original segments as sacred and use the new points primarily for display. The decision usually depends on whether the implementer trusts the floating point arithmetic, or has been burned in the past.

Our approach is to break segments, but only when the endpoints are representable in single precision. Specifically, if an endpoint lies on a segment, which we can test exactly in double precision, then we can break the segment. Thus, the only remaining degeneracy is the case of overlapping segments with same start and end points.

## 4   Conclusion

We have given an algorithm building an arrangement of red/blue line segments that is optimal in not only time and space but also in the algebraic degree of its predicates. This makes it possible to guarantee correct results using only double the precision of the input coordinates. Our initial Java implementation, in Figure 3, can be found at `http:\\www.cs.unc.edu\~snoeyink\demos\rbseg`.

Intersections in the resulting arrangement are represented by pointers to their segments, as in [7]. Representing their coordinates requires rational numbers with numerators of degree three and denominators of degree 2. In future work, we will apply snap rounding ideas [17, 14, 16] to consistently round the output back to single precision.

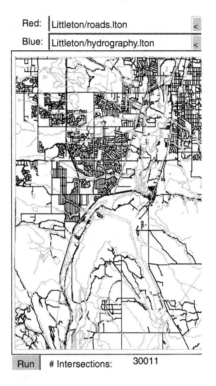

**Fig. 3.** Intersection of roads and hydrography data in our Java applet

## Acknowledgements

We thank Emo Welzl for asking whether intersections could be counted in batches, and thank Sergei Bespamyatnikh for discussions on this algorithm. We gratefully acknowledge the support of NSF Grant 9988742.

## References

1. Ivan J. Balaban. An optimal algorithm for finding segment intersections. In *Proc. 11th Annu. ACM Sympos. Comput. Geom.*, pages 211–219, 1995.
2. U. Bartuschka, K. Mehlhorn, and S. Näher. A robust and efficient implementation of a sweep line algorithm for the straight line segment intersection problem. In *Proc. Workshop on Algorithm Engineering*, pages 124–135, 1997.
3. J. L. Bentley and T. A. Ottmann. Algorithms for reporting and counting geometric intersections. *IEEE Trans. Comput.*, C-28(9):643–647, September 1979.
4. J. F. Blinn. A trip down the graphics pipeline: Line clipping. *IEEE Comput. Graph. Appl.*, 11(1):98–105, 1991.
5. J.-D. Boissonnat and J. Snoeyink. Efficient algorithms for line and curve segment intersection using restricted predicates. In *Proc. 15th Annu. ACM Sympos. Comput. Geom.*, pages 370–379, 1999.

6. Jean-Daniel Boissonnat and Franco P. Preparata. Robust plane sweep for intersecting segments. *SIAM J. Comp.*, 29(5):1401–1421, 2000.

7. Andreas Brinkmann and Klaus Hinrichs. Implementing exact line segment intersection in map overlay. In *Proc. 8th Intl. Symp. Spatial Data Handling*, pages 569–579. International Geographical Union, 1998.

8. T. M. Chan. A simple trapezoid sweep algorithm for reporting red/blue segment intersections. In *Proc. 6th Canad. Conf. Comput. Geom.*, pages 263–268, 1994.

9. Bernard Chazelle. Reporting and counting segment intersections. *J. Comput. Syst. Sci.*, 32:156–182, 1986.

10. Bernard Chazelle and H. Edelsbrunner. An optimal algorithm for intersecting line segments in the plane. *J. ACM*, 39(1):1–54, 1992.

11. Bernard Chazelle, H. Edelsbrunner, Leonidas J. Guibas, and Micha Sharir. Algorithms for bichromatic line segment problems and polyhedral terrains. *Algorithmica*, 11:116–132, 1994.

12. Olivier Devillers and Andreas Fabri. Scalable algorithms for bichromatic line segment intersection problems on coarse grained multicomputers. *Internat. J. Comput. Geom. Appl.*, 6:487–506, 1996.

13. J. R. Driscoll, N. Sarnak, D. D. Sleator, and R. E. Tarjan. Making data structures persistent. *J. Comput. Syst. Sci.*, 38:86–124, 1989.

14. M. Goodrich, L. J. Guibas, J. Hershberger, and P. Tanenbaum. Snap rounding line segments efficiently in two and three dimensions. In *Proc. 13th Annu. ACM Sympos. Comput. Geom.*, pages 284–293, 1997.

15. Michael T. Goodrich and Roberto Tamassia. *Data Structures and Algorithms in Java*. John Wiley & Sons, New York, NY, 1998.

16. Leonidas Guibas and David Marimont. Rounding arrangements dynamically. *Internat. J. Comput. Geom. Appl.*, 8:157–176, 1998.

17. J. D. Hobby. Practical segment intersection with finite precision output. *Comput. Geom. Theory Appl.*, 13(4):199–214, October 1999.

18. H. G. Mairson and J. Stolfi. Reporting and counting intersections between two sets of line segments. In R. A. Earnshaw, editor, *Theoretical Foundations of Computer Graphics and CAD*, volume 40 of *NATO ASI Series F*, pages 307–325. Springer-Verlag, Berlin, West Germany, 1988.

19. Victor J. Milenkovic. Practical methods for set operations on polygons using exact arithmetic. In *Proc. 7th Canad. Conf. Comput. Geom.*, pages 55–60, 1995.

20. D. M. Mount. Storing the subdivision of a polyhedral surface. *Discrete Comput. Geom.*, 2:153–174, 1987.

21. L. Palazzi and J. Snoeyink. Counting and reporting red/blue segment intersections. *CVGIP: Graph. Models Image Process.*, 56(4):304–311, 1994.

22. D. D. Sleator and R. E. Tarjan. A data structure for dynamic trees. *J. Comput. Syst. Sci.*, 26(3):362–381, 1983.

23. R. B. Tilove and A. A. G. Requicha. Closure of boolean operations on geometric entities. *Comput. Aided Design*, 12:219–220, 1980.

# Tight Error Bounds of Geometric Problems on Convex Objects with Imprecise Coordinates

Takayuki Nagai [*1] and Nobuki Tokura[2]

[1] Research Fellow of the Japan Society for the Promotion of Science
email:nagai@ics.es.osaka-u.ac.jp
[2] Graduate School of Engineering Science, Osaka University,Japan
email:tokura@ics.es.osaka-u.ac.jp

**Abstract.** We study accuracy guaranteed solutions of geometric problems defined on convex region under an assumption that input points are known only up to a limited accuracy, that is, each input point is given by a convex region that represents the possible locations of the point. We show how to compute tight error bounds for basic problems such as convex hull, Minkowski sum of convex polygons, diameter of points, and so on. To compute tight error bound from imprecise coordinates, we represent a convex region by a set of half-planes whose intersection gives the region. Error bounds are computed by applying rotating calipers paradigm to this representation.

## 1 Introduction

The accuracy of computed results is determined by two factors: arithmetic error and input error. In a sense they are similar because both affect computed values. However, they are essentially different. We can reduce arithmetic error as much as we need by a sophisticated algorithm that involves less intermediate arithmetic errors or a high-precision arithmetic. On the other hand, any improvement of algorithm or arithmetic can not remove the entire effect of input error. The best we can do is to carefully analyze the effect of input error and guarantee the accuracy of computed results.

Geometric computation with imprecise data has been studied in problems such as convex hull[4], visibility information inside a simple polygon [3], and so on. *Epsilon Geometry*[5] gives a framework to implement geometric algorithms from approximate primitives. An algorithm that computes approximate convex hull based on the framework is known[6].

In this paper, we show accuracy-guaranteed solutions of geometric problems defined on convex object with imprecise coordinates. We assume that each input point is known only up to a limited accuracy, which means that a point is given by a convex region that represents the possible locations of the point. Such points are considered to contain errors in coordinates, or to move arbitrarily in

---

[*] This work is supported by the Grant in Aid for Scientific Research of the Ministry of Education, Science and Cultures of Japan.

J. Akiyama et al. (Eds.): JCDCG 2000, LNCS 2098, pp. 252-263, 2001.
© Springer-Verlag Berlin Heidelberg 2001

restricted areas. Our concern is efficient computation of tight error bound in each problem instance. For example, a tight error bound of convex hull is given by the intersection and the union of all possible convex hulls as we proposed in [8]. In this paper we generalize input model and show how to compute accuracy guaranteed solutions for problems: convex hull, Minkowski sum of convex polygons, the diameter of a set of points, and so on.

To derive tight error bound, we first consider to represent a convex region by a set of half-planes whose intersection gives the region instead of by its boundary. Under this representation, which we call *half-plane representation*, the convexity of a region is always established. Then, based on *rotating calipers paradigm*[11], we characterize the solution of a problem by half-plane representation, and derive accuracy guaranteed solution of it.

The organization of this paper is as follows. In Section 2, basic geometric terms are introduced. In Section 3, we introduce *half-plane representation* of a convex region. In Section 4, we consider geometric problems with imprecise input, and show how to compute accuracy guaranteed solutions based on half-plane representation.

## 2    Preliminaries

For a directed line $l$, $h_{\mathcal{L}}(l)$ and $h_{\mathcal{R}}(l)$ denote the closed left and the closed right half-plane of $l$ respectively. By $h_{\mathcal{L}}^-(l), h_{\mathcal{R}}^-(l)$ we denote the open left and right half-planes. For a region $R$ in the plane, a directed support $l$ of $R$ is called *left support* if it contains $R$ in $h_{\mathcal{L}}(l)$, or called *right support* if it contains $R$ in $h_{\mathcal{R}}(l)$. Directed lines $t_{\mathcal{L}}(R, \theta)$ and $t_{\mathcal{R}}(R, \theta)$ are the left and the right support of $R$ whose angles with respect to x-axis are $\theta$ respectively. For a real-valued function $d(\theta)$ and an angle $\omega$, $l(d, \omega)$ denotes directed line : $(\sin \omega, -\cos \omega) \cdot (x, y) = d(\omega)$, where $\cdot$ represents inner product. The convex hull $conv(P)$ of a set $P$ of points in the plane is the smallest convex region that contains all points in $P$. The inverse of Ackermann's function is denoted by $\alpha(n)$.

## 3    Half-plane Representation of a Convex Region

In this section, we introduce *half-plane representation(HPR)* of a convex region in the plane. We show that operations on HPR give intersection, convex hull, Minkowski sum, and Minkowski difference between convex regions.

**Definition 1. (Half-Plane Representation(HPR) of a Convex Region)**
    Let $R$ be a convex region in the plane. A real-valued function $f(\theta)$ is a half-plane representation (HPR) of $R$ if it satisfies both $f(\theta + 2\pi) = f(\theta)$ and $R = \cap_{0 \le \theta < 2\pi} h_{\mathcal{L}}(l(f, \theta))$.

The idea is that we represent a convex region by a set of half-planes whose intersection gives the region; computation of a convex region is reduced to that of a corresponding set of half-planes. Any convex region $R$ has an HPR of it: $R$

is equal to $\cap_{0\leq\theta<2\pi}h_{\mathcal{L}}(t_{\mathcal{L}}(R,\theta))$. In addition, a convex region has many HPRs corresponding to it. For example, see the regular triangle in Fig. 1. Since a triangle can be represented by the intersection of three half-planes, the periodic discrete function in Fig. 2 is an HPR of the triangle. On the other hand, the triangle has infinitely many left supports shown as dashed lines in Fig. 1. Thus the periodic continuous function shown in Fig. 3, that is composed of three pieces of sinusoidal curves, is also an HPR of the triangle. Among HPRs of a convex region $R$ the HPR $d(\theta)$ that satisfies $l(d,\theta) = t_{\mathcal{L}}(R,\theta)$ for all $\theta$ is called *the normalized half-plane representation* of $R$, which is denoted by $d(R,\theta)$.

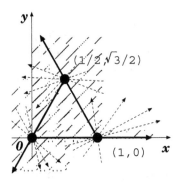

**Fig. 1.** A regular triangle given by the intersection of half-planes.

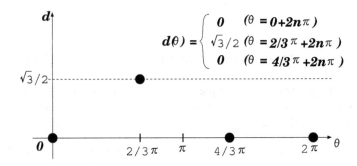

**Fig. 2.** A half-plane representation of a triangle by a discrete function.

The normalized HPR of a convex region $R$ gives right supports as well as left supports; right support $t_{\mathcal{R}}(R,\theta)$ is given by $(\sin\theta, -\cos\theta)\cdot(x,y) = -d(R,\theta+\pi)$

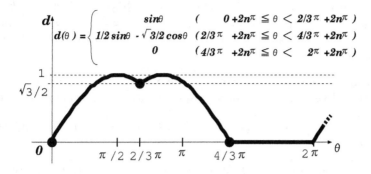

$$d(\theta) = \begin{cases} \sin\theta & ( \quad 0 +2n\pi \leqq \theta < 2/3\pi +2n\pi ) \\ 1/2 \sin\theta - \sqrt{3}/2 \cos\theta & ( 2/3\pi +2n\pi \leqq \theta < 4/3\pi +2n\pi ) \\ 0 & ( 4/3\pi +2n\pi \leqq \theta < 2\pi +2n\pi ) \end{cases}$$

**Fig. 3.** A half-plane representation of a triangle by a continuous function.

because $t_{\mathcal{L}}(R, \theta + \pi)$, which has the opossite direction to $t_{\mathcal{R}}(R, \theta)$, is given by $(\sin(\theta+\pi), -\cos(\theta+\pi)) \cdot (x, y) = d(R, \theta + \pi)$. The property is essential to apply rotating calipers paradigm to HPR.

**Lemma 2.** *Let $l_1, l_2, l_3$ and $l_4$ be directed lines $x \sin\theta - y \cos\theta = a$, $x \sin\theta - y \cos\theta = b$, $x \sin\theta - y \cos\theta = \min\{a, b\}$, $x \sin\theta - y \cos\theta = \max\{a, b\}$ respectively. Then, $h_{\mathcal{L}}(l_1) \cap h_{\mathcal{L}}(l_2)$ equals $h_{\mathcal{L}}(l_3)$ and $h_{\mathcal{L}}(l_1) \cup h_{\mathcal{L}}(l_2)$ equals $h_{\mathcal{L}}(l_4)$.*

*Proof.* Omitted.                                                                    □

**Theorem 3.** *Let $f(\theta)$ and $g(\theta)$ be HPRs of $P$ and $Q$ respectively. Then, the pointwise minimum function $\min\{f(\theta), g(\theta)\}$ is an HPR of $P \cap Q$.*

*Proof.* Note that $P \cap Q$ equals $\left( \bigcap_{0 \leq \theta < 2\pi} h_{\mathcal{L}}(l(f, \theta)) \right) \cap \left( \bigcap_{0 \leq \theta < 2\pi} h_{\mathcal{L}}(l(g, \theta)) \right)$, that corresponds to $\bigcap_{0 \leq \theta < 2\pi} (h_{\mathcal{L}}(l(f, \theta)) \cap h_{\mathcal{L}}(l(g, \theta)))$. Hence, by Lemma 2, $\bigcap_{0 \leq \theta < 2\pi} (h_{\mathcal{L}}(l(\min\{f, g\}, \theta)))$ is an HPR of $P \cap Q$.                              □

**Theorem 4.** *For convex regions $P$ and $Q$, let $R$ be $conv(P \cup Q)$. The pointwise maximum function $\max\{d(P, \theta), d(Q, \theta)\}$ is the normalized HPR of $R$.*

*Proof.* Since $R$ is the convex hull of $P \cup Q$, $h_{\mathcal{L}}(t_{\mathcal{L}}(R, \theta))$ equals $h_{\mathcal{L}}(t_{\mathcal{L}}(P, \theta)) \cup h_{\mathcal{L}}(t_{\mathcal{L}}(Q, \theta))$. Hence, by Lemma 2, $\max\{d(P, \theta), d(Q, \theta)\}$ gives $d(R, \theta)$.       □

For regions $P$ and $Q$, the Minkowski sum of $P$ and $Q$, denoted by $P \oplus Q$, is $\bigcup_{q \in Q}\{p + q | p \in P\}$. Because Minkowski sum of convex regions is also convex, we can represent the sum by HPR.

**Theorem 5.** *For convex regions $P$ and $Q$, the normalized HPR of $P \oplus Q$ is $d(P, \theta) + d(Q, \theta)$.*

*Proof.* First, consider two points $p, q$. By a simple arithmetic we have:

$$d(p \oplus q, \theta) = d(p, \theta) + d(q, \theta). \tag{1}$$

Next, consider $P \oplus q$. From Theorem 4, $\max_{p \in P}\{d(p \oplus q, \theta)\}$ gives $d(P \oplus q, \theta)$, which equals $\max_{p \in P}\{d(p, \theta) + d(q, \theta)\}$ by (1). Because $d(P, \theta)$ is $\max_{p \in P}\{d(p, \theta)\}$, we have $d(P \oplus q, \theta) = d(P, \theta) + d(q, \theta)$. Similarly, we can show that $d(P \oplus Q, \theta) = d(P, \theta) + d(Q, \theta)$.  □

For region $P$ and $Q$, the Minkowski difference of $P$ and $Q$, denoted by $P \ominus Q$, is $\cap_{q \in Q}\{p - q | p \in P\}$. We can prove a similar theorem on Minkowski difference.

**Theorem 6.** *For convex regions $P$ and $Q$, the function $d(P, \theta) - d(Q, \theta)$ is an HPR of $P \ominus Q$.*

*Proof.* From the observation that $d(p \ominus q, \theta) = d(p, \theta) - d(q, \theta)$, where $p$ and $q$ are points, we can prove the theorem by a similar argument in Theorem 5.  □

We give examples of HPRs for basic geometric objects, and show how the HPR in Fig. 3 is obtained.

*Example 1.* The normalized HPR of a point P(a,b)
The left support $t_{\mathcal{L}}(P, \theta)$, which passes through $P$, is given by $x \sin \theta - y \cos \theta = a \sin \theta - b \cos \theta$. Hence, $d(P, \theta) = a \sin \theta - b \cos \theta$.

*Example 2.* The normalized HPR of a circle C of radius r with center (a,b)
Observe that the left support $t_{\mathcal{L}}(C, \theta)$ contacts $C$ at point $(a + r \cos(\theta - \pi/2), b + r \sin(\theta - \pi/2))$. Then, the normalized HPR of $C$ is given as follows.

$$d(C, \theta) = (a + r \cos(\theta - \pi/2)) \sin \theta - (b + r \sin(\theta - \pi/2)) \cos \theta$$
$$= a \sin \theta - b \cos \theta + r$$

*Example 3.* The convex hull of a set P of points $p_1, \ldots, p_n$
From Theorem 4, we have $d(conv(P), \theta) = \max_{1 \le i \le n}\{d(p_i, \theta)\}$. Let $(a_i, b_i)$ be the coordinates of $p_i$, then the HPR $d(conv(P), \theta)$ is expressed as follows:

$$d(conv(P), \theta) = \max_{1 \le i \le n}\{a_i \sin \theta - b_i \cos \theta\}. \tag{2}$$

Observe that the normalized HPR in Fig. 3 satisfies the above equation.

## 3.1  Region Construction from Half-plane Representation

Once we have the boundary representation of a convex polygon $P$, its HPR is constructed from (2). Now consider to translate an HPR to its corresponding convex region.

In the paper, we consider only a class of HPRs : HPRs derived from HPRs of circles by operations $\min, \max, +,$ and $-$. This class of HPR is enough for our purpose since it can represent variety of convex regions: circle, point (which is a circle of radius zero), convex polygon (which is the convex hull of its vertices), the intersection and the Minkowski sum/difference of these.

As long as this class of HPR is concerned, each piece of an HPR is given by a sinusoidal curve of the form: $d(\theta) = a \sin \theta - b \cos \theta + r$, which we call *primitive HPR* (if $r \geq 0$, this is the normalized HPR of a circle of radius $r$ with center $(a, b)$). Hence, the construction of corresponding convex region is reduced to the construction of region $\cap_{\alpha \leq \theta \leq \beta} h_{\mathcal{L}}(l(a \sin \theta - b \cos \theta + r, \theta))$, which we call *primitive region*. Without loss of generality, we assume $\beta - \alpha \leq \pi$. In case of $\beta - \alpha > \pi$, we can divide the construction of primitive region into two steps: one for $\alpha < \theta \leq (\alpha + \beta)/2$ and the other for $(\alpha + \beta)/2 < \theta \leq \beta$.

The shape of primitive region is classified according to the sign of $r$. In case of $r \geq 0$, the region is unbounded and surrounded by two half-lines and a circular arc (see the left case in Fig. 4). In case of $r < 0$, the region is unbounded and surrounded by two half-lines. (see the right case in Fig. 4).

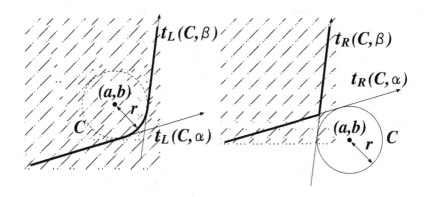

**Fig. 4.** Two types of primitive regions.

Hence, the corresponding convex region of an HPR in this class can be constructed by intersecting two types of primitive regions. Since half-planes in an HPR are sorted in increasing order of angles, this construction can be done in $O(k)$ time, where $k$ is the number of primitives in an HPR.

**Theorem 7.** *The corresponding convex region of an HPR that is composed of $k$ pieces of primitive HPRs can be constructed in $O(k)$ time.*

## 4    Accuracy Guaranteed Solutions

In this section, we show accuracy-guaranteed solutions of geometric problems on convex object with imprecise coordinates. Error bounds are introduced for problems: convex hull of points, Minkowski sum of convex polygons, diameter of points. Tight error bounds of these problems are represented by HPR. We are based on the idea that once we have an HPR of convex hull, we can apply rotating calipers paradigm[11] to it.

## 4.1   Input Model

An *ε-Point* is a convex region that represents the possible locations of a point, which is a model of a point whose coordinates are imprecise but known up to some accuracy. Input is given by a set $S$ of $ε$-Points. They are allowed to have different shapes, and to intersect each other. We use the notation $\hat{\ }$ (e.g. $\hat{p},\hat{q}$) for $ε$-Points.

Intuitively, a set of $ε$-Points represents all possible sets of points. The formal definition is as follows.

**Definition 8 (Possible sets of points).** Let $S$ be a set of $ε$-Points $\hat{p}_1, \ldots, \hat{p}_n$. A set $T$ of points $q_1, \ldots, q_n$ is a *possible set of points of $S$* if there is a one-to-one function $f_T : S \to T$ s.t. $f_T(\hat{p}_i) = q_j$ iff $q_j \in \hat{p}_i$. We call each $q_j$ *a possible point of $\hat{p}_i$*. $A(S)$ denotes the set of all possible sets of points.

Note that each point is considered to occupy exactly one location in any possible set of points, which makes a sense to derive tight error bounds. A *possible geometric object* refers to a geometric object that can be realized in a possible set of points.

## 4.2   Convex Hull of Points

Let $CH(S)$ denote the set of all possible convex hulls of $S$, that is, $CH(S) = \{conv(P)|P \in A(S)\}$. Consider region $\underline{conv}(S)$ given by $\cap_{C \in CH(S)} C$, that is the intersection of all possible convex hulls, and $\overline{conv}(S)$ given by $\cup_{C \in CH(S)} C$, that is the union of all possible convex hulls. Because any possible convex hull $C$ satisfies $\underline{conv}(S) \subseteq C \subseteq \overline{conv}(S)$, $\underline{conv}(S)$ and $\overline{conv}(S)$ give tight error bound of any possible convex hulls; this is the same formalization as in our previous work[8].

**Theorem 9.** *The function* $\max_{\hat{p} \in S}\{-d(\hat{p}, \theta + \pi)\}$ *is an HPR of* $\underline{conv}(S)$.

*Proof.* We refer the elements of $S$ by $\hat{p}_1, \ldots, \hat{p}_n$. First, observe that

$$\underline{conv}(S) = \bigcap_{C \in CH(S)} C = \bigcap_{p_i \in \hat{p}_i} conv\left(\bigcup_{i=1}^{n} \{p_i\}\right). \tag{3}$$

From Theorem 3 and Theorem 4, an HPR $d(\theta)$ of $\underline{conv}(S)$ is given by:

$$d(\theta) = \min_{p_i \in \hat{p}_i}\{\max_{1 \leq i \leq n}\{d(p_i, \theta)\}\}. \tag{4}$$

Since $p_1, \ldots, p_n$ are independent each other, we can exchange the order of *min* and *max* in (4). In addition, $\min_{p_i \in \hat{p}_i}\{d(p_i, \theta)\}$ equals $-d(\hat{p}_i, \theta + \pi)$ because $h_{\mathcal{L}}(\min_{p_i \in \hat{p}_i}\{d(p_i, \theta)\}, \theta)$ corresponds to $h_{\mathcal{L}}(t_{\mathcal{R}}(\hat{p}_i, \theta))$. Hence, $d(\theta)$ can be simplified to $\max_{1 \leq i \leq n}\{-d(\hat{p}_i, \theta + \pi)\}$. $\square$

**Theorem 10.** *The function* $\max_{\hat{p} \in S}\{d(\hat{p}, \theta)\}$ *is the normalized HPR of* $\overline{conv}(S)$.

*Proof.* We refer the elements of $S$ by $\hat{p}_1$, ..., $\hat{p}_n$. Since $\overline{conv}(S)$ is convex as we proved in [8], it is expressed as follows:

$$\overline{conv}(S) = conv(\overline{conv}(S)) = conv\left(\bigcup_{p_i \in \hat{p}_i} conv\left(\bigcup_{i=1}^{n}\{p_i\}\right)\right). \tag{5}$$

From Theorem 4, the normalized HPR $d(\theta)$ of $\overline{conv}(S)$ is given by:

$$d(\theta) = \max\{\max_{p_i \in \hat{p}_i \; 1 \le i \le n}\{d(\hat{p}_i, \theta)\}\}. \tag{6}$$

Since $p_1, \ldots, p_n$ are independent each other, we can exchange the order of maxs in (6). In addition, $\max_{p_i \in \hat{p}_i}\{d(p_i, \theta)\}$ equals $d(\hat{p}_i, \theta)$ because half-plane $h_{\mathcal{L}}(\max_{p_i \in \hat{p}_i}\{d(p_i, \theta)\})$ corresponds to $h_{\mathcal{L}}(t_{\mathcal{L}}(\hat{p}_i, \theta))$. Hence, $d(\theta)$ can be simplified to $\max_{1 \le i \le n}\{d(\hat{p}_i, \theta)\}$. □

We refer the HPRs introduced in Theorem 9 and Theorem 10 by $\underline{conv}(S, \theta)$ and $\overline{conv}(S, \theta)$ respectively.

**Corollary 11.** *For a set $S$ of $\varepsilon$-Points given by convex polygons that have $n$ vertices, the boundary of $\underline{conv}(S)$ and $\overline{conv}(S)$ can be constructed in $O(n \log n)$ time. In case that $\varepsilon$-Points are given by $n$ circles, the same running time applies.*

*Proof.* Because $\underline{conv}(S, \theta)$ and $\overline{conv}(S, \theta)$ are envelopes of HPRs of $\varepsilon$-Points, their complexities depend on geometric object we use as $\varepsilon$-Point. In case that $\varepsilon$-Point is given by convex polygon, their complexities are $O(n\alpha(n))$ (by a similar argument in [2]) and $O(n)$ respectively. Hence, the HPRs can be constructed in $O(n \log n)$ time [7], and the boundaries of $\underline{conv}(S)$ and $\overline{conv}(S)$ can be constructed in additional $O(n\alpha(n))$ and $O(n)$ time respectively.

In case $\varepsilon$-Points are given by $n$ circles, the HPRs have complexity of $O(n)$, and can be constructed in $O(n \log n)$ time. Hence, $\underline{conv}(S)$ and $\overline{conv}(S)$ can be constructed in additional $O(n)$ time. □

*Remark.* If accuracy guaranteed convex hull is going to be used in following computations, it is better to keep it in HPR form. Theorem 9 suggests that if we have $\underline{conv}(V, \theta)$ and $\underline{conv}(W, \theta)$, we can construct $\underline{conv}(V \cup W, \theta)$ by computing $\max\{\underline{conv}(V, \theta), \underline{conv}(W, \theta)\}$ while the convex hull of $\underline{conv}(V)$ and $\underline{conv}(W)$ does not necessarily give $\underline{conv}(V \cup W)$ (for example, consider a case that $W$ has possible points that are collinear). Translation of an HPR to its corresponding boundary will cause additional error.

## 4.3   Minkowski Sum of Convex Polygons

Let $P$ be a convex polygon with vertices $V$ that are specified by $\varepsilon$-Points, and let $Q$ be one with $W$. We define the set $MS(P, Q)$ of all possible Minkowski sums of $P$ and $Q$ by $MS(P, Q) = \{P' \oplus Q' | P' \in CH(V), Q' \in CH(W)\}$, where $\oplus$ denotes Minkowski sum. In this definition, we use possible convex hulls of

$P$ and $Q$ instead of possible polygons of them to preserve the convexity of the Minkowski sums.

Let $P \oplus Q$ denote $\cap_{R \in MS(P,Q)} R$, the intersection of all possible Minkowski sums, and let $P \overline{\oplus} Q$ denote $\cup_{R \in MS(P,Q)} R$, the union of all possible Minkowski sums. As in convex hull problem, $P \oplus Q$ and $P \overline{\oplus} Q$ give a tight error bound of $P \oplus Q$.

**Lemma 12.** *Let $P$ be a convex polygon with vertices $V$ that are specified by $\varepsilon$-Points, and let $Q$ be one with $W$. The union of all possible Minkowski sums $P \overline{\oplus} Q$ is convex.*

*Proof.* Observe that $P \overline{\oplus} Q$ equals $\overline{conv}(V) \oplus \overline{conv}(W)$. From the convexity of $\overline{conv}(V)$ and $\overline{conv}(W)$ (for detail, see [8]), $P \overline{\oplus} Q$ is also convex.    □

**Theorem 13.** *Let $P$ be a convex polygon with vertices $V$ that are specified by $\varepsilon$-Points, and let $Q$ be one with $W$. Functions $d(\underline{conv}(V), \theta) + d(\underline{conv}(W), \theta)$ and $d(\overline{conv}(V), \theta) + d(\overline{conv}(W), \theta)$ are HPRs of $P \oplus Q$ and $P \overline{\oplus} Q$ respectively.*

*Proof.* By Lemma 12, there exists an HPR of $P \overline{\oplus} Q$. From Theorem 3, Theorem 4, and Theorem 5, an HPR $d(\theta)$ of $P \oplus Q$ is calculated as follows.

$$
\begin{aligned}
d(\theta) &= \min_{R \in MS(P,Q)} \{d(R, \theta)\} \\
&= \min_{P' \in CH(V), Q' \in CH(W)} \{d(P', \theta) + d(Q', \theta)\} \\
&= \min_{P' \in CH(V)} \{d(P', \theta)\} + \min_{Q' \in CH(W)} \{d(Q', \theta)\} \\
&= d(\underline{conv}(V), \theta) + d(\underline{conv}(W), \theta)
\end{aligned}
$$

Similarly, we have $d(P \overline{\oplus} Q, \theta) = d(\overline{conv}(V), \theta) + d(\overline{conv}(W), \theta)$.    □

**Corollary 14.** *Let $P$ be a convex polygon with vertices $V$ that are specified by $\varepsilon$-Points, and let $Q$ be one with $W$. From HPRs $d(\underline{conv}(V), \theta)$, $d(\underline{conv}(W), \theta)$, $d(\overline{conv}(V), \theta)$ and $d(\overline{conv}(W), \theta)$ whose complexities are $K, L, M, N$ respectively, the boundaries of $P \oplus Q$ and $P \overline{\oplus} Q$ can be constructed in $O(K + L)$ time and $O(M + N)$ time respectively.*

### 4.4   Diameter of Points

By applying rotating calipers paradigm[11] to HPR, the diameter $diam(R)$ of a convex region $R$ can be represented by $diam(R) = \max_{0 \le \theta < 2\pi} \{w(R, \theta)\}$, where $w(R, \theta)$ is $d(R, \theta) - (-d(R, \theta + \pi))$, the distance between $t_L(R, \theta)$ and $t_R(R, \theta)$. Let $\underline{width}(S, \theta)$ denote $\min_{C \in CH(S)} \{w(C, \theta)\}$ and let $\overline{width}(S, \theta)$ denote $\max_{C \in CH(S)} \{w(C, \theta)\}$. Error bound of diameter can be computed similarly.

**Theorem 15.** *The minimum $\underline{diam}(S)$ and the maximum $\overline{diam}(S)$ of possible diameters are given as follows:*

$$
\overline{diam}(S) = \max_{0 \le \theta < 2\pi} \{\overline{width}(S, \theta)\}, \underline{diam}(S) = \max_{0 \le \theta < 2\pi} \{\underline{width}(S, \theta)\}. \tag{7}
$$

*Proof.* The first equation is obviously true. We give a proof of the second equation. Let $\underline{d}(\hat{p}, \hat{q})$ denote the minimum of possible distances between $\hat{p}$ and $\hat{q}$, that is, $\min_{p \in \hat{p}, q \in \hat{q}}\{d(p, q)\}$. We denote by $r$ the value of $\max_{\hat{p}, \hat{q} \in S}\{\underline{d}(\hat{p}, \hat{q})\}$. Note that $\underline{diam}(S)$ equals $r$ because $\underline{diam}(S)$ could not be smaller than $r$ and could not be greater than it, which can be proved by contradiction, but details are omitted here.

The rest we have to prove is $\max_{0 \le \theta < 2\pi}\{\underline{width}(S, \theta)\} = r$. First, observe that each local maximum of $\underline{width}(S, \theta)$ in range $[0, 2\pi]$ corresponds to minimum of possible distances between $\varepsilon$-Points. We prove that the maximum of $\underline{width}(S, \theta)$ equals $r$, which establishes our claim, by showing that there is a possible convex hull $C^*$ that satisfies $w(C^*, \alpha) = \underline{width}(S, \alpha) = r$ at some angle $\alpha$.

Let $\hat{v}, \hat{w}$ be $\varepsilon$-Points such that $\underline{d}(\hat{v}, \hat{w}) = r$, and $v, w$ be the possible points of them that satisfy $d(v, w) = \underline{d}(\hat{v}, \hat{w})$. We denote by $l_v$ the left support of $\hat{v}$ perpendicular to segment $\overline{vw}$, by $l_w$ the right support of $\hat{w}$ similarly.

We show that no $\varepsilon$-Point is in $h_{\mathcal{L}}^-(l_v)$ by contradiction. Assume that an $\varepsilon$-Point $\hat{p}$ is in $h_{\mathcal{L}}^-(l_v)$, and let $p, w'$ be the possible points of $\hat{p}, \hat{w}$ such that $d(p, w') = \underline{d}(\hat{p}, \hat{w})$. Because segment $\overline{pw'}$ acrosses the strip between $l_v$ and $l_w$, $d(p, w')$ is greater than $d(v, w)$. It means that $\underline{d}(\hat{p}, \hat{w})$ is greater than $\underline{d}(\hat{v}, \hat{w})$, which is a contradiction. Similarly, we can show that no $\varepsilon$-Point is in $h_{\mathcal{R}}^-(l_w)$. By these argument, there is a possible set $A$ of points where all possible points except $v, w$ lie between $l_v$ and $l_w$. The convex hull $C^*$ of $A$ satisfies $w(C^*, \alpha) = \underline{width}(S, \alpha) = r$ at some angle $\alpha$. $\qquad\square$

In the literature of convex hull algorithm for discs[9], D. Rappaport essentially mentioned about $\max_{0 \le \theta \le 2\pi}\{\underline{width}(S, \theta)\}$ as a different measure of the diameter of $S$ in case that $S$ is given by circles. From our analysis, the geometric meaning of the measure is proved to be the minimum of possible diameters.

**Theorem 16.** $\underline{width}(S, \theta) = \max\{\underline{conv}(S, \theta) + \underline{conv}(S, \theta + \pi), 0\}$.

*Proof.* We denote by $f(\theta)$ the function $\underline{conv}(S, \theta)$, that is, $\max_{\hat{p} \in S}\{-d(\hat{p}, \theta + \pi)\}$, and by $g(\theta)$ the function $-\underline{conv}(S, \theta + \pi)$, that is, $\min_{\hat{p} \in S}\{d(\hat{p}, \theta)\}$. For an angle $\omega$, $l_0$ and $l_1$ denote $l(f, \omega)$ and $l(g, \omega)$ respectively.

By Lemma 2, $l_0$ is a right support of some $\varepsilon$-Point $\hat{v}$, and $l_1$ is a left support of $\hat{w}$. Note that $l_0$ is the rightmost among the right supports of $\varepsilon$-Points with angle $\omega$; similarly $l_1$ is the leftmost among left supports. It means that all $\varepsilon$-Points $\hat{p}$ satisfies:

$$\hat{p} \cap h_{\mathcal{L}}(l_0) \ne \emptyset \wedge \hat{p} \cap h_{\mathcal{R}}(l_1) \ne \emptyset. \qquad (8)$$

Hence, $\underline{width}(S, \omega)$ is determined by $\hat{v}$ and $\hat{w}$.

In case of $l_1 \subset h_{\mathcal{L}}(l_0)$ i.e. $g(\omega) < f(\omega)$, $\hat{v}$ and $\hat{w}$ are seperated by $l_0$ and $l_1$. The value of $\underline{width}(S, \omega)$ equals the distance between $l_0$ and $l_1$, that is, $f(\omega) - g(\omega)$. In case of $l_0 \subset h_{\mathcal{L}}(l_1)$ i.e. $g(\omega) \ge f(\omega)$, any line $l$ that lies in $h_{\mathcal{R}}(l_0) \cap h_{\mathcal{L}}(l_1)$ stabs all $\varepsilon$-Points, which implies $\underline{width}(S, \omega) = 0$. $\qquad\square$

Note that to find an angle $\omega$ s.t. $\underline{width}(S,\omega) = 0$ is equivalent to find a common transversal [2]. Our approach is also related to *the thinnest strip transversal problem*. The problem is equivalent to compute $\min_{0 \leq \theta < 2\pi}\{\underline{width}(S,\theta)\}$, the possible minimum width. J.Robert and G.Toussaint solved this problem by using *rotating calipers* paradigm and the dynamic maintenance of convex hulls[10]. Our analysis shows that the same problem can be solved by envelope construction.

In the following, we denote the $k$th maximum value as $\max^{(k)}$ (if the $k$th maximum value does not exist, the value of $\max^{(k)}$ is defined to be $-\infty$).

**Theorem 17.** *For $n \geq 2$, consider a set $S$ of $\varepsilon$-Points $\hat{p}_1,\ldots,\hat{p}_n$, and functions $f(\theta) = \max_{1 \leq i \leq n}\{d(\hat{p}_i,\theta)\}$, $f_2(\theta) = \max_{1 \leq i \leq n}^{(2)}\{d(\hat{p}_i,\theta)\}$, and the mapping $M(\theta) : \Re \mapsto 2^N$ s.t. $M(\theta) = \{i|d(\hat{p}_i,\theta) = f(\theta)\}$.*
*Then, $\overline{width}(S,\theta)$ is given by the following equation:*

$$\overline{width}(S,\theta) = \begin{cases} f(\theta) + f(\theta + \pi) \\ \quad (\text{if } M(\theta) \neq M(\theta + \pi) \vee |M(\theta)| \geq 2) \\ \max\{f(\theta) + f_2(\theta + \pi), f(\theta) + f_2(\theta + \pi)\} \\ \quad (\text{if } M(\theta) = M(\theta + \pi) \wedge |M(\theta)| = 1). \end{cases}$$

*Proof.* Remember that $\overline{width}(S,\theta)$ should be determined by two distinct $\varepsilon$-Points. From the observation of $\overline{width}(S,\theta) = \max_{1 \leq i,j \leq n, i \neq j}\{d(\hat{p}_i,\theta) + d(\hat{p}_j,\theta + \pi)\}$, the value of $\overline{width}(S,\theta)$ is given as claimed.    □

**Corollary 18.** *For a set $S$ of $\varepsilon$-Points given by convex polygons that have $n$ vertices, $\underline{diam}(S)$ can be computed in $O(n\log n)$ time and $\overline{diam}(S)$ can be computed in expected $O(n\alpha(n)\log n)$ time. The same running time applies for a case that $\varepsilon$-Points are given by $n$ circles.*

*Proof.* We can compute $\underline{diam}(S)$ once we have $\underline{conv}(S,\theta)$ that can be constructed in $O(n\log n)$ time. To compute $\overline{diam}(S)$, we need $\overline{conv}(S,\theta)$ and the 1-level in the arrangement of normalized HPRs of $\varepsilon$-Points, that can be computed in expected $O(n\alpha(n)\log n)$ time [1].    □

## 4.5    Other Problems

If an HPR of a convex object is given, we can apply rotating calipers paradigm because support lines, which is calipers, can be computed from the HPR. As we have seen in computation of diameter, we can consider that HPRs of $\underline{conv}(S)$ and $\overline{conv}(S)$ give accuracy guaranteed calipers.

For example, consider to compute the minimum distance $dist(P,Q)$ from convex polygon $P$ to $Q$. The distance is given by $\max_{0 \leq \theta < 2\pi}\{-d(Q,\theta+\pi) - d(P,\theta)\}$ (observe that if $dist(P,Q)$ is negative, $P$ intersects $Q$). Hence, the maximum of possible minimum distances between convex polygons whose vertices are specified by sets $T,U$ of $\varepsilon$-Points is given by $\max_{0 \leq \theta < 2\pi}\{-\underline{conv}(U,\theta + \pi) - \underline{conv}(T,\theta)\}$. Similarly, other problems can be solved in accuracy guaranteed form by applying rotating calipers paradigm. However, we must be careful when multiple calipers are used at the same time. Consider $\min_{0 \leq \theta \leq 2\pi} \{width(S, \theta) \cdot width(S, \theta + \pi/2)\}$. Although it gives a lower bound of the area of the smallest enclosing rectangle, the same $\varepsilon$-Point would contribute to both $width(S, \omega)$ and $width(S, \omega + \pi/2)$ at some angle $\omega$; this results in underestimation of the area.

## 5   Conclusion

In this paper, we have considered geometric problems defined on convex objects under an assumption that input points are known only up to a limited accuracy. The main tools that we use are *half-plane representation* and rotating calipers paradigm. The problems are solved in accuracy guaranteed form by them. Error bounds presented in this paper can be computed efficiently in each problem instance. They are tight even when input contains different sizes of error.

## References

1. P.K.Agarwal, M.de Berg, J.Matoušek, and O.Schwarzkopf: Constructing Levels in Arrangements and Higher Order Voronoi Diagrams, *Proceedings of 10th Annual ACM Symposium on Computational Geometry*, pages 67-75, 1994.
2. M.Atallah and C.Bajaj: Efficient Algorithms for Common Transversals, *Information Processing Letters*, vol.25, pages 87 – 91, 1987.
3. L.Cai, J.M.Keil: Computing Visibility Information in an Inaccurate Simple Polygon, *International Journal of Computational Geometry & Applications*, vol.7, No. 6, pages 515-537, 1997.
4. P.G.Franciosa, C.Gaibisso, G.Gambosi, M.Talamo: A convex hull algorithm for points with approximately known positions, *International Journal of Computational Geometry & Applications*, vol. 4, No.2, pages 153-163, 1994.
5. L.Guibas, D.Salein, J.Stolfi: Epsilon geometry: building robust algorithm from imprecise computations, *Proceedings of 5th Annual ACM Symposium on Computational Geometry*, pages 208-217, 1989.
6. L.Guibas, D.Salein, J.Stolfi: Constructing Strongly Convex Approximate Hulls with Inaccurate Primitives, *Algorithmica*, vol. 9, pages 534-560, 1993.
7. J.Hershberger: Finding the Upper Envelope of $n$ Line Segments in $O(n \log n)$ Time, *Information Processing Letters*, vol. 33, pages 169-174, 1989.
8. T.Nagai, S.Yasutome and N.Tokura: Convex Hull Problem with Imprecise Input, *Lecture Notes in Computer Science*, vol. 1763, pages 207 – 219, 1998.
9. David Rappaport: A convex hull algorithm for discs, and applications, *Computational Geometry: Theory and Applications*, vol. 1, No. 3, pages 171-187, 1992.
10. J.Robert and G.Toussaint: Computational Geometry and Facility Location, *Proceedings of International Conference on Operations Research and Management Science*, pages B-1 – B-19, 1990.
11. G.Toussaint: Solving Geometric Problems with the "Rotating Calipers", *Proceedings of IEEE MELECON'83*, pages A10.02/1 – 4, 1983.

This article was processed using the LaTeX macro package with LLNCS style

# Triangle Contact Systems, Orthogonal Partitions, and Their Hit Graphs

Atsuhiro Nakamoto

Department of Mathematics, Osaka Kyoiku University
4-8-1 Asahigaoka, Kashiwara, Osaka 582-8582 Japan

**Abstract.** A *triangle contact system* is a geometrical object on the plane consisting of a finite number of triangular disks $\{c_1 \dots c_n\}$ triangle bounded by the segments $AB$ and $C$ such that every vertex of each $c_i$ coincides with an interior point of either $AB$ or a segment of some $c_j$. An *orthogonal partition* is a geometrical object on the plane consisting of a finite number of horizontal segments $q_1, \dots, q_n$ and vertical segments, which are contained in a rectangle $R$ bounded by segments $AB, \dots$ the horizontal segments such that every endpoint of each coincides within an interior point of either $R$ or some $q_i$ that every endpoint of each $q_i$ coincides within an interior point of either $R$ or some $a_j$. From $S$ we can construct a planar triangulation and a planar quadrangulation according to the adjacency of elements in $S$ or $P$ respectively. In this paper, we give a new theory for triangle contact or orthogonal partitions and their graphs constructed from them.

## 1 Introduction

Consider a set $S = \{S_1, \dots, S_n\}$ of circular disks arranged in the plane in which any two circular disks intersect at most one point. According to the adjacency of the circular disks we can construct the graph $G_S$ whose vertex set corresponds to $S$. It is easy to see that $G_S$ can be drawn in the plane without edge crossings. For any arbitrarily given graph $G$ in the plane, does there exist an arrangement $S$ of disks such that $G = G_S$? This problem has been affirmatively solved in [1, 6]. There are several researches concerning the problem whether a given graph on the plane can be represented as a geometrical object arranged in the plane [5, 4, 10]. In this paper, we consider this problem.

Let $G$ be a *simple* graph (i.e., without loops and multiple edges) which is *embedded* in the plane so that any two edges do not intersect except their ends. Such a graph is called a *plane graph*. We denote the vertex set, the edge set and the face set by $V(G)$, $E(G)$ and $F(G)$ respectively. A vertex $v$ or an edge $e$ of a plane graph $G$ is said to be *outer* (resp., *inner*) if $v$ or $e$ is (resp., is not) incident to the infinite region. A *k-cycle* means a cycle of length exactly $k$. A plane *triangulation* is a plane graph with each face bounded by a 3-cycle, while a plane *quadrangulation* is a plane graph with each face bounded by a 4-cycle. It is easy to see that every quadrangulation on the plane is *bipartite*, i.e., its vertices

J. Akiyama, M. Kano, and M. Urabe (Eds.): JCDCG 2000, LNCS 2098, pp. 264–273, 2001.
© Springer-Verlag Berlin Heidelberg 2001

can be colored black white so that any two adjacent vertices had different colors. We deal only with triangulations that can admit such a coloring.

In this paper we introduce geometric objects arranged in the plane called triangle contact systems and orthogonal plane partitions corresponding to plane triangulations and describe their relations from a graph theoretic point of view. They have already been dealt in the papers [3, 7, 8] independently. This paper is to put these results together and add a further observation for them.

## 2  Triangle Contact Systems and Plane Triangulations

Let $T$ be a triangle in the plane with outer three straight segments $A_1$, $A_2$ and $A_3$. Suppose that the triangular disk $T_1, \ldots T_n$ are placed in the interior of $T$ so that each vertex of every $T_i$ coincides with an endpoint of a segment of some $T_j$ with $j \neq i$ or an endpoint of some $A_k$ (In this case we say that $T_i$ hits $T_j$ or $T_i$ hits $A_k$). Let $\mathcal{T} = \{A_1, A_2, A_3, T_1, \ldots T_n\}$. Then, such arrangement of $\mathcal{T}$ is said to be a *triangle contact system*. We often regard the set as triangle contact system. We regard two triangle contact systems as the same if they have the same combinatorial structure with respect to the adjacency of $A_1, A_2, A_3, T_1, \ldots T_n$.

A partially oriented graph $G_{\mathcal{T}}$ with vertex set $\{a_1, a_2, a_3, t_1, \ldots t_n\}$ is said to be the *hit graph* of $\mathcal{T}$ if there exists a bijection $b : V(G_{\mathcal{T}}) \to \{A_1, A_2, A_3, T_1, \ldots T_n\}$ such that

(i) $b(a_j) = A_j$ for $j = 1, 2, 3$ and $b(t_i) = T_i$ for $i = 1, \ldots n$,
(ii) $G_{\mathcal{T}}$ has exactly three unoriented edges $\{a_1 a_2, a_2 a_3, a_1 a_3\}$ for the outer triangle and for any $x, y \in V(G_{\mathcal{T}})$, $G_{\mathcal{T}}$ has an oriented edge $xy$ if and only if $b(x)$ hits $b(y)$.

**Proposition 1. (Fraysseix, Mendez and Rosenstiehl [3])** *The hit graph of a triangle contact system $\mathcal{T}$ is a plane triangulation in which each inner edge is oriented so that every inner vertex has outdegree exactly 3. For example, see Fig. 1.*

An orientation of a plane triangulation with each inner vertex of outdegree exactly 3 an *inner 3-orientation*. A plane triangulation associated with an inner 3-orientation is called *inner 3-oriented plane triangulation*.

The hit graph $G_{\mathcal{T}}$ is a plane graph with exactly 3 outer unoriented edges and exactly $3(|V(G_{\mathcal{T}})| - 3)$ inner oriented edges (since each inner vertex has outdegree exactly 3). Hence $G_{\mathcal{T}}$ has exactly $3 + 3(|V(G_{\mathcal{T}})| - 3) = 3|V(G_{\mathcal{T}})| - 6$ edges. On the other hand, we know by Euler's formula that every plane graph $G$ has at most $3|V(G)| - 6$ edges and the equality holds when $G$ is a triangulation. Thus $G_{\mathcal{T}}$ is an inner 3-oriented plane triangulation.

The converse of Proposition 1 holds as the following theorem.

**Theorem 1. (Fraysseix, Mendez and Rosenstiehl [3])** *For every triangle contact system $S$, there exists a unique inner 3-oriented plane triangulation*

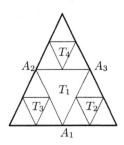
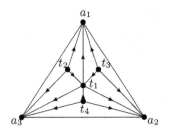

**Fig.1** Triangle contact system and triangle contact graph $G_T$

which is the hit graph of $S$. Conversely, for every inner 3-orientated plane tri-angulation $G$, there exists a unique triangle contact system $S$ whose triangle contract graph is $G$.

We give the proof of the first part since the $\cdots$ has been desried Proposition.

Spose that an inner 3-oriented plane triangulation $G$ has ter vertices $a, b, c$ and inner vertices. We label their vertices by $v_1, v_2, \ldots v_n$, as follows. Let $\overline{xy}$ denote an edge oriented from $x$ to $y$, or this symbol means that it is oriented to $y$. Let $G_n$ be a partially oriented graph obtained from $G$ by removing $c$. For $i = n, n-1, \ldots, 1$, we successively define the graph $G \supset G_n \supset \cdots \supset G_0$. To do so we observe the following claim.

*Claim.* Let $R_k$ be a 2-cell region of $G$ bounded by (not necessary oriented) closed walk $W$ of length $k$. Then $R_k$ has exactly $k-3$ edges oriented from vertices on $W$ toward the interior of $R_k$.

By Euler's formula, we can conclude that if there are $m$ vertices in the interior of $R_k$, then $R_k$ has exactly $3m + k - 3$ edges. Since each inner vertex in $R_k$ has total degree 3, there are $k - 3$ oriented edges going into $W$. Thus the claim holds.

Spose that $G_i$ has a boundary walk $s_1 \cdots s_t$ where $a = s_1$ and $b = s_t$. Since the vertices $a, b$ have no outgoing edges, we have $\overrightarrow{s_2 s_1}$ and $\overrightarrow{s_{t-1} s_t}$. Hence there exists a vertex $s_j$ with $2 \le j \le t-1$ such that $\overrightarrow{s_j s_{j-1}}$ and $\overrightarrow{s_j s_{j+1}}$. Let $v_i$ be $s_j$ and let $G_{i-1} = G_i - v_i$ for $i = n, n-1, \ldots, 1$. Note that all edges incident to $v_i$ in $G_i$ except $\overrightarrow{v_i s_{j-1}}$ and $\overrightarrow{v_i s_{j+1}}$ are oriented to $v_i$. Clearly each $G_i$ has precisely $i + 2$ vertices.

Now we construct a triangle contact system $S$ with the three segments $A, B, C$ corresponding to $a, b, c$ and triangles disk $T_1, \ldots T_n$ corresponding to $v_1, \ldots v_n$ respectively. Draw two half lines $A : y = x$ and $B : y = -x$ in $R_+^2 = \{(x, y) \in R^2 : x, y \ge 0\}$ from the origin. The two lines $A$ and $B$ correspond to the vertices $a$ and $b$ respectively. Clearly this corresponds to $G_0$. Spose that the vertex $v_i$ has neighbors $v_{s_1} \ldots v_{s_t}$ $(t \ge 2)$ in $G_i$ in the clockwise order ( where $\overrightarrow{v_i v_{s_1}}$, $\overrightarrow{v_i v_{s_t}}$ and $\overrightarrow{v_{s_j} v_i}$ for $j = 2, \ldots, t-1$) and $v_i$ is oriented $v_k \notin V(G_i)$ in $G$.

Suppose that the arrangement $S_{G_{i-1}}$ consisting of $A$, $B$, $T_1, \cdots, T_{i-1}$ has already been constructed. Particular, it is $T_i$ as shown in Fig 2 to construct $S_G$ so that

(i) one segment $L$ of $T_i$ is on the line $y = i$,

(ii) both two ends are of $S_{i-1}$ and $S$ which correspond to $s_1$ and $s_t$,

(iii) the third vertex (called the top) of $X_i$ on top of $S_{i-1}$ and $S_n$ has $y$-coordinate $k$ where $v_i$ is oriented $v_k$ in $G$ with $v_k \notin V(G_i)$.

Since the top of $S_2, \ldots, S_t$ have $y$-coordinate exactly and the top of $T_{s_t}$ and $T_{s_t}$ have $y$-coordinate strictly greater than $i$, we can place $T_i$ as in (i). Finally the horizontal zone has segment $G'$ on a line $y$ corresponding to the top vertex $c$.

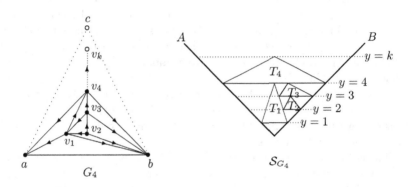

**Fig. 2** Construction of $S_G$ from $G$

**Corollary 1.** *Let $\Omega$ be the set of all triangle contact systems. Let $\Gamma$ be the set of all inner 3-orientated plane triangulations. Then, there exists a bijection $b : \Omega \to \Gamma$ such that for any $T \in \Omega$, $b(T)$ is the hit graph of $T$.*

**Theorem 2 (Fraysseix, Mendez and Rosenstiehl [3]).** *Every plane triangulation has an inner 3-orientation.*

The authors in [3] have proved Theorem 2 by construction. There is another proof of Theorem 2 by induction on the number of vertices [7].

Let $T$ be a triangulation and $D_1$ be an inner 3-orientation of $T$. If $D_1$ has an oriented cycle $C$ then we obtain another inner 3-orientation of $T$ denoted by $D_2$ by changing the direction of the edges of $C$. We call this operation a *cycle reversal*. Actually, if $D_1$ and $D_2$ are different inner 3-orientations of the same triangulation $T$ then we define the *subtraction* $D = D_1 - D_2$ to be the oriented subgraph of $D_1$ such that $V(D) = V(T)$ and $E(D) = \{xy \in E(D_1) : yx \in E(D_2)\}$. Then for any vertex $w \in V(D)$, the total degree of $v$ in $D$ is its degree in $D$. One each vertex has the same out degree in $D_1$ and $D_2$. Here $D$ an edge-disjoint union of several bar torus. This means that $D_1$ and $D_2$ can be transformed into another repeatedly by applications of cycle reversals. Here we would like to consider whether or not $D_1$ and $D_2$ can be transformed into each other by the reversals only by 3-cycles. A cycle reversal of a cycle of length $n$ is called an *n-cycle reversal*.

**Theorem 3 (Nakamoto, Ota and Tanuma [7]).** *Let $D_1$ and $D_2$ be inner 3-orientations of the same plane triangulation with $n$ vertices. Then, $D_1$ and $D_2$ are equivalent to each other.*

In order to prove Theorem it suffices to show that are reversible all $k$-cycle with $k \geq 4$ in an inner 3-oriented triangulation $D$ an be obtained by a sequence of 3-cycle reversals.

Using Euler's formula we can find that there is an oriented path $P$ through the interior of from $x \in V(C)$ to $y \in V(C)$ as follows.

Let $R_k$ be the 2-cell region bounded by $C$. By Claim 2 let $v$ be a vertex in $C$ which has an oriented edge toward the interior of $C$. Consider oriented paths from $v$ going through only oriented edges of $R_k$, and cut $R_k$ along all such oriented paths from $v$. If there is such an oriented path reaching to a vertex on $C$ other than $v$ toward, and hence we may suppose that there is no such path. Thus all the other vertices of $R_k$ lie on the same region say $R$, among the several regions obtained from $R_k$ by cutting. Clearly $R$ is bounded by (not necessarily oriented) closed walk of length strictly greater than $k$. Having no edge toward the interior of from the boundary. This contradicts Claim 2 and hence there exists a required $P$.

Then $P$ and one of oriented paths, say $P_1$ on $C$ between $x$ and $y$ form an oriented cycle bounding less number of faces than $C$. By induction hypothesis, the oriented cycle $P \cup P_1$ can be reversed. After the reverse $P$ and then other oriented path on $C$ between $x$ and $y$ also form an oriented cycle bounding less number of faces than $C$. This can be reversed similarly. In the resulting oriented graph only the direction of $P$ has been reversed since the oriented path is reversed twice. Therefore we are done.

Theorem 2 clarifies the relation between triangle contact systems and inner 3-oriented triangulations. By Theorem this relation can be translated into the relation between some equivalence classes of triangle contact systems and inner oriented triangulations as follows.

Let $T$ and $T'$ be two triangle contact systems. Suppose that their height graphs $G_T$ and $G_{T'}$ (which are inner 3-orientated triangulations by Proposition 1) can be transformed into another by reversing a 3-cycle $t_i t_j t_k$ into $t_k t_j t_i$. This means that $T$ has three triangle disk $T_i$, $T_j$ and $T_k$ such that $T_i$, $T_j$, $T_j$, $T_k$, $T_i$, respectively, and changing the adjacency of $T_i$, $T_j$ as shown in Fig 3 (fixing the adjacency of the triangle disk ) transform $T$ into $T'$. (Note that we don't know whether $T$ and $T'$ can is isotopically be transformed into each other.) We say that two triangle contact systems $T$ and $T'$ are *equivalent* and denoted by $T \sim T'$ if they can be transformed into each other by one of the two deformations among three triangle disk

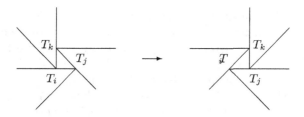

**Fig. 3** Changing the place of $T_i$, $T_j$ and $T_k$

Let $T$ be an oriented triangulation and $D$ an inner 3-orientation of $T$. If $D$ is the hit graph of a triangle contact system then $T$ is said to be a *hit nonoriented graph* of $T$.

**Corollary 2.** *Let $\Omega$ be the set of all triangle contact systems. Let $\Pi$ be the set of all nonoriented plane triangulations. Then, there exists a bijection $c : \Omega/\sim \longrightarrow \Pi$ such that for any $T \in \Omega/\sim$, $c(T)$ is the hit nonoriented graph of $T$.*

Now consider how many 3-cycle reversals are needed to transform two inner 3-orientations of a triangulation. Let $G$ be a triangulation and let $D_1$ and $D_2$ be two inner 3-orientations of $G$. Let $d(D_1, D_2)$ be the minimum number of 3-cycle reversals needed to transform $D_1$ and $D_2$. Let $d(G)$ be the minimum of $d(D_1, D_2)$, where $D_1$ and $D_2$ range all inner 3-orientations of $G$.

**Theorem 4 (Nakamoto, Ota and Tanuma [7]).**

$$d(G) \leq \frac{1}{2}n^2 - 5n + \frac{25}{2}.$$

**Proposition 2 (Nakamoto, Ota and Tanuma [7]).** *There exists a plane triangulation $T$ with two inner 3-orientations $D_1$ and $D_2$ such that*

$$d(D_1, D_2) = \begin{cases} \dfrac{1}{3}n^2 - 3n + \dfrac{17}{3} & (n = 9, 12, 15, \ldots) \\[2mm] \dfrac{1}{3}n^2 - 3n + \dfrac{20}{3} & \text{otherwise} \end{cases}$$

By Theorem 4 and Proposition 2 the order of $d(G)$ cannot be improved. We conjecture that the coefficient of the primer bound is $\frac{1}{3}$.

## 3 Orthogonal Plane Partitions and Plane quadrangulations

We introduce a geometrical object called an *orthogonal plane partition* which has a relation to the quadrangulation associated with their inner partial orientation. As results for these objects obtain parallel to triangle contact systems and their corresponding partial orientations.

A *orthogonal plane partition* $\mathcal{P} = \{A, B, C, D, a_1, \ldots a_p, b_1, \ldots b_q\}$ is a geometric object on the plane consisting of a set of item or of horizontal segments $a_1, \ldots a_p$ and vertical segments $b_1, \ldots b_q$ in a rectangle $R$ bounded by 4 segments $A, B$ and $D$ with 4 horizontal such that every end point of each $a_i$ coincides with an interior point of either $B, D$ or some $b_j$ and that every end point of each $b_i$ coincides with an interior point of either $A, C$ or some $a_j$. For two segments $a$ and $b$, if an end point of $a$ coincides with an interior point of $b$ the the segment $a$ is said to *hit* $b$. Observe that for any $i$ and $j$ $a_i$ does not hit $a_j$ and $b_i$ does not hit $b_j$. We regard two orthogonal plane partitions $\mathcal{P}$ and $\mathcal{P}'$ as the same if they have the attached adjacency of segments which induces a bijection between the set of rectangles of $R$ and that of $\mathcal{P}'$.

there exists a bijection between the set of rectangles of $R$ and that of $\mathcal{P}'$.

A *partially oriented graph* $G_{\mathcal{P}}$ with vertex set $\{a, b, c, d, x_1, \ldots x_p, y_1, \ldots y_q\}$ is said to be the *hit graph* of $\mathcal{P}$ if there exists a bijection $\phi : V(G_{\mathcal{P}}) \to \mathcal{P}$ such that

(i) $\phi(a) = A$, $\phi(b) = B$, $\phi(c) = C$, $\phi(d) = D$, $b(x_i) = a_i$ for $i = 1 \ldots p$, and $b(y_i) = b_i$ for $i = 1, \ldots q$,

(ii) $G$ has exactly $n$ non-oriented edges $ab, bc, cd$ and for forming the outer quadri-lateral and for any $x, y \in V(G_T)$, $G$ has an oriented edge $xy \in E(G_T)$ if and only if $b(x)$ hits $b(y)$,

(iii) if $G$ has a face bounded by a 4-cycle $xyx'y'$ where $x, x' \in \{a, c, x_1, \ldots x_p\}$ and $y, y' \in \{b, d, y_1, \ldots y_q\}$ then there is a rectangle which is bounded by 4 segments $\phi(x), \phi(y), \phi(x'), \phi(y')$ in this order and whose interior is empty.

The third conditions is actually needed though the corresponding condition for triangle contacts systems was not needed. It is what a 3-connected planar graph is uniquely embeddable in the plane [11]. A triangulation is always 3-connected while planar quadrangulation is not necessarily 3-connected and might have different embeddings. That is important that one special fixed embedding so that planar bipartite graph corresponds to a fixed orthogonal plane partition.

The hit graph $G$ is the planar quadrangulation in which each inner edge is oriented so that every inner vertex has outdegree exactly 2 [7]. We call this orientation an *inner 2-orientation* (For example see Fig 4) A planar triangulation associated with 2-orientation is an inner 2-orientation is called *inner 2-oriented plane triangulation*. The following theorem has been proved.

**Theorem 5 (Nakamoto and Watanabe [7]).** *For every orthogonal plane partition* $\mathcal{P}$, *there exists a unique inner 2-oriented plane quadrangulation which is the hit graph of* $\mathcal{P}$. *Conversely, for every inner 2-orientated plane quadrangulation* $G$, *there exists a unique orthogonal plane partition* $\mathcal{P}$ *whose hit graph is* $G$.

**Corollary 3.** *Let* $\Psi$ *be the set of all orthogonal plane partitions. Let* $\Sigma$ *be the set of all inner 2-orientated plane quadrangulations. Then, there exists a bijection* $b : \Psi \to \Sigma$ *such that for any* $\mathcal{T} \in \Psi$, $b(\mathcal{T})$ *is the hit graph of* $\mathcal{T}$.

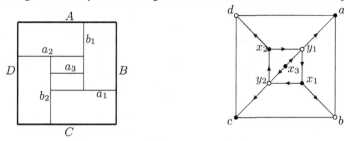

**Fig. 4** Orthogonal plane partition and its hit graph $G_{\mathcal{P}}$

---

**Theorem 6 (Nakamoto and Watanabe [8]).** *Every plane quadrangulation has an inner 2-orientation.*

Let $G$ be a plane quadrangulation and $D_1$ be an inner 2-orientation in $G$. By reversal, we can obtain a different inner 2-orientation of $G$. Similarly to the case for plane quadrangulations, we can obtain the following theorem.

**Theorem 7 (Nakamoto and Watanabe [8]).** *Let $D_1$ and $D_2$ be inner 2-orientations of the same plane quadrangulation. Then, $D_1$ and $D_2$ are equivalent to each other.*

Theorem 6 clarifies the relation between orthogonal plane partitions and inner 2-orientations of plane quadrangulations.

Let $\mathcal{P}$ and $\mathcal{P}'$ be two orthogonal plane partitions. Suppose that their inner 2-orientations of $G_{\mathcal{P}}$ and $G_{\mathcal{P}'}$ can be transformed into each other by a 4-cycle reversal changing $x_i y_j x_k y_l$ into $y_l x_k y_j x_i$. This means that the inner segments $a_i, b_j, a_k, b_l$ such that $a_i, b_j, a_k, b_l$ hit $b_j, a_k, b_l, a_i$ respectively, and changing the adjacency of $a_i, b_j, a_k, b_l$ as shown in Fig 5 (the adjacency of other triangular disks) transform $\mathcal{P}$ into $\mathcal{P}'$. Notice that we don't know whether $T_1$ and $T_2$ can be simply transformed into each other. We say two triangle contact systems $T$ and $T'$ are *equivalent* and denoted by $T \sim T'$ if they can be transformed into each other by one of the above reversals among four inner segments.

Let $G$ be an oriented plane quadrangulation and let $D$ be an inner 2-orientation of $G$. If $G$ is the hit graph of an orthogonal plane partition $\mathcal{P}$, then $G$ is said to be a *hit nonoriented graph* of $\mathcal{P}$.

By Theorem 7, 6 and the uniqueness of orthogonal plane partitions, the relation between $G$ and $\Sigma$ in Corollary 3 can be translated into the relation between the equivalence classes of orthogonal plane partitions and plane quadrangulations.

**Corollary 4.** *Let $\Psi$ be the set of all orthogonal plane partitions. Let $\Lambda$ be the set of all nonoriented plane quadrangulations. Then, there exists a bijection $c : \Psi/\sim \to \Lambda$ such that for any $\mathcal{P} \in \Omega/\sim$, $c(\mathcal{P})$ is the hit nonoriented graph of $\mathcal{P}$.*

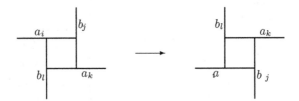

**Fig. 5** Changing the adjacency of $a_i$, $b_j$, $a_k$ and $b_l$

## 4  Remarks

In this paper we mentioned the three orthogonal tactical systems and orthogonal partitions. And we described that these correspond to inner 3-orientations of triangulations and inner 2-orientations of quadrangulations and that every inner triangulation and every inner quadrangulation has an inner 3-orientation and an inner 2-orientation, respectively.

Observe that every plane graph $G$ can be transformed into a triangulation $T$ by joining vertices to each non-triangular face. Then $G$ also has an orientation such that each inner vertex of $G$ has outdegree at most 3, though the outdegree for each vertex cannot be specified. In general, a necessary and sufficient condition for a graph $G$ to have an orientation of outdegree specified for each vertex is known in the following proposition.

**Proposition 3 (Frank [2]).** *Let $G$ be a graph and let $f : V(G) \to \{0, 1, \ldots\}$ be a function such that for each $v \in V(G)$, $0 \le f(v) \le \deg(v)$. Then $G$ has an orientation such that each vertex $v$ of $G$ has outdegree $f(v)$ if and only if for any $S \subset V(G)$,*

$$\sum_{v \in S} f(v) \ge |E([S])|,$$

*where $[S]$ denotes the subgraph of $G$ induced by $S$.*

Theorems 5 and 6 can be proved by Proposition 3 as follows. We show that a triangulation $G$ has an orientation such that each inner vertex has outdegree exactly 3 and each outer vertex has outdegree exactly 1. Consider any $S \subset V(G)$ and the subgraph $[S]$ of $G$ induced by $S$. By Euler's formula we have $E([S]) \le 3|S| - 6$ since $[S]$ is a plane graph. On the other hand, we have $\sum_{v \in S} f(v) \ge 3(|S| - 3) + 3 \times 1 = 3|S| - 6$ since $S$ contains at most 3 outer vertices. Thus by Proposition 3 the required orientation exists. Finally reversing the orientation from the ordered edges we can obtain an inner 3-orientation of $G$. We similarly prove Theorem 6 using the fact that every inner quadrangulation of $n$ vertices has at most $2n - 4$ edges.

Since every plane graph $G$ can be included in a plane graph of some triangulation and a double cover of $G$ can be represented by a barrage, an inner triangle contact system. But we want this representation to have a *free* vertex of a triangle disk (i.e. a vertex not touching others) since it has been obtained from a triangle contact system by removing several triangular disks and other edges.

Now we turn attention to bipartite graphs. Every bipartite graph G and indeed its graph of some bipartite quadrangulation. Hence G has an orientation with each inner vertex of out-degree at most 2 (though the out-degree for each vertex cannot be specified) and G can be presented as a sub-arrangement of an orthogonal plane partition which might have reed so far several segments.

When we want to consider geometric arrangements with the same graphs and with no fewer vertices by segments (i.e., an orthogonal plane partition) and triangles (i.e., a triangle contact system) represents. If there is an arrangement consisting of $n$ $k$-gons $(k \geq 4)$ in the interior of a $k'$-gon with no free vertex, then the corresponding partially oriented bipartite graph has exactly $n + k'$ vertices and $nk + k' \geq 4n + k'$ edges. Also we have a graph with $m$ vertices has at most $3m - 6$ edges. Thus $k'$ must depend on $n$ and we want $k'$.

# References

1. G. R. Brightwell and E. R. Scheinerman, SIAM Discrete Math. 6 (1993), 214–229.

2. A. Fraysseix, Orientation of graphs and boundary flow, *Graphs*, *Num.* 113 (1996), 1-142.

3. H. de Fraysseix, P. O. de Mendez, and P. Rosenstiehl, Combin. Probab. Comput. 3 (1994), 233–246.

4. H. de Fraysseix, P. O. de Mendez, and J. Pach. Representation of planar graphs by segments. *Intuitive Geometry* (Szeged, 1991), 109–117. Colloq. Math. Soc. János Bolyai 63, North-Holland and Amsterdam, 1994.

5. A. Kaneko and M. Watanabe, String and composition of graphs, 49 (1991), 231–236.

6. P. Koebe, Kontaktprobleme der konformen Abbildung, Ber. Verh. Sächs. Akademie der Wissenschaften zu Leipzig, Math. Phys. Klasse 88 (1936), 141–164.

7. A. Nakamoto, K. Ota and T. Tanuma, The e-cycle reversion in oriented triangulations, Yokohama Math. J. 44 (1997), 123–139.

8. A. Nakamoto and M. Watanabe, Cycle reversals in oriented plane quadrangulations and orthogonal plane partitions, J. Geometry 68 (2000), 200–208.

9. S. Negami and A. Nakamoto, Diagonal transformations of graphs on closed surfaces, Sci. Rep. Yokohama Nat. Univ, Sec. I No. 40 (1993), 71–97.

10. R. Tamassia and I. G. Tollis, Tessellation representations of planar graphs, Twenty-Sixth Annual Allerton Conference on Communication, Control and Computing, 48–57.

11. H. Whitney, A theorem on graphs, Ann. of Math. 32 (1931), 378–390.

# Note on Diagonal Flips and Chromatic Numbers of Quadrangulations on Closed Surfaces

Atsuhiro Nakamoto¹ and Seiya Negami²

Department of Mathematics, Osaka Keizai University, 1-2-8 Asahigaoka, Higashi-Osaka, Osaka 858 58 2, Japan.
² Department of Mathematics, Faculty of Education and Human Sciences, Yokohama National University, 79-2 Tokiwadai, Hodogaya-ku, Yokohama 208 50 Japan

**Abstract.** We shall show that if two quadrangulations on a closed surface with sufficiently large representativity can be transformed into each other by a sequence of diagonal flips, then they have the same chromatic number.

## 1 Introduction

A *quadrangulation* $G$ on a closed surface $F$ is a simple graph embedded on $F$ so that each face of $G$ is bounded by a closed walk of length 4. A *diagonal flip* of $G$ is to replace one diagonal $u_1u_3$ (or $u_2u_4$) in a hexagonal region $u_1u_2u_3u_4$ with the other $u_2u_4$, where $u_1u_2u_3u_4$ is obtained around vertex-edge, etc.

We call this set of all alternations a *diagonal flip* applied collectively. Two quadrangulations on $F$ are said to be *equivalent under diagonal flips* if they can be transformed into each other by a sequence of diagonal flips. There have been a lot of studies [2–8] on the equivalence over quadrangulations on closed surfaces under diagonal flips. However, there has never been a paper written on the relationship between diagonal flips and "colorings" of quadrangulations on closed surfaces.

A *coloring* of a graph $G$ is an assignment $f : V(G) \to \{1, 2, 3 \ldots\}$ of colors to vertices such that $f(u) \neq f(v)$ for each edge $uv \in E(G)$ and is called a $k$-coloring if $|f(V(G))| \leq k$. The chromatic number is defined as the minimum $k$ such that $G$ admits a $k$-coloring. Consider a triangulation $G$ on a closed surface $F$ which is colored by suitable $k$-coloring $f : V(G) \to \{1, 2, 3, \ldots\}$ and look at the situation around an edge $u_1u_2$. Then $f(u_2) \neq f(u_3)$ while $f(u_1)$ may be equal to $f(u_4)$. If $f(u_1) = f(u_4)$, then $f$ cannot be used as a coloring of the quadrangulation $G'$ obtained from $G$ by a diagonal flip of $u_2u_3$. Even if $G'$ admits a $k$-coloring, it is still possible to get it into forming $f$ locally.

Someone might not expect the connection between diagonal flips and the chromatic number of quadrangulations on closed surfaces. However, we shall show a surprising fact, as follows. A graph $G$ embedded on a closed surface $F$ is said to be $r$-representative if any essential simple closed curve on $F$ meets $G$ in at least $r$ points. The maximum $r$ such that $G$ is $r$-representative is called the *representativity* of $G$ (See [9] for the details of "representativity").

**Theorem 1.** *For any closed surface $F$ there exists a natural number $R = R(F)$ such that if two $R$-representative triangulations $G_1$ and $G_2$ of $F$ are equivalent under diagonal flips then $\chi(G_1) = \chi(G_2)$.*

Figure 1 presents two 3-representative triangulations on the torus, each of which can be obtained from the above rectangle by identifying two pairs of parallel sides. It is clear that they cannot be transformed into each other by using the diagonal flip. However, the former has chromatic number 3 while the latter one has 4, as the number of alternatives at the time suggests that the assumption on representativity in the above theorem is actually needed.

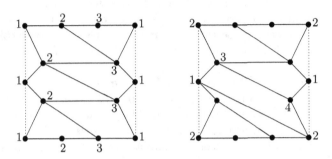

**Fig. 1** Two triangulations on the torus

For finitely many combinations of numbers, the assumption of representativity in the above two quadrangulations under diagonal flipping in general even in the case of sufficiently large representativity. However, if we confine the surfaces to be orientable then we can prove the following theorem:

**Theorem 2.** *For an orientable closed surface $S_g$ of genus $g$ there exists a natural number $R_+ = R_+(g)$ such that two $R_+$-representative triangulations $G_1$ and $G_2$ on $S_g$ are equivalent under diagonal flips if and only if $\chi(G_1) = \chi(G_2)$.*

On the other hand, the same statement for non-orientable closed surfaces is believed not to hold actually. We can prove the following with a little more than:

**Theorem 3.** *For a non-orientable closed surface $N_{2k+\varepsilon}$ of genus $2k + \varepsilon$ with $\varepsilon = 0$ or 1, there exists a natural number $R_- = R_-(2k+\varepsilon)$ such that two $R_-$-representative triangulations $G_1$ and $G_2$ on $N_{2k+\varepsilon}$ are equivalent under diagonal flips if and only if $\chi(G_1) \neq \chi(G_2) = 0$ or $= 4 - \varepsilon$.*

Both diagonal flips and chromatic numbers are said to be combinatorial objects. However, we shall introduce an algebraic invariant called the "edge parity" or "edge parity" which will mediate between those combinatorial objects in our proof of the theorems, which will be given in Section 3. These theorems must be very serious for the reader, which we shall not know with the deep results in [4] and [7].

## 2  Cycle Parities

The parity was defined first in [8] as studies on the quadrangulations and diagonal flips and has been developed in [7] as studies on colorings of quadrangulations. We shall give now whether theory of cycle parities going in the latter. We denote the fundamental group by $\pi_1(\cdot)$ and the 1-dimensional homology groups with the coefficient ring $R$ by $H_1(\cdot; R)$, assuming that the readers know those basic concepts in algebraic topology.

Let $F^2$ be a closed surface. We call a homomorphism $\tilde{\rho} : \pi_1(F^2) \to \mathbb{Z}_2$ a cycle parity over $F$. A closed curve $\ell$ is said to be even (and odd) under a cycle parity $\tilde{\rho}$ if $\tilde{\rho}([\ell]) = 0$ (or $= 1$), where $[\ell]$ denotes the homotopy class to which $\ell$ belongs. In particular, a cycle parity $\tilde{\rho}$ is said to be trivial if $\tilde{\rho}([\ell]) = 0$ for all $\ell$. Two parities $\tilde{\rho}$ and $\tilde{\rho}'$ are said to be equivalent if there is a homeomorphism $h : F \to F$ which induces an automorphism $h_* : \pi_1(F) \to \pi_1(F)$ with $\tilde{\rho} h_* = \tilde{\rho}'$.

A graph $G$ embedded on a closed surface $F$ is said to be evenly embedded on $F$ if each face is among a $2$-cell and is bounded by a closed walk of even length. It is easy to see that two closed walks in $G$ are evenly embedded graph with the same length modulo 2 if they are homotopic to each other on $F$ and that a closed walk in $G$ generates $\pi_1(F)$. So a cycle parity $\tilde{\rho}_G$ can be defined by $\tilde{\rho}_G([W] = |W| \pmod 2$ for a closed walk $W$ in $G$, where $|W|$ stands for the length of $W$. We call the $\tilde{\rho}_G$ the cycle parity of $G$. Two cycle parities $\tilde{\rho}_G$ and $\tilde{\rho}_{G'}$ of two embeddings $G$ and $G'$ are congruent if there is a homeomorphism $h : F \to F$ with $\tilde{\rho}_{h(G)} = \tilde{\rho}_{G'}$. It is clear that $\tilde{\rho}_G$ is trivial if and only if $G$ is bipartite.

Let $p : \pi_1(F) \to H_1(F; \mathbb{Z})$ and $\bar{p} : H_1(F; \mathbb{Z}) \to H_1(F; \mathbb{Z}_2)$ be the canonical reductions. Any cycle parity $\tilde{\rho} : \pi_1(F) \to \mathbb{Z}_2$ factors through $H_1(F; \mathbb{Z})$ and $H_1(F^2; \mathbb{Z}_2)$ naturally. That is, there exist two homomorphisms $\rho : H_1(F^2; \mathbb{Z}) \to \mathbb{Z}_2$ and $\bar{\rho} : H_1(F; \mathbb{Z}_2) \to \mathbb{Z}_2$ such that $\tilde{\rho} = \rho \cdot p$ and $\rho = \bar{\rho} \cdot \bar{p}$. Thus, $\tilde{\rho}$ and $\rho$ can be determined uniquely by the reduction $\bar{\rho} : H_1(F; \mathbb{Z}_2) \to \mathbb{Z}_2$, which belongs to the $\mathbb{Z}_2$-cohomology group $H^1(F; \mathbb{Z}_2)$ finally. It may be preferable to call $\rho$ or $\bar{\rho}$ the cycle parity rather than $\tilde{\rho}$ when we discuss them roughly.

Let $S_g$ denote the orientable closed surface of genus $g$ which can be obtained as the sphere with $g$ handles. Choose $g$ pairs of simple closed curves $\{a_i, b_i\}$ on $S_g$ corresponding to one of the handles, so that $a_i$ and $b_i$ cross exactly once transversely for $i = 1, \ldots, g$ and both of $a_i$ and $b_i$ cross no other $a_j$ and $b_j$ with $i \neq j$. The system $\{a_1, b_1, \ldots, a_g, b_g\}$ presents a base of $H_1(S_g; \mathbb{Z})$ and the parity $\rho$ (and also $\tilde{\rho}$ and $\bar{\rho}$) can be determined by $(\rho([a_1]), \rho([b_1]), \ldots, \rho([a_g]), \rho([b_g]))$. That is, we denote $\rho = (\rho([a_1]), \rho([b_1]), \ldots, \rho([a_g]), \rho([b_g]))$, which is a sequence of $0$'s and $1$'s.

**Theorem 4.** *Any non-trivial cycle parity $\rho$ on $S_g$ with $g \geq 1$ is equivalent to* $(1, 0, \ldots, 0)$.

Let $N_k$ denote the non-orientable closed surface of genus $k$ which can be obtained from the sphere by adding $k$ cross-caps. Below we shall classify the non-orientable closed surfaces into two classes, depending on the parity of their general "type" $N_{2k+1}$ can be regarded as the orientable closed surface $S_k$ with one crosscap added. We choose a 1-sided simple closed curve along the crosscap and pairs of simple closed curves $\{a_i, b_i\}$ on $S_k$ as well as in the previous, so that crosses no $a_i$ and no $b_i$. Then the system $\{x, a_1, b_1, \ldots, a_k, b_k\}$ forms a base of $H_1(N_{2k+1}; \mathbf{Z})$ and $[x]$ is an unique element of order 2 in $H_1(N_{2k+1}; \mathbf{Z})$. We express a cycle parity $\rho$ as $(\rho[x]), \rho[a_1]), \rho[b_1]), \ldots, \rho[a_k]), \rho[b_k])$.

**Theorem 5.** Any non-trivial cycle parity $\rho$ on $N_{2k+1}$ with $k \geq 1$ is given exactly one of $A = (1, 0, 0, \ldots, 0)$, $B = (1, 1, 0, \ldots, 0)$ and $C = (0, 1, 0, \ldots, 0)$.

The "even type" $N_{2k}$ can be generated into one Klein bottle and their orientable closed surface $S_{k-1}$. Choose a pair of simple closed curves $\{m, \ell\}$ on the Klein bottle so that $m$ is 2-sided and cuts open the Klein bottle into an annulus and $\ell$ crosses $m$ transversely exactly once. Then we take a system $\{m, \ell, a_1, b_1, \ldots, a_{k-1}, b_{k-1}\}$ of simple closed curves on $N_{2k}$. This forms a base of $H_1(N_{2k}; \mathbf{Z})$ and $[m]$ is an unique element of order 2 in $H_1(N_{2k}; \mathbf{Z})$. We write $\rho = (\rho[m]), \rho[\ell]), \rho[a_1]), \rho[b_1]), \ldots, \rho[a_{k-1}]), \rho[b_{k-1}])$.

**Theorem 6.** Any non-trivial cycle parity $\rho$ on $N_{2k}$ with $k \geq 2$ is given exactly one of $D = (0, 0, 1, 0, \ldots, 0)$, $E = (0, 1, 0, 0, \ldots, 0)$ and $F = (1, 0, 0, 0, \ldots, 0)$. For the Klein bottle $N_2$, $\rho$ is given exactly one of $E = (0, 1)$ and $F = (1, 0)$.

A cycle parity over $N_k$ is said to be of type A, B, C, D, E or F if it is congruent to each of the cycle parities A to F in Theorems 5 and 6.

# 3  Proof of Theorems

We shall prove Theorems 1, 2 and 3 though this section introducing some known results we need in order.

Following the theorem presents a fundamental result in the theory of diagonal flips in triangulations on closed surfaces, shown in Nakamoto [4]. It's implies that the cycle parity is complete invariant for the equivalence over quadrangulations with sufficiently large representativity.

**Theorem 7 (Nakamoto [4]).** For any closed surface $\overline{F}$, there exists a natural number $M = M(\overline{F})$ such that two quadrangulations $G_1$ and $G_2$ of $\overline{F}$ with $|V(G_1)| = |V(G_2)| \geq M$ are equivalent under diagonal flips if and only if they have the same cycle parities.

On the other hand, Hutchinson [1] has proved that there exists a natural number $H_+ = H_+(g)$ such that every $H_+$-representative quadrangulation on $S_g$ is 3-colorable. Her result on quadrangulations coincides with the bipartite theorem for 3-chromatic. Since the are only two cycle parity classes over an orientable closed surface, these two alternatives must correspond their trivial cycle parity and a non-trivial cycle parity respectively. Thus, her formula holds with $H_+(g) = $

$\max \{M(S_g), H_+(g)\}$ si mean $R_+$- ep sentati veg has at least $R$ veti es.
Theorem 1 follows from it.

For non orientable close surfaces, we need the following theorem which characterizes the maximum number of adding ulations with sufficiently large rep-
esentativity in terms of the parities

**Theorem 8 (Nakamoto, Negami and Ota [7]).** *The exists a number* $H_- = H_-(k)$ *such that any* $H_-$-*representation quadrangulation* $G$ *on* $N_k$ *is 4-bi colorable:*

  (i) $G$ 2-chromatically odd only if $\rho_G$ is true.

  (ii) $G$ 3-chromatically odd only if $\rho_G$ is of a type $C$, $D$ or $E$.

  (iii) $G$ 4-chromatically odd only if $\rho_G$ is of a type $A$, $B$ or $F$.

Since we quasi identity quadrangulations by recognizing cycle parities For-
em 1 holds with $R(N_k) = H_-(k)$. By theorem then $H_-$- representive quad
angulation on $N_{2k+1}$ has cycle parity type Ci ftis 3-chromatic $\rho$ while
sub quadrangulation on $N_{2k}$ has cycle parity of type F ftis 4-chromatic
It is simple that two $H_-(2k+\varepsilon)$- representive quadrangulations $G_1$ and $G_2$
have common cycle parities if $\chi(G_1) = \chi(G_2) = 4-\varepsilon$ Furthermore if a
$R_-$-representation with $R_-(k) = \max\{M(N_k), H_-(k)\}$ then they at least
$M(N_k)$ vertices and hence they are equi valent under diagonal flips Theorem
7 Hs Theorem 3 follows. □

*Remark.* We have concluded that two $R_+$- or $R_-$- representation quadrangulations $G_1$ and $G_2$ can be transformed into each other by sequence of diagonal
flips. However the quadrangulations appearing results here may not be
$R_+$ or $R_-$- representative and the chromatic numbers may not be equal to
those of $G_1$ and $G_2$.

# References

1. J P Hutchinson "The edge-coloring number embedded on surface which all face even side d J. Combin. Theory, Ser. B 65 (1995), 139–155.
2. A. Nakamoto "Triangulations and quadrangulations of surfaces", Doctoral thesis, Kobe university, 1996
3. A Nakamoto, Diagonal transformations quadrangulations of surfaces J. Graph Theory 21 (1996), 289–299.
4. A Nakamoto Diagonal transformations the properties of quadrangulations on surfaces J. Combin. Theory, Ser. B 67 (1996), 202–211.
5. A Nakamoto and K. Ota, Diagonal transformations quadrangulations and Dehn twists preserving properties J. Combin. Theory, Ser. B 69 (1997), 13–41.
6. A. Nakamoto and K Ota, Diagonal transformations graphs and Dehn twists of surfaces J. Combin. Theory, Ser. B 70 (1997), 292–300.
7. A, Nakamoto Negami and K Ota Chromatic numbers and cycle parities of quadrangulations on nonorientable closed surfaces admitted
8. S Negami and Nakamoto, Diagonal transformation of graphs on closed sur-faces Sci. Rep. Yokohama Nat. Univ., Sec. I 40 (1993), 71–97.

9. N. Bertand R. Vitra, Representativity of surface embedding, *Paths, Flows, and VLSI-Layout ("Algorithm and Combinatorics" Vol. 9* (Kte, Lovás, H. J. Prömel and A. Shijver eds.) Spier-Verlag, Berlin (1990), 293–328.

# An Extension of Cauchy's Arm Lemma with Application to Curve Development

Joseph O'Rourke*

Dept. Comput. Sci., Smith College
Northampton, MA 01063, USA
orourke@cs.smith.edu

**Abstract.** Cauchy's "Arm Lemma" may be generalized to permit non-convex "openings" of a planar convex chain. Although this (and further extensions) were known, no proofs have appeared in the literature. Here two induction proofs are offered. The extension can then be employed to establish that a curve that is the intersection of a plane with a convex polyhedron "develops" without self-intersection.

## 1 Introduction

Cauchy used his famous "Arm Lemma" to establish the rigidity of convex poly-hedra [Cro97, p. 228]. Cauchy's lemma says that if $n - 2$ consecutive angles of a convex polygon of $n$ vertices are opened but not beyond $\pi$, keeping all but one edge length fixed and permitting that "missing" edge $e$ to vary in length, then $e$ lengthens (or retains its original length). This lemma has numerous applications, including at least one to curve development [OS89]. Here we describe an appar-ently little-known generalization of Cauchy's lemma to permit opening of the angles beyond $\pi$, as far reflex as they were originally convex. The conclusion re-mains the same: $e$ cannot shorten. We then apply this extension to another result on curve development, that "slice curves" develop without self-intersection.

The first part of this paper (Section 2) concentrates on the generalization of Cauchy's lemma. The issue of self-intersection is addressed in Section 3, and the curve development result is discussed in Section 4.

## 2 Cauchy's Arm Lemma Extended

Let $A = (a_0, a_1, \ldots, a_n)$ be an $n$-link polygonal chain in the plane with $n$ fixed edge lengths $\ell_i = |a_i a_{i+1}|$, $i = 0, \ldots, n - 1$. We call the vertices $a_i$ the *joints* of the chain, $a_0$ (which will always be placed at the origin) the *shoulder*, and $a_n$ the *hand*. Define the *turn angle* $\alpha_i$ at joint $a_i$, $i = 1, \ldots, n - 1$ to be the angle in $[-\pi, \pi]$ that turns the vector $a_i - a_{i-1}$ to $a_{i+1} - a_i$, positive for left (counterclockwise) and negative for right (clockwise) turns.

---

* Supported by NSF grant CCR-9731804.

J. Akiyama, M. Kano, and M. Urabe (Eds.): JCDCG 2000, LNCS 2098, pp. 280–291, 2001.
© Springer-Verlag Berlin Heidelberg 2001

Define an open polygonal chain $A$ to be *convex* if its joints determine a (nondegenerate) convex polygon, i.e., all joints are distinct points (in particular, $a_n \neq a_0$), all joints lie on the convex hull of $A$ and they do not all lie on a line. Note there is no chain link between $a_n$ and $a_0$. The turn angles for a convex chain all lie in $[0, \pi)$; but note this is not a sufficient condition for a chain to be convex, for it is also necessary that the angles at $a_0$ and $a_n$ be convex.

We can view the configuration of a polygonal chain $A$ to be determined by two vectors: the fixed edge lengths $L = (\ell_0, \ldots, \ell_{n-1})$ and the variable turn angles $\alpha = (\alpha_1, \ldots, \alpha_{n-1})$, with the convention that $a_0$ is placed at the origin and $a_n$ horizontally left of $a_0$. Let $\mathcal{C}_L(\alpha) = A$ be the configuration so determined. We use $\alpha$ to represent the angles of the initial configuration, and $\beta$ and $\gamma$ to represent angles in a reconfiguration.

Let $D(A)$ be the open disk of radius $|a_n a_0|$ (the length of the "missing" link) centered on the shoulder joint $a_0$. We will call $D(A)$ the *forbidden (shoulder) disk*. We may state Cauchy's arm lemma in the following form:

**Theorem 0.** Cauchy's Arm Lemma. *If $A = \mathcal{C}_L(\alpha)$ is a convex chain with fixed edge lengths $L$, and turn angles $\alpha$, then in any reconfiguration to $B = \mathcal{C}_L(\beta)$ with turn angles $\beta = (\beta_1, \ldots, \beta_{n-1})$ satisfying*

$$\beta_i \in [0, \alpha_i] \tag{1}$$

*we must have $|b_n b_0| \geq |a_n a_0|$, i.e., the hand cannot enter the forbidden disk $D(A)$.*

Cauchy's lemma is sometimes known as Steinitz's lemma, because Steinitz noticed and corrected an error in the proof a century after Cauchy [Cro97, p. 235]. Many proofs of Cauchy's lemma are now known, e.g., [SZ67,Sin97] and [AZ98, p. 64].

Our generalization of Cauchy's lemma replaces the 0 in Eq. (1) by $-\alpha_i$, and is otherwise identical:

**Theorem 1.** Cauchy Extension. *If $A = \mathcal{C}_L(\alpha)$ is a convex chain with fixed edge lengths $L$, and turn angles $\alpha$, then in any reconfiguration to $B = \mathcal{C}_L(\beta)$ with turn angles $\beta = (\beta_1, \ldots, \beta_{n-1})$ satisfying*

$$\beta_i \in [-\alpha_i, \alpha_i] \tag{2}$$

*we must have $|b_n b_0| \geq |a_n a_0|$, i.e., the hand cannot enter the forbidden disk $D(A)$.*

The intuition, illustrated in Fig. 1, is perhaps best captured by Chern [Che67, p. 36]: "if an arc is 'stretched,' the distance between its endpoints becomes longer." See also our Java applet illustrating the theorem and its corollaries.[1] Although the chain may become nonconvex, Eq. (2) ensures that the movement constitutes a form of straightening. Note that Theorem 1 makes no claim about

---

[1]  http://cs.smith.edu/~orourke/Cauchy/ .

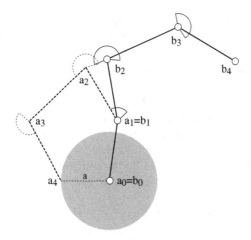

**Fig. 1.** A reconfiguration of a 4-link convex chain $A$ to chain $B$, satisfying Eq. (2), leaves $b_n$ outside the forbidden disk (shaded). The valid angle ranges are marked by circular arcs.

steadily increasing hand-shoulder separation during some continuous movement to $B$; indeed a continuous opening could first increase and later decrease the separation. Rather the claim is that a final configuration satisfying Eq. (2) cannot place the hand in the forbidden disk.

As with Cauchy's arm lemma, one may expect many different proofs of such a fundamental result. We sketch three proofs of Theorem 1 here; for full details, see [O'R00]. First we discuss a proof by S. S. Chern, followed by two induction proofs.

## 2.1   Chern's Proof of Theorem 1

There are three possible generalizations of Cauchy's Theorem 0: to smooth curves, to space curves, and to nonconvex angles. All three generalizations hold, and there is some confusion in the literature on when each was established. Certainly the primary source is acknowledged to be a 1921 paper by Axel Schur [Sch21]. Connelly mentions in [Con82, p. 30] that Schur generalized Cauchy's lemma to the smooth case. Some other mentions of Schur's theorem in the literature, e.g., in [Gug63, p.31], phrase it as the smooth, planar equivalent of Cauchy's lemma, which does not capture the nonconvexity permitted in the statement of Theorem 1. But others, notably the exposition by Chern [Che67], state it as a generalization also to space curves, and employ the absolute value of curvature, implicitly permitting nonconvexity.

Schur states his main result as follows:

> SATZ: "Verbiegt man eine ebene doppelpunktlose Kurve, die mit ihrer Sehne einen konvexen geschlossenen doppelpunktlosen Linienzug bildet, so wird dabei die Sehne länger."

Roughly: "If one deforms a simple curve in the plane that (together with its chord) forms a convex simple closed curve, then the chord becomes longer."

Later he makes clear that his "deformation" is deformation into space without changing the curvature: "By deformation of a curve, we mean that after deforming the tangent planes the lengths and angles of the line segments should remain unchanged. If the curvature of the curve is continuous, then this is equal to demanding that the curvature doesn't change." It seems, then, that Schur did not generalize the opening aspect of Cauchy's lemma (to which he does not refer), but rather extended to deformations into space without angle changes. Moreover, it appears his proof (and claim) are for polygonal curves, not smooth curves.

Chern states Schur's theorem as follows [Che67, p. 36]:

> THEOREM: Let $C$ be a plane arc with the curvature $k(s)$ which forms a convex curve with its chord $AB$. Let $C^*$ be an arc of the same length referred to the same parameter $s$ such that its curvature $k^*(s) \leq k(s)$. If $d^*$ and $d$ are the lengths of the chords joining their endpoints, then $d \leq d^*$.

He makes clear that by "curvature" he means the absolute value of the curvature, thus explicitly including nonconvex openings. Although he attributes the theorem to Schur, neither his theorem statement, and certainly not his proof, match Schur's. Let us then refer to this full generalization of Cauchy's theorem as the Chern-Schur theorem.

Chern's proof is carried out largely in the domain of the "tangent indicatrixes" of $C$ and $C^*$, and relies on smoothness of the curves. He claims the theorem for "sectionally smooth curves" [p. 39], e.g., for polygonal curves, but without proof. Thus, as far as I can determine, no proof of Theorem 1 has appeared in the literature.

However, Chern's differential geometry proof may be specialized to our planar, nonsmooth instance, and although some of its elegance is lost, it can be carried through to establish Theorem 1 [O'R00]. Therefore Theorem 1 follows in spirit if not in fact from the Chern-Schur theorem.

The next section proves the theorem via an induction proof that makes explicit many relationships only implicit in Chern's proof.

## 2.2   First Induction Proof of Theorem 1

Although we impose no restriction on self-intersection of the chain, we will show in Theorem 3 that the chain remains simple. Note that, because we fix $a_0$ to the origin, and the first turn angle is at joint $a_1$, in any reconfiguration the first edge of the chain is fixed.

The first induction proof of Theorem 1 requires a few preparatory lemmas, whose proofs may be found in [O'R00]. We start with the simple observation that negating the turn angles reflects the chain.

**Lemma 1.** *If a chain $A = C_L(\alpha)$ is reconfigured to $B = C_L(\beta)$ with $\beta_i = -\alpha_i$, $i = 1, \ldots, n-1$, then $B$ is a reflection of $A$ through the line $M$ containing $a_0 a_1$, and $|b_n b_0| = |a_n a_0|$.*    □

Call a reconfiguration $B = C_L(\beta)$ of a convex chain $A = C_L(\alpha)$ which satisfies the constraints of Eq. (2) a *valid reconfiguration*, and call the vector of angles $\beta$ *valid angles*. Define the *reachable region* $R_L(\alpha)$ for a convex chain $A = C_L(\alpha)$ to be the set of all hand positions $b_n$ for any valid reconfiguration $B = C_L(\beta)$. One can view Theorem 1 as the claim that $R_L(\alpha) \cap D(A) = \emptyset$. (Again, see our applet[1] for an illustration of this claim.) It is well known [HJW84][O'R98, p. 326] that the reachable region for a chain with no angle constraints is a shoulder-centered closed annulus, but angle-constrained reachable regions seem unstudied.

For the first proof we need two technical lemmas.

**Lemma 2.** *The configuration of a chain $A = C_L(\alpha)$ is a continuous function of its turn angles $\alpha$.*    □

**Lemma 3.** *$R_L(\alpha)$ is a closed set.*    □

We use this lemma to help identify, among potential counterexamples, the "worst" violators. Define a configuration $B = C_L(\beta)$ to be *locally minimal* if there is a neighborhood $N$ of $\beta$ such that, for all $\beta' \in N$, the determined hand position $b_n'$ is no closer to the shoulder: $|b_n' a_0| \geq |b_n a_0|$. Thus the hand's distance to the shoulder is locally minimal.

**Lemma 4.** *Let $B = C_L(\beta)$ be a reconfiguration of convex chain $A = C_L(\alpha)$ with $b_n \in D(A)$. Then either $b_n = a_0$, or there is some locally minimal configuration $B' = C_L(\beta')$ with $b_n' \in D(A)$.*    □

The above lemma will provide a "hook" to reduce $n$ in the induction step. We separate out the base of the induction in the next lemma.

**Lemma 5.** *Theorem 1 (Cauchy Extension) holds for $n = 2$.*
**Proof:** A 2-link chain's configuration is determined by single angle at $a_1$. The reachable region is a single circular arc exterior to $D(A)$, centered on $a_1$, of radius $\ell_1$. See Fig. 2.    □
We now prove Theorem 1 by induction.
**Proof:** Lemma 5 establishes the theorem for $n = 2$. Assume then that the theorem holds for all chains of $n-1$ or fewer links. We seek to establish it for an $n$-link chain $A = C_L(\alpha)$, $n > 2$. Assume, for the purposes of contradiction, that $A$ may be reconfigured so that the hand falls inside the forbidden disk $D(A)$. We seek a contradiction on a shorter chain. By Lemma 4, one of two cases holds: the hand reaches $a_0$, or there is a locally minimal configuration.

1. Suppose $B = C_L(\beta)$ is such that $b_n = a_0$, as illustrated in Fig. 3(b). We know that $\ell_{n-1} = |a_{n-1} a_n| < |a_{n-1} a_0|$ by the triangle inequality: By definition, $A$ forms a nondegenerate convex polygon, so the triangle $\triangle(a_0, a_{n-1}, a_n)$ is

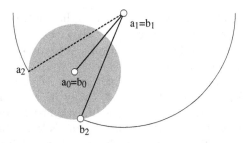

**Fig. 2.** $R_L(\alpha)$ for a 2-link chain is a circle arc centered on $a_1 = b_1$.

nondegenerate (see Fig. 3(a)). Now consider the chains $A'$ and $B'$ obtained by removing the last links $a_{n-1}a_n$ and $b_{n-1}b_n$. First, $A'$ is a convex chain of $n - 1$ links, so the induction hypothesis applies and says that $A'$ cannot be validly reconfigured to place $b_{n-1}$ closer to $a_0$ than $a' = |a_{n-1}a_0|$. $B'$ places $b_{n-1}$ at distance $\ell_{n-1}$ from $a_0$, which we just observed is less than $a'$. It remains to argue that $B'$ is a valid reconfiguration of $A'$, i.e., that it satisfies Eq. (2). However, this is satisfied for $i = 1, \ldots, n - 2$ because these angles are not changed by the shortening, and after shortening there is no constraint on $\beta_{n-1}$. Thus $B'$ is a valid reconfiguration of $A'$ but places the hand in the forbidden disk $D(A')$, a contradiction.

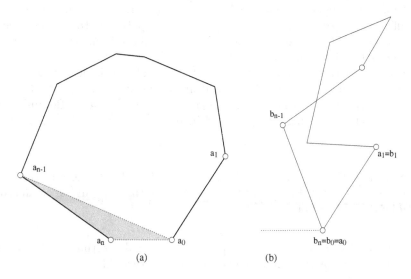

(a)    (b)

**Fig. 3.** Case 1: $b_n = a_0$. (Drawing (b) is not accurate.)

2. We may henceforth assume, by Lemma 4, that there is a locally minimal configuration $B = \mathcal{C}_L(\beta)$ that places $b_n \in D(A)$. Again we seek to shorten the chain and obtain a contradiction.

First we establish that at least one[2] $\beta_i$ is at the limit of its valid turn range: $\beta_i = \pm\alpha_i$. Suppose to the contrary that all $\beta_i$, $i = 1, \ldots, n-1$, are strictly interior to their allowable turn ranges: $\beta_i \in (-\alpha_i, \alpha_i)$. Let $M$ be the line containing $b_0 b_n$. Consider two cases:

(a) Some $b_i$, $i = 1, \ldots, n-1$, does not lie on $M$. Then because $\beta_i$ is not extreme, the subchain $(b_{i+1}, \ldots, b_n)$ may be rotated about $b_i$ in both directions. Because $b_i$ is off $M$, one direction or the other must bring $b_n$ closer to $b_0$, contradicting the fact that $b_n$ is locally minimal.

(b) All $b_i$ lie on $M$. Then there must be some $b_i$ which is extreme on $M$. For this $b_i$, $\beta_i = \pm\pi$. But $\alpha_i \in [0, \pi)$: the nondegeneracy assumption bounds $\alpha_i$ away from $\pi$, and so bounds $\beta_i$ away from $\pm\pi$.

Henceforth let $b_i$ be a joint whose angle $\beta_i$ is extreme. If $\beta_i = -\alpha_i$, then reflect $B$ about $b_0 b_n$ so that $\beta_i = \alpha_i$ is convex. By Lemma 1, this does not change the distance from $b_n$ to the shoulder, so we still have $b_n \in D(A)$.

We are now prepared to shorten the chains. Let $A'$ and $B'$ be the chains resulting from removing $a_i$ and $b_i$ from $A$ and $B$ respectively:

$$A' = (a_0, a_1, \ldots, a_{i-1}, a_{i+1}, \ldots, a_n) \tag{3}$$
$$B' = (b_0, b_1, \ldots, b_{i-1}, b_{i+1}, \ldots, b_n) \tag{4}$$

A crucial point to notice is that $|b_{i-1} b_{i+1}| = |a_{i-1} a_{i+1}|$ because $\beta_i = \alpha_i$; this was the reason for focusing on an extreme $\beta_i$. Therefore $B'$ is a reconfiguration of $A'$. Of course both $A'$ and $B'$ contain $n-1$ links, so the induction hypothesis applies. Moreover, because $i \leq n-1$, the $b_i$ removed does not affect the position of $b_n$. So $b_n \in D(A)$ by hypothesis. To derive a contradiction, it only remains to show that $B'$ is a valid reconfiguration of $A'$, i.e., one that satisfies the turn constraints (2).

Let $\alpha'_{i+1}$ be the turn angle at $a_{i+1}$ in $A'$. We analyze this turn angle in detail, and argue later that the situation is analogous at $a_{i-1}$. Let $\theta$ be the angle of the triangle $\triangle_i = \triangle(a_i, a_{i+1}, a_{i-1})$ at $a_{i+1}$; see Fig. 4(a). Because $A$ is a convex chain, cutting off $\triangle_i$ from $A$ increases the turn angle at $a_{i+1}$ in $A'$:

$$\alpha'_{i+1} = \theta + \alpha_{i+1} \tag{5}$$

Now consider the turn angle $\beta'_{i+1}$ at $b_{i+1}$ in $B'$. Although here the turn could be negative, as in Fig. 4(b), it is still the case that the turn is advanced by $\theta$ by the removal of $\triangle_i$:

$$\beta'_{i+1} = \theta + \beta_{i+1} \tag{6}$$

We seek to prove that $\beta'_{i+1} \in [-\alpha'_{i+1}, \alpha'_{i+1}]$. Substituting the expressions from Eqs. (5) and (6) into the desired inequality yields:

$$-\alpha_{i+1} - 2\theta \leq \beta_{i+1} \leq \alpha_{i+1} \tag{7}$$

And this holds because $\theta > 0$ and $\beta_{i+1} \in [-\alpha_{i+1}, \alpha_{i+1}]$ (because $B$ is a valid reconfiguration of $A$). The intuition here is that the nesting of the turn angle

---

[2] In fact I believe that all must be extreme, but the proof only needs one.

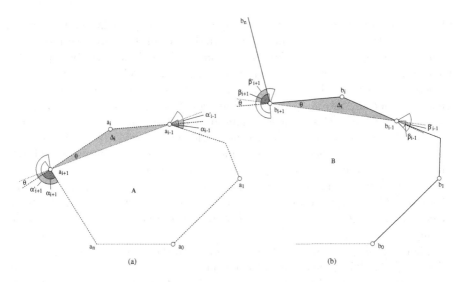

**Fig. 4.** (a) Shortening the chain $A$ by removal of $a_i$ determines new, larger turn angles $\alpha'_{i+1}$ and $\alpha'_{i-1}$ at $a_{i+1}$ and $a_{i-1}$ respectively. (b) Here the turn angles $\beta_{i+1}$ and $\beta'_{i+1}$ are negative.

ranges at $a_{i+1}$ in $A$ and $A'$ (evident in Fig. 4(a)) carries over, rigidly attached to $\triangle_i$, to $B$, so that satisfying the tighter constraint in $B$ also satisfies the looser constraint in $B'$.

Although the situation is superficially different at $a_{i-1}$ because our definition of turn angle depends on the orientation of the chain, it is easily seen that the turn constraint is identical if the orientation is reversed.

We have thus established that $B'$ is a valid reconfiguration of $A'$. By the induction hypothesis, the hand $b_n$ of $B'$ cannot enter the forbidden disk $D(A')$. By assumption $b_n \in D(A)$. But note that $D(A') = D(A)$, because, as mentioned above, deleting $a_i$ and $b_i$ does not affect the positions of $a_n$ and $b_n$ respectively. This contradiction shows that our assumption that $b_n \in D(A)$ cannot hold, and establishes the theorem.

$\square$

The following corollary extends the distance inequality to every point of the chain.

**Corollary 2** *Let $A = \mathcal{C}_L(\alpha)$ be a convex chain as in Theorem 1, and let $p_1$, $p_2 \in A$ be any two distinct points of the chain. Then in any valid reconfiguration $B$, the points $q_1$, $q_2 \in B$ corresponding to $p_1$ and $p_2$ satisfy $|q_1 q_2| \geq |p_1 p_2|$, i.e., they have not moved closer to one another.* $\square$

## 2.3  Second Induction Proof of Theorem 1

We now sketch a second proof, which avoids reliance on locally minimal configurations. Unlike the previous proof, this one employs Cauchy's Arm Lemma

(Theorem 0), rather than proving it along the way. The proof is again inductive, by contradiction from a shortened chain, and relies on the same detailed argument concerning the turn angle ranges. None of those details will be repeated.

**Proof:** Let $A = \mathcal{C}_L(\alpha)$ be the given convex chain, and $C = \mathcal{C}_L(\gamma)$ a valid reconfiguration that places $c_n \in D(A)$, in contradiction to the theorem. We first construct an "intermediate" configuration $B = \mathcal{C}_L(\beta)$ with $\beta_i = |\gamma_i|$ for all $i = 1, \ldots, n-1$, i.e., $B$ is a convex chain formed by flipping all turns in $C$ to be positive. Note that, because $\gamma$ is a valid angle vector for $A$, $\gamma_i \in [-\alpha_i, \alpha_i]$, and so $\beta_i \in [0, \alpha_i]$. As this is exactly the Cauchy arm opening condition, Eq. (1), we may apply Theorem 0 to conclude that $b = |b_n b_0| \geq |a_n a_0| = a$.

Now we focus attention on chains $B$ and $C$. Because $\gamma_i = \pm \beta_i$, $\gamma_i \in [-\beta_i, \beta_i]$. Therefore, $C$ is a valid reconfiguration of $B$. But here is the point: every angle $\gamma_i$ of $C$ is extreme with respect to $B$, and so there is no need to invoke local minimality.

Choose an $i$ and remove $b_i$ from $B$ and $c_i$ from $C$, obtaining shorter chains $B'$ and $C'$. Applying the argument from the previous section verbatim, we conclude that $C'$ is a valid reconfiguration of $B'$. But because $B'$ has $n-1$ links, the induction hypothesis applies and shows that $c_n$ cannot enter the forbidden disk $D(b)$, with $b = |b_n b_0|$. Because $b \geq a$, $c_n$ cannot be in $D(A)$ either. This contradicts the assumption and establishes the theorem.    □

## 3    Noncrossing of Straightened Curve

Define a polygonal chain to be *simple* if nonadjacent segments are disjoint, and adjacent segments intersect only at their single, shared endpoint. By our nondegeneracy requirement, convex chains are simple. In particular, any opening of a convex chain via Cauchy's arm lemma (Theorem 0) remains simple because it remains convex. We now establish a parallel result for the generalized straightening of Theorem 1. We generalize slightly to permit the convex chain to start with the hand at the shoulder.

**Theorem 3.** *If $A = (a_0, \ldots, a_n) = \mathcal{C}_L(\alpha)$ is a closed convex chain with $n$ fixed edge lengths $L$ and turn angles $\alpha$, closed in the sense that $a_n = a_0$, then any valid reconfiguration to $B = \mathcal{C}_L(\beta)$ is a simple polygonal chain.*

**Proof:** Suppose to the contrary that $B$ is nonsimple. Let $q_2$ be the first point of $B$, measured by distance along the chain from the shoulder $b_0$, that coincides with an earlier point $q_1 \in B$. Thus $q_1$ and $q_2$ represent the same point of the plane, but different points along $B$. See Fig. 5. Because $B$ is nonsimple, these "first touching points" exist, and we do not have both $q_1 = b_0$ and $q_2 = b_n$ (because that would make $B$ a simple, closed chain). Let $p_1$ and $p_2$ be the points of $A$ corresponding to $q_1$ and $q_2$.

Corollary 2 guarantees that $|q_1 q_2| \geq |p_1 p_2|$. But $|q_1 q_2| = 0$, and because the $q$'s do not coincide with the original hand and shoulder, $|p_1 p_2| > 0$. This contradiction establishes the claim.    □

One could alternatively prove this theorem by induction on the length of the chain, showing that in a continuous motion to $B$, the first violation of simplicity is either impossible by the induction hypothesis, or directly contradicts Theorem 1.

**Corollary 4** *A valid reconfiguration of an open convex chain remains simple.*
**Proof:** The proof of Theorem 3 guarantees that even the final missing edge is not crossed, so the corollary is obtained by simply ignoring that last edge.    □

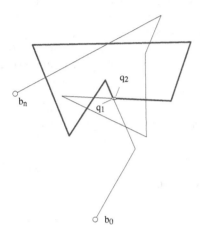

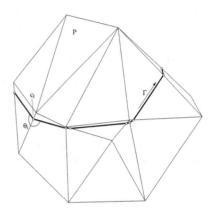

**Fig. 5.** Violation of Theorem 1. $q_1 = q_2$ is the first point of self-contact; the initial portion of $B$, up to $q_2$, is highlighted.

**Fig. 6.** $\Gamma$ is the intersection of a plane (not shown) with polyhedron $P$.

## 4   Application to Curve Development

Although the intersection of a plane $\Pi$ with a convex polyhedron $P$ is a convex polygon $Q$ within that plane, on the surface of $P$, this *slice curve* (Fig. 6) can be nonconvex, turning both left and right. The "development" of a curve on a plane is determined by its turning behavior on the surface. Thus slice curves develop (in general) to nonconvex, open chains on a plane. An earlier result is that a closed convex polygonal curve on a convex polyhedron, i.e., one whose turns are *all* leftward on the surface, develops to a simple polygonal chain [OS89]. That proof relied on Cauchy's Arm Lemma (Theorem 0). We claim that slice curves also develop without self-intersection, despite their nonconvexity. The proof relies on Theorem 1, the extension of Cauchy's lemma.

Let $\Gamma$ be an oriented slice curve on the surface of $P$. Let $c_0, c_1, \ldots, c_n$ be the *corners* of $\Gamma$, the points at which $\Gamma$ crosses a polyhedron edge with a dihedral angle different from $\pi$, or meets a polyhedron vertex. Define the *right surface angle* $\theta(p)$ at a point $p \in \Gamma$ to be the total incident face angle at $p$ to the right of the directed curve $\Gamma$ at $p$. Only at a corner $c_i$ of $\Gamma$ is the right surface angle $\theta_i$ different from $\pi$. Note that $\theta_i$ could be greater or less than $\pi$, i.e., the slice curve could turn right or left on the surface.

Define the *right development* of $\Gamma$ to be a planar drawing of the polygonal chain $\Gamma$ as the chain $B = (b_0, b_1, \ldots, b_n)$ with the same link lengths, $|b_i b_{i+1}| = |c_i c_{i+1}|$ for $i = 0, \ldots, n - 1$, and with exterior angle $\theta_i$ to the right of $b_i$ the same as the surface angle to the right of $\Gamma$ at $c_i$ on $P$, for all $i = 1, \ldots, n - 1$. Left development is defined similarly, with any curve "between" considered a development of $\Gamma$. If $\Gamma$ avoids all polyhedron vertices, then the left and right developments are identical, and so the development of $\Gamma$ is unique.

In [O'R01] I prove that the developed chain $B$ is a valid reconfiguration (in the sense used in Section 2.2, i.e., satisfies Eq. (2)) of the chain $A$ representing $Q$ in $\Pi$. Via Corollary 4, this leads to:

**Theorem 5.** *A slice curve $\Gamma = P \cap \Pi$, the intersection of a convex polyhedron $P$ with a plane $\Pi$, develops on a plane to a simple (noncrossing) polygonal curve.*

Because the Chern-Schur Theorem encompasses smooth curves, Theorem 5 should generalize to slice curves for any convex body $B$.

**Acknowledgments**

I thank Erik Demaine, Martin Demaine, Anna Lubiw, and Godfried Toussaint for a clarifying discussion. Anna Lubiw suggested the proof in Section 2.3. I thank Michael Albertson for a technical suggestion, Therese Biedl for help with German, and Veronica Morales for writing the Java applet.[1]

# References

AZ98.    M. Aigner and G.M. Ziegler. *Proofs from THE BOOK*. Springer-Verlag, Berlin, 1998.

Che67.   S.S. Chern. Curves and surfaces in Eucidean space. In S. S. Chern, editor, *Studies in Global Geometry and Analaysis*, volume 4 of *MAA Studies in Mathmatics*, pages 16–56. Math. Assoc. Amer., 1967. Also in vol. 27 of the *MAA Studies in Mathmatics, Global Differential Geometry*, 1989, pp. 99–139.

Con82.   R. Connelly. Rigidity and energy. *Invent. Math.*, 66:11–33, 1982.

Cro97.   P. Cromwell. *Polyhedra*. Cambridge University Press, 1997.

Gug63.   H.W. Guggenheimer. *Differential Geometry*. McGraw-Hill, 1963.

HJW84.   J.E. Hopcroft, D.A. Joseph, and S.H. Whitesides. Movement problems for 2-dimensional linkages. *SIAM J. Comput.*, 13:610–629, 1984.

O'R98.   J. O'Rourke. *Computational Geometry in C (Second Edition)*. Cambridge University Press, 1998.

O'R00.   J. O'Rourke. On the development of the intersection of a plane with a polytope. Technical Report 068, Dept. Comput. Sci., Smith College, Northampton, MA, June 2000. LANL arXiv cs.CG/0006035 v3, http://cs.smith.edu/~orourke/papers.html.

O'R01.   J. O'Rourke. On the development of the intersection of a plane with a poly-
         tope. *Comput. Geom. Theory Appl.*, 2001. Submitted.
OS89.    J. O'Rourke and C. Schevon. On the development of closed convex curves on
         3-polytopes. *J. Geom.*, 13:152–157, 1989.
Sch21.   A. Schur. Über die Schwarzche Extremaleigenschaft des Kreises unter den
         Kurven konstantes Krümmung. *Math. Ann.*, 83:143–148, 1921.
Sin97.   D. Singer. *Geometry: Plane and Fancy.* Springer-Verlag, Berlin, 1997.
SZ67.    I. J. Schoenberg and S.K. Zaremba. On Cauchy's lemma concerning convex
         polygons. *Canad. J. Math.*, 19:1062–1077, 1967.

# On the complexity of the union
# of geometric objects

János Pach*

Courant Institute, New York University
and Hungarian Academy of Sciences

**Abstract.** Given a family $\mathcal{C}$ of regions bounded by simple closed curves in the plane, the *complexity* of their union is defined as the number of points along the boundary of $\cup\mathcal{C}$, which belong to more than one curve. Similarly, one can define the complexity of the union of 3-dimensional bodies, as the number of points on the boundary of the union, belonging to the surfaces of at least *three* distinct members of the family. We survey some upper bounds on the complexity of the union of $n$ geometric objects satisfying various natural conditions. These problems play a central role in the design and analysis of many geometric algorithms arising in motion planning and computer graphics.

## 1 Introduction

Given a family $\mathcal{C}$ of polygons in the plane (or, a family of polyhedra in $\mathbb{R}^d$), the number of sides of their union (resp., the total number of its faces of all dimensions) is called the *combinatorial complexity*, or, simply, the *complexity* of $\cup\mathcal{C}$. This notion can be easily generalized to families of other simply shaped geometric objects with piecewise smooth boundaries. The complexity of a set is closely related to its *description size*, i.e., the number of parameters needed for its description.

Many basic problems in computational geometry related to motion planning [39, 40, 41], range searching [22, 18], computer graphics [1], and geographic information systems [6] lead to questions about the complexity of the boundary of the union of certain geometric objects. This was the motivation behind a lot of research during the past 15 years, establishing upper bounds for the complexity of the union of various objects.

We mention three simple examples.

*1.1. Linear programming.* Given a family $\mathcal{C}$ of $n$ half-spaces in $\mathbb{R}^d$, we want to maximize a linear function on the boundary of their union, $\mathrm{Bd}(\cup\mathcal{C})$. The running time of the simplex algorithm, as well as many other naive solutions to this problem, is proportional to the total number of vertices of $\mathrm{Bd}(\cup\mathcal{C})$. According

---

* Supported by the National Science Foundation (USA) and the National Fund for Scientific Research (Hungary). e-mail: pach@cims.nyu.edu

J. Akiyama, M. Kano, and M. Urabe (Eds.): JCDCG 2000, LNCS 2098, pp. 292–307, 2001.

to McMullen's Upper Bound Theorem [31], this number cannot exceed

$$\binom{n - \lceil d/2 \rceil}{\lfloor d/2 \rfloor} + \binom{n - 1 - \lceil (d-1)/2 \rceil}{\lfloor (d-1)/2 \rfloor},$$

with equality for cyclic polytopes and for all other simplicial neighborly polytopes.

*1.2. Motion planning amidst obstacles.* Let $R$ be a "robot," i.e., a convex polygonal object with a small number $r$ of sides, which is free to translate amidst a collection of $n$ convex polygonal obstacles, $C_1, C_2, \ldots, C_n$. Fix a reference point $O$ (the origin) within $R$. In order to decide whether the robot can be moved from a fixed position to another without colliding with any of the obstacles, and to plan such a motion if it exists, we want to describe the space $F$ of *free placements* of $R$, i.e., the locus of all positions of the reference point which correspond to placements of the robot, in which it does not intersect any $C_i$ (see Fig. 1). It is easy to see that $R$ intersects $C_i$ if and only if the corresponding reference point belongs to the "expanded obstacle" $C'_i = C_i \oplus (-R)$, where $\oplus$ stands for the *Minkowski sum*, i.e.,

$$C'_i = \{c - r \mid c \in C_i, r \in R\}.$$

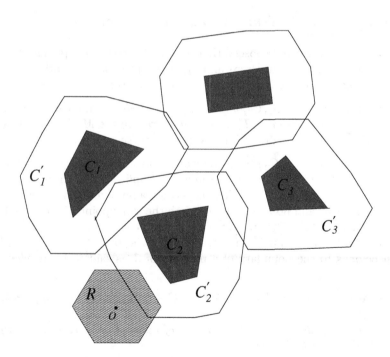

**Figure 1:** The space of free placements of the robot $R$.

In other words, the space $F$ of free placements of the robot is equal to the complement of $\cup_{i=1}^{n} C'_i$. Therefore, the running time of any algorithm for the description of $F$ is at least the number of vertices of the union of the expanded obstacles, and there are a number of efficient algorithms which almost achieve this time bound.

The above approach was initiated by Lozano-Pérez and Wesley [29], and it can also be applied to the situation when the robot is allowed to rotate [25], moves in three dimensions [4], has many arms [17], etc.

*1.3. Overlay of maps.* In most geographic information systems the data is stored in several thematic maps, each representing only one kind of information. E.g., there are separate maps for average temperature, for average precipitation, altitude, etc. A *face* of a map corresponds to a region where the given parameter is roughly constant. One often has to combine the information, i.e., to compute the overlay of two or more maps. This procedure results in a new map, more complex than the original ones. Consider, for example, two maps, and put their faces in a single list $F_1, F_2, \ldots, F_n$. Let $F_i'$ denote a region obtained from $F_i$ by very slightly shrinking it. It is easy to see now that the total number of vertices of the new combined map is proportional to the complexity of the boundary of $\cup_{i+1}^n F_i'$. Indeed, each vertex of the overlay will give rise to a "hole" determined by the $F_i'$.

# 2    An example – Translational motion planning

First we make some simple observations related to the motion planning problem stated above (1.2). Two curves, $\gamma_1$ and $\gamma_2$, are said to *cross* each other at a point, if at this point $\gamma_1$ passes from one side of $\gamma_2$ to the other.

**Lemma 2.1.** ([24]) *Let $C_1, C_2$, and $R$ be convex bodies in the plane, and assume that $C_1$ and $C_2$ are disjoint. Then the boundaries of the Minkowski sums $C_1' = C_1 \oplus (-R)$ and $C_2' = C_2 \oplus (-R)$ cross at most twice.*

Let $\mathcal{C} = \{C_1, C_2, \ldots, C_n\}$ be a family of simply connected regions in the plane bounded by simple closed curves. Assume, for simplicity, that these curves are in *general position*, i.e., any two of them cross only a finite number of times, no two touch each other, and no three pass through the same point. If any two distinct curves $\mathrm{Bd}(C_i)$ and $\mathrm{Bd}(C_j)$ cross at most twice, then $\mathcal{C}$ is usually called a family of *pseudo-disks*. A maximal connected piece of the boundary of $\cup \mathcal{C} = \cup_{i=1}^n C_i$, which belongs to the boundary of a member of $\mathcal{C}$, is said to be an *elementary arc*.

**Theorem 2.2.** ([24]) *Let $\mathcal{C} = \{C_1, C_2, \ldots, C_n\}$ be a family of $n \geq 3$ pseudo-disks in the plane.*

*Then the boundary of $\cup_{i=1}^n C_i$ consists of at most $6n - 12$ elementary arcs. This bound cannot be improved.*

In the case when each $C_i$ is a disk, the proof of Theorem 2.2 is straightforward (see Fig. 2). Assign to each $C_i$ its center, $p_i$, and connect $p_i$ to $p_j$ by a straight-line segment if and only if $\mathrm{Bd}(C_i)$ and $\mathrm{Bd}(C_j)$ cross each other, and at least one of their crossing points belongs to $\mathrm{Bd}(\cup \mathcal{C})$. It is easy to verify that no two segments cross each other, i.e., the resulting graph $G$ is *planar*. Therefore, $G$ has at most $3n - 6$ edges, each of which corresponds to at most two crossings between the circles that belong to $\mathrm{Bd}(\cup\mathcal{C})$. Therefore, the number of crossings on $\mathrm{Bd}(\cup\mathcal{C})$, and hence the number of elementary arcs is at most $6n - 12$. Using the terminology introduced in the first paragraph of this paper, we obtain that the *complexity* of $\cup\mathcal{C}$ is at most linear in $n$.

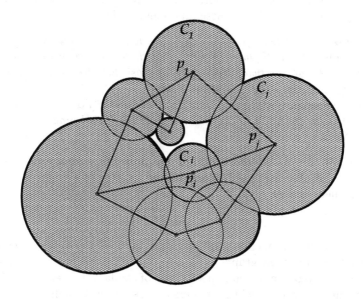

**Figure 2:** The proof of Theorem 2.2 for disks.

One can combine a well known algorithm of Bentley and Ottmann [7] for reporting all intersections in a collection of line segments with some standard divide-and-conquer techniques to obtain the following

**Corollary 2.3.** *There is an $O(nr\log^2 n)$-time algorithm for describing the space of free placements of a convex polygonal robot with $r$ sides among a set of polygonal obstacles with a total of $n$ sides, and for computing a collision-free path between any two given positions of the robot, if such a path exists.*

Leven and Sharir [27] reduced the $\log^2 n$ factor in the last result to $\log n$.

## 3    Allowing 3 intersections – Ackermann's function

What happens if we somewhat weaken the condition in Theorem 2.2, by assuming that the boundaries of any two members of $\mathcal{C}$ cross at most *three* times, rather than twice? At first glance this problem seems to be foolish, because if two closed curves are in general position, then they can cross only an *even* number of times. However, by a slight modification we obtain a meaningful question with a surprising answer.

Construct recursively an infinite sequence of integer-valued functions $A_1(n)$, $A_2(n), \ldots$ on the set of positive integers, as follows. Let $A_1(n) = 2n$ for every $n$. If $A_k$ has already been defined for some $k \geq 1$, then let $A_{k+1}(n) = A_k^{(n)}(1)$. In other words, we $n$ times iterate the function $A_k$, and take its value at 1. The function $A(n) = A_n(n)$ is called *Ackermann's function*. It grows so fast that $A(4)$ is a tower of 65536 2's! Consequently, the inverse of Ackermann's function is an extremely slowly growing funtion, whose all "practical" values are smaller than 4. For basic properties of these functions, see [43].

**Theorem 3.1.** ([11]) *Let $\{\gamma_1, \gamma_2, \ldots, \gamma_n\}$ be a family of simple curves in general position in the upper half-plane. Assume that the endpoints of each curve are on the x-axis, and that any two curves cross at most three times. Let $C_i$ denote the bounded region enclosed by $\gamma_i$ and the x-axis.*

*Then the boundary of $\cup_{i=1}^n C_i$ consists of at most $O(n\alpha(n))$ elementary arcs, where $\alpha(n)$ is the inverse of Ackermann's function. This bound is asymptotically tight.*

Let $\{f_1, f_2, \ldots, f_n\}$ be a collection of real-valued functions, each defined on a subinterval of $\mathbb{R}$. For any $x \in \mathbb{R}$, consider the set $I(x)$ of all indices $i$, for which $f_i(x)$ is defined, and let

$$f(x) = \max_{i \in I(x)} f_i(x).$$

This partially defined function (and its graph) is called the *upper envelope* of the functions $f_i$ (and of their graphs).

Perhaps the most important special case of the last theorem is the following result of Hart and Sharir, which answers a question of Atallah [5].

**Theorem 3.2.** ([20]) *The upper envelope of n non-vertical straight-line segments in the plane consists of at most $O(n\alpha(n))$ linear pieces.*

To verify that this result follows from Theorem 3.1, it is sufficient to notice that by attaching to each segment two vertical rays pointing downwards, one at each of its endpoints, we obtain a family of two-way infinite curves, any pair of which cross at most three times.

Wiernik and Sharir [46] showed that Theorem 3.2 is asymptotically tight.

A sequence of integers $a(i) \in \{1, 2, \ldots, n\}$ ($i = 1, 2, \ldots$) is called a *Davenport-Schinzel sequence* (of order 3) if no two consecutive elements are the same, and there is no alternating subsequence of length 5, i.e., there are no indices $i_1 < i_2 < \ldots < i_5$ such that $a(i_1) = a(i_3) = a(i_5)$, $a(i_2) = a(i_4)$, but $a(i_1) \neq a(i_2)$.

In fact, Hart and Sharir proved that the length of any such sequence is at most $O(n\alpha(n))$. Theorem 3.2 can be easily deduced from this result. Indeed, if we order the linear pieces of the upper envelope of the segments $s_1, s_2, \ldots, s_n$ from left to right, and replace each piece by the index $i$ of the segment $s_i$ it belongs to, we obtain a Davenport-Schinzel sequence (see Fig. 3). To see this, we have to check only that this sequence has no alternating subsequence of length 5. However, this immediately follows from the fact that any two segments cross at most once.

1 2    1    3    1    3    2    4    2 5 2 5 4 5    3

**Figure 3:** The upper envelope of segments and the corresponding Davenport-Schinzel sequence.

## 4    Well-behaved intersections – The role of parity

If we try to further weaken the condition in Theorem 2.2, assuming only that the boundaries of any two members of $\mathcal{C}$ cross at most *four* times, then the situation completely deteriorates. Consider a collection of $n$ pairwise crossing line segments, no three of which pass through the same point, and enclose each of them in a very narrow triangle whose width is at most $\varepsilon > 0$. If $\varepsilon$ is small enough, then every pair of triangles intersect in precisely four points, and all $4\binom{n}{2}$ intersection points belong to the boundary of their union (Fig. 4).

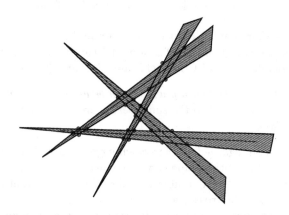

**Figure 4:** $n$ pairwise crossing triangles with $\Omega(n^2)$ intersections on the boundary of their union.

As Whitesides and Zhao [45] discovered, if we exclude certain types of crossings between the members of $\mathcal{C}$, it is possible to give a linear upper bound on the complexity of $\cup\mathcal{C}$, even if two members of $\mathcal{C}$ may intersect in more than two points. A family $\mathcal{C}$ of simply connected regions bounded by simple closed curves in general position in the plane is called $k$-*admissible*, if any two members $C_1, C_2 \in \mathcal{C}$ have at most $k$ boundary points in common and $C_1 \setminus C_2$ is connected (see Fig. 5). Clearly, we can restrict our attention to the case when $k$ is even, because the members of $C$ are in general position, i.e., their boundary curves cannot touch each other, so any two of them intersect in an even number of points.

**Theorem 4.1.** ([45]) *Let $C = \{C_1, C_2, \ldots, C_n\}$ be a $k$-admissible family of $n \geq 3$ simply connected regions in general position in the plane.*

*Then the boundary of $\cup_{i=1}^{n} C_i$ consists of at most $k(3n-6)$ elementary arcs, and this bound cannot be improved.*

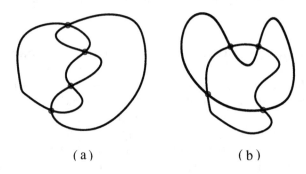

( a )                         ( b )

**Figure 5:** A pair of regions belonging to some
(a) 4-admissible, (b) non-admissible family.

In case $k = 2$ Theorem 4.1 reduces to Theorem 2.2.

It was pointed out in [35] that Theorem 4.1 can be easily deduced from the following remarkable result of Chojnacki (alias Hanani) [9] (see also [44, 28]).

**Lemma 4.2.** ([9]) *Suppose that a graph $G$ can be drawn in the plane so that any two of its edges not incident to the same vertex cross an even number of times. Then $G$ is planar.*

This result can be regarded as a far-reaching generalization of the elementary fact that if two points of the plane can be connected by an arc crossing a fixed simple closed curve $\gamma$ an even number of times, then they can also be connected by an arc which does not cross $\gamma$ at all.

To see that Lemma 4.2 implies Theorem 4.1, it is enough to bound the number of points on $\mathrm{Bd}(\cup C)$, which belong to the boundary of more than one member of $C$. For every $C_i$ which contributes at least one arc to $\mathrm{Bd}(\cup C)$, fix a point $p_i$ in the interior of such an arc. For any pair $C_i, C_j \in C$ which has an intersection point $q \in \mathrm{Bd}(\cup C)$, draw an edge (but only one!) between $p_i$ and $p_j$, as follows. Starting from $p_i$, follow $\mathrm{Bd}(C_i)$ to $q$ in clockwise direction, and from there follow $\mathrm{Bd}(C_j)$ to $p_j$ in counter-clockwise direction. It is not hard to verify that any two edges of this graph not incident to the same vertex cross an even number of times. Thus, the graph has at most $3n - 6$ edges. That is, there are at most $3n - 6$ pairs $\{C_i, C_j\}$ contributing intersection points to $\mathrm{Bd}(\cup C)$, and each of them can contribute at most $k$ points.

## 5   Counting special intersections

Theorem 2.2 can be regarded as an upper bound on the number all intersection points between the curves $\mathrm{Bd}(C_i)$, which lie on the boundary of $\cup_{i=1}^n C_i$. Another way to generalize this result is to drop the condition that any pair of boundary curves intersect in at most two points, but to count only those intersections on $\mathrm{Bd}(\cup_{i=1}^n C_i)$, which belong to such pairs.

If two members of $\mathcal{C}$ have precisely two boundary points in common, then these points are called *regular intersection points*. All other intersection points between boundary curves are called *irregular*.

**Theorem 5.1** ([35]) *Given a family $\mathcal{C}$ of $n \geq 3$ convex regions in general position in the plane, let $R$ and $I$ denote the number of regular and irregular intersection points of their boundaries, resp., which belong to $\mathrm{Bd}(\cup\mathcal{C})$.*

*Then we have $R \leq 2I + 6n - 12$.*

The last result is sharper than Theorem 2.2 in the sense that to obtain a $6n - 12$ upper bound on the number of elementary arcs (the number of intersection points) on $\mathrm{Bd}(\cup\mathcal{C})$, we do not have to exclude all irregular intersections. It is sufficient to assume that no such intersection occurs on $\mathrm{Bd}(\cup\mathcal{C})$. However, for some technical reasons, we have been unable to establish Theorem 5.1 without the additional assumption that all members of $\mathcal{C}$ are *convex*. We conjecture that this assumption can be dropped.

It is not hard to show that the coefficient of $I$ in Theorem 5.1 cannot be replaced by any constant smaller than 2.

If we want to get a non-trivial (i.e., subquadratic) upper bound on $R$, we have to limit the number of times two boundary curves are allowed to cross each other. But even under such an assumption we cannot expect a *linear* upper bound. There is a family of $n$ disks and rectangles in general position in the plane satisfying $R = \Omega(n^{4/3})$. The best positive result in this direction is the following.

**Theorem 5.2.** ([3]) *Let $\mathcal{C}$ be a family of $n$ simply connected regions in the plane. Suppose that they are bounded by simple closed curves in general position, any two of which intersect in at most $s$ points, where $s$ is a constant.*

*Then there exists $\delta = \delta(s) > 0$ such that the number $R$ of regular intersection points on $\mathrm{Bd}(\cup\mathcal{C})$ satisfies*

*(i) $R = O(n^{2-\delta})$;*

*(ii) $R = O(n^{1.5+\varepsilon})$ for any $\varepsilon > 0$, provided that every member of $\mathcal{C}$ is convex.*

## 6   The union of fat triangles – Counting holes

The construction at the beginning of Section 4, showing that the union of $n$ triangles may have *quadratic* complexity, uses extremely narrow triangles. It was proved in [30] that if we restrict how narrow the triangles can be, we can still establish a nearly linear upper bound on the complexity of their union. For any $\delta > 0$, a triangle is said to be *$\delta$-fat* if each of its angles is at least $\delta$.

**Theorem 6.1.** ([30]) *For any fixed $\delta > 0$, the boundary of the union of $n$ $\delta$-fat triangles in the plane consists of at most $O(n \log \log n)$ elementary arcs.*

By a slight modification of the construction of Wiernik and Sharir [46] cited after Theorem 3.2, one can easily give an example of $n$ equilateral ($\pi/3$-fat) triangles, whose union has a slightly superlinear boundary complexity ($\Omega(n\alpha(n))$).

Given a family $\mathcal{C}$ of simply connected regions in the plane, a connected component of the complement of $\cup\mathcal{C}$ is called a *hole* determined by $\mathcal{C}$. The proof of Theorem 6.1 is based on the fact that every family of $n$ $\delta$-fat triangles in the plane determines at most a linear number of holes. The strongest known bound of this type is the following.

**Theorem 6.2.** ([36]) *Any family of $n$ $\delta$-fat triangles in the plane determines $O\left(\frac{n}{\delta} \log \frac{2}{\delta}\right)$ holes. This bound is tight up to the logarithmic factor.*

This result can be used to establish a more general upper bound for the number of holes determined by a family of triangles with given angles.

**Theorem 6.3.** ([36]) *Let $\mathcal{C} = \{C_1, C_2, \ldots, C_n\}$ be a family of $n > 1$ triangles in the plane, and let $\alpha_i$ denote the smallest angle of $C_i$ ($1 \leq i \leq n$). Suppose $0 < \alpha_1 \leq \alpha_2 \leq \cdots \leq \alpha_n$, and let $k \leq n$ be the largest integer satisfying $\sum_{i=1}^{k} \alpha_i < \pi$.*
*Then $\mathcal{C}$ determines $O(nk \log k)$ holes. Furthermore, there exists a family $\mathcal{C}' = \{C'_1, C'_2, \ldots, C'_n\}$, where $C'_i$ is isomorphic to $C_i$ and $\mathcal{C}'$ determines $\Omega(nk)$ holes.*

If we consider *infinite wedges* (i.e., convex cones) rather than triangles, then the same bound holds not only for the number of *holes*, but also for the *complexity* of the boundary of the union. The following result strengthens some earlier bounds in [14].

**Theorem 6.4.** ([36]) *Let $\mathcal{C}$ be a family of $n$ wedges in the plane with angles $0 < \alpha_1 \leq \alpha_2 \leq \cdots \leq \alpha_n < \pi$. Let $k \leq n$ be the largest integer satisfying $\sum_{i=1}^{k} \alpha_i < \pi$.*
*If $k \geq 2$, then the boundary complexity of $\cup\mathcal{C}$ is $O(nk \log k)$. Furthermore, there exists a family of $n$ wedges with angles $\alpha_1, \alpha_2, \ldots, \alpha_n$, which determines $\Omega\left((\pi - \alpha_n)nk\right)$ holes.*

The concept of $\delta$-fatness, as well as Theorem 6.2, has been extended to arbitrary polygons by van Kreveld [26]. For other extensions and generalizations, see [38],[37],[15], [13], and [12].

## 7    Fat objects in space

Many of the theorems in previous sections have natural generalizations to higher dimensions. In this section, we mention only a few 3-dimensional results. Given a family $\mathcal{C}$ of 3-dimensional bodies, an *edge* of their union is defined as a maximal connected arc on $\mathrm{Bd}(\cup\mathcal{C})$, which belongs to two distinct members of $\mathcal{C}$. A point of $\mathrm{Bd}(\cup\mathcal{C})$, belonging to *three* distinct members of $\mathcal{C}$ is called a *vertex*. A maximal connected 2-dimensional piece of $\mathrm{Bd}(\cup\mathcal{C})$, which belongs to a *single* member of $\mathcal{C}$, is a *face*. The *complexity* of the boundary of $\cup\mathcal{C}$ is defined as the total number of vertices, edges, and faces of the boundary. These numbers are related to one another via Euler's Formula.

Let $C$ be a collection of topological balls in $\mathbb{R}^3$ such that the intersection of any two of their surfaces is either empty or is a simple closed curve, and the intersection of any three surfaces consists of at most two points. Then $C$ is called a family of *pseudo-balls*. Taking the intersection of the surface of each member of $C$ with all the other members, and applying Theorem 2.2 to the resulting 2-dimensional arrangements, we obtain

**Corollary 7.1.** ([24]) *The complexity of the boundary of the union of $n$ pseudo-balls in $\mathbb{R}^3$ is $O(n^2)$. This bound is asymptotically tight.*

Another generalization of Theorem 2.2 provides an upper bound on the complexity of the space of free placements of a convex polyhedral robot which is allowed to translate amidst polyhedral obstacles in 3-space.

**Theorem 7.2.** ([4]) *Let $\{C_1, C_2, \ldots, C_n\}$ be a family of pairwise disjoint convex polyhedral "obstacles" in $\mathbb{R}^3$ with a total of $N$ faces, and let $R$ be a convex polyhedral "robot," whose number of faces is a constant.*

*Then the complexity of the union of the "expanded obstacles" $C_1 \oplus (-R)$, $C_2 \oplus (-R), \ldots, C_n \oplus (-R)$ is $O(nN \log n)$. In the worst case this bound cannot be improved, apart from the logarithmic factor.*

For the case, when the robot is a ball, we have a similar result.

**Theorem 7.3.** ([2]) *Let $\{C_1, C_2, \ldots\}$ be a family of pairwise disjoint convex polyhedral "obstacles" in $\mathbb{R}^3$ with a total of $N$ faces, and let $R$ be a ball-shaped "robot."*

*Then the complexity of the union of the "expanded obstacles" $C_1 \oplus (-R)$, $C_2 \oplus (-R), \ldots, C_n \oplus (-R)$ is $O(N^{2+\varepsilon})$ for every $\varepsilon > 0$.*

In particular, the last result shows that the complexity of the union of $n$ congruent infinite cylinders in $\mathbb{R}^3$ is only at most slightly superquadratic. No non-trivial (subcubic) upper bound is known for infinite cylinders with arbitrary radii.

Theorem 3.2 has the following analogue.

**Theorem 7.4.** ([34]) *The complexity of the upper envelope of $n$ non-vertical triangles in $\mathbb{R}^3$ is $O(n^2\alpha(n))$. This bound is asymptotically tight.*

A somewhat weaker form of the last result is true in a much more general setting: the complexity of the upper envelope of $n$ $(d-1)$-dimensional algebraic surface patches in $\mathbb{R}^d$, satisfying some natural conditions, is $O(n^{d-1+\varepsilon})$ for every $\varepsilon > 0$ (see [19, 42]).

It seems likely that the results of Section 6 also generalize to higher dimensions. The intersection of two (three) half-spaces in general position in 3-space is called a *dihedral (resp., trihedral) wedge*. A wedge (tetrahedron) is called $\delta$-*fat* if its dihedral angle (resp., each of its solid angles) is at least $\delta$.

We conjecture that the complexity of the union of any family of $\delta$-fat tetrahedra in $\mathbb{R}^3$ is at most slightly superquadratic in $n$. To prove this conjecture for *congruent* tetrahedra, it would be sufficient to show that the same statement holds for $\delta$-fat trihedral wedges. We can prove only a weaker result.

**Theorem 7.5.** ([33]) *The complexity of the boundary of the union of n $\delta$-fat dihedral wedges in $\mathbb{R}^3$ is $O(n^{2+\varepsilon})$ for every $\varepsilon > 0$.*

We do not know the answer to the following simple question: is it true that the complexity of the union of $n$ cubes in $\mathbb{R}^3$ is $o(n^3)$? In fact, we do not even know whether $n$ cubes in $\mathbb{R}^3$ always determine at most $o(n^3)$ holes.

It is not hard to show that the complexity of the union of *axis-parallel* cubes is $O(n^2)$ and that this bound is asymptotically tight. In the case of *congruent* axis-parallel cubes, this bound can be improved to linear [8]. For congruent, but not necessarily axis-parallel cubes, we have the following recent result.

**Theorem 7.6.** ([33]) *The complexity of the boundary of the union of n congruent cubes in $\mathbb{R}^3$ is $O(n^{2+\varepsilon})$ for every $\varepsilon > 0$.*

# 8    Fat wedges – Extremal hypergraph theory

It was shown by Katona and Kovalev [21, 23] that for any family $\mathcal{C}$ of convex sets in $\mathbb{R}^d$, the number of *holes*, i.e., connected components of $\mathbb{R}^d \setminus \bigcup \mathcal{C}$, is at most $\sum_{i=0}^{d} \binom{n}{i}$, with equality only if $\mathcal{C}$ consists of hyperplanes or parallel strips in general position. In fact, if any $d$ members of $\mathcal{C}$ have only a bounded number, $s$, of boundary points in common, then the *complexity* of $\bigcup \mathcal{C}$ is also $O(n^d)$, because each vertex $p$ of the union is determined by $d$ members, whose boundaries pass through $p$.

The aim of this section is to sketch a proof of the following weak (but non-trivial) version of Theorem 7.5.

**Proposition 8.1.** *There is an $\varepsilon > 0$ such that the complexity of the boundary of the union of n $\delta$-fat wedges in $\mathbb{R}^d$ is $O(n^{3-\varepsilon})$.*

Let $K^{(3)}(m)$ denote a complete 3-uniform hypergraph with 3 disjoint $m$-element vertex classes, consisting of all triples containing exactly one element from each class. As in [34], our basic tool is Erdös's result from extremal hypergraph theory.

**Lemma 8.2.** ([16]) *Let H be a 3-uniform hypergraph on n vertices containing no subhypergraph isomorphic to $K^{(3)}(m)$. Then H has at most $n^{3-1/m^2}$ triples.*

Three $l$-membered families of half-spaces, $\mathcal{H}_1, \mathcal{H}_2, \mathcal{H}_3 \subset \mathbb{R}^3$ are said to *meet regularly* if their arrangement, restricted to the convex hull of the points

$$\{\text{Bd}(h_1) \cap \text{Bd}(h_2) \cap \text{Bd}(h_3) \mid h_1 \in \mathcal{H}_1, h_2 \in \mathcal{H}_2, h_3 \in \mathcal{H}_3\},$$

is combinatorially isomorphic to the arrangement

$$\mathcal{H}_1^0 = \{\{(x, y, z) \in \mathbb{R}^3 : x \le i\} \mid 1 \le i \le l\},$$
$$\mathcal{H}_2^0 = \{\{(x, y, z) \in \mathbb{R}^3 : y \le j\} \mid 1 \le j \le l\},$$

$$\mathcal{H}_3^0 = \{\{(x,y,z) \in \mathbb{R}^3 : z \le k\}, | 1 \le k \le l\},$$

restricted to the cube $[1,l]^3$. (See Fig. 6.)

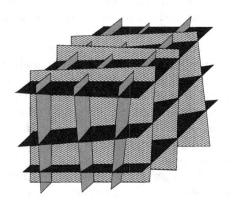

**Figure 6:** The boundary planes of three 3-membered families of regularly meeting half-spaces.

**Lemma 8.3.** ([34]) *For any $l$, there exists $L = L(l)$ such that any three $L$-membered families of half-spaces in $\mathbb{R}^3$ contain three $l$-membered subfamilies which meet regularly.*

Obviously, it is sufficient to prove Proposition 8.1 in the special case when the angle of every wedge is precisely $\delta$, because every wedge can be obtained as the union of at most $\lceil \pi/\delta \rceil$ such wedges. Color the wedges with a constant number of colors so that the directions of the (infinite) edges and the directions of the corresponding faces of any two wedges of the same color differ from each other by less than 1 degree.

Assume, in order to obtain a contradiction, that there is a family $\mathcal{W}$ of $n$ dihedral wedges whose union has at least $n^{3-\varepsilon}$ vertices on its boundary, for some $\varepsilon > 0$. Combine the last two lemmas. We obtain that if $\varepsilon = \varepsilon(l) > 0$ is small enough, there exist three pairwise disjoint $l$-membered monochromatic subfamilies $\mathcal{W}_1, \mathcal{W}_2, \mathcal{W}_3 \subset \mathcal{W}$ such that, expressing every $w_{st} \in \mathcal{W}_s$ as the intersection of two half-spaces, $h_{st} \cap h'_{st}$ $(1 \le s \le 3, 1 \le t \le l)$, the families of half-spaces

$$\mathcal{H}_1 := \{h_{1i} \mid 1 \le i \le l\},$$
$$\mathcal{H}_2 := \{h_{2j} \mid 1 \le j \le l\},$$
$$\mathcal{H}_3 := \{h_{3k} \mid 1 \le k \le l\}$$

meet regularly, and every point of the set

$$S := \{\text{Bd}(h_{1i}) \cap \text{Bd}(h_{2j}) \cap \text{Bd}(h_{3k}) \mid (1 \le i,j,k \le l)\}$$

belongs to the boundary of $\cup\mathcal{W}$. Note that any two half-spaces belonging to the same family $\mathcal{H}_s$ are almost parallel. We can also assume without loss of generality that the isomorphism between $\mathcal{H}_1, \mathcal{H}_2, \mathcal{H}_3$ and $\mathcal{H}_1^0, \mathcal{H}_2^0, \mathcal{H}_3^0$ takes each half-space $h_{1i}$ (resp., $h_{2j}, h_{3k}$) to $\{(x, y, z) \in \mathbb{R}^3 : x \leq i\}$ (resp., $\{(x, y, z) \in \mathbb{R}^3 : y \leq j\}$, $\{(x, y, z) \in \mathbb{R}^3 : z \leq k\}$).

It follows from the fact that any two wedges belonging to the same $\mathcal{W}_s$ have the same color, that their edges are almost parallel. Therefore, we can choose a plane $P$ such that the angle between $P$ and the edge of every wedge $w \in \mathcal{W}_1 \cup \mathcal{W}_2 \cup \mathcal{W}_3$ is larger than, say, 30 degrees. This implies, for example, that $w \cap P$ is $\delta/2$-fat (in the plane). Translating $P$ parallel to itself, if necessary, we can assume without loss of generality that it does not pass through any vertex of $\mathrm{Bd}(\cup \mathcal{W})$, and that $P$ cuts the set $S$ into two parts as equally as possible. Let $R$ (and $B$) denote the set of elements of $S$ on one side of $P$ (on the other, resp.). Color any point $(i, j, k) \in [1, l]^3$ red or blue according to whether $\mathrm{Bd}(h_{1i}) \cap \mathrm{Bd}(h_{2j}) \cap \mathrm{Bd}(h_{3k})$ belongs to $R$ or $B$, and denote the sets of red and blue points by $R'$ and $B'$, resp.

A set $X$ of points with integer coordinates is called *convex in a given direction* if, for every segment $xy$ parallel to this direction, both of whose endpoints belong to $X$, all other integer points of $xy$ also belong to $X$. We obviously have

**Claim 8.4.** *$R'$ and $B'$ are convex in the directions of all three coordinate axes.*

Connect a point $(i, j, k) \in R'$ to $(i', j', k') \in B'$ by a *directed edge*, whenever their distance is 1, i.e., when they differ only in one of their coordinates, and in this coordinate their difference is 1. Using the fact that $|R'|, |B'| \geq \lfloor l^3/2 \rfloor$, it follows by standard isoperimetric inequalities that the number of directed edges is at least $l^2/2$. We can assume, by symmetry, that at least $l^2/12$ of them are parallel to the $z$-axis and are pointed upwards. Let $D$ denote the orthogonal projection of these edges to the $(x, y)$-plane, i.e.,

$$D := \{(i, j) \in [1, l]^2 \mid \exists k \text{ such that } (i, j, k) \in R' \text{ and } (i, j, k + 1) \in B'\}.$$

In view of Claim 8.4, we have

**Claim 8.5.** *The set $D \subseteq [1, l]^2$ is convex in the directions of both coordinate axes, and $|D| \geq l^2/12$.*

The last claim easily implies that $D$ contains all integer points within an axis-parallel square, whose side length is at least $l/50$.

Switching back to the original picture, this means that there are two subfamilies $\mathcal{H}_1' \subset \mathcal{H}_1, \mathcal{H}_2' \subset \mathcal{H}_2$, each of size $l' := \lceil l/50 \rceil$, whose cross-sections on the plane $P$ "meet regularly," i.e., are combinatorially isomorphic to the arrangement

$$\{\{(x, y) \in \mathbb{R}^2 : x \leq i\} \mid 1 \leq i \leq l'\},$$
$$\{\{(x, y) \in \mathbb{R}^2 : y \leq j\} \mid 1 \leq j \leq l'\},$$

restricted to $[1, l']^2$. Consider now the $l'$-membered families of wedges, $\mathcal{W}_1' \subset \mathcal{W}_1$ and $\mathcal{W}_2' \subset \mathcal{W}_2$ corresponding to the members of $\mathcal{H}_1'$ and $\mathcal{H}_2'$, resp.

It follows from the definitions that, for every $h_1 \in \mathcal{H}_1'$, $h_2 \in \mathcal{H}_2'$, the point $\mathrm{Bd}(h_1) \cap \mathrm{Bd}(h_2) \cap P$ belongs to the boundary of

$$P \cap ((\cup \mathcal{W}_1') \cup (\cup \mathcal{W}_2')).$$

Consequently, the complexity of the union of all planar wedges $w \cap P$ ($w \in (\cup \mathcal{W}_1') \cup (\cup \mathcal{W}_2')$) is at least $(l')^2$.

Recall that, by the choice of the direction of $P$, the intersection of every element of $\mathcal{W}_1' \cup \mathcal{W}_2'$ with $P$ is $(\delta/2)$-fat planar wedge. Thus, according to Theorem 6.1, the complexity of their union cannot exceed $O(l' \log l')$. This contradicts the conclusion of the last paragraph, provided that $l' = \lceil l/50 \rceil$ is sufficiently large (and $\varepsilon = \varepsilon(l) > 0$ is sufficiently small). This completes the proof of Proposition 8.1.

# References

1. P. Agarwal, M. Katz, and M. Sharir: Computing depth order and related problems, *Comput. Geom. Theory Appls.* **5** (1995), 187–206.
2. P. Agarwal and M. Sharir: Pipes, cigars, and kreplach: The union of Minkowski sums in three dimensions, *Discrete Comput. Geom.* 24 (2000), 645–685.
3. B. Aronov, A. Efrat, D. Halperin, and M. Sharir: On the number of regular vertices of the union of Jordan regions, in: *Algorithm Theory, SWAT'98 (Stockholm), Lecture Notes in Comput. Sci.* **1432**, Springer-Verlag, Berlin, 1998, 322–334.
4. B. Aronov and M. Sharir: On translational motion planning of a convex polyhedron in 3-space, *SIAM J. Comput.* **26** (1997), 1785–1803.
5. M. Atallah: Some dynamic computational geometry problems, *Computers and Mathematics with Applications* **11** (1985), 1171–1181.
6. M. de Berg, M. Katz, F. van der Stappen, and J. Vleugels: Realistic input models for geometric algorithms, in: *Proc. 13th Annual Symposium on Computational Geometry*, ACM Press, 1997, 294–303.
7. J. L. Bentley and T. A. Ottmann: Algorithms for reporting and counting geometric intersections, *IEEE Trans. Comput.* **C-28** (1979), 643–647.
8. J.-D. Boissonnat, M. Sharir, B. Tagansky, and M. Yvinec: Voronoi diagrams in higher dimensions under certain polyhedral distance functions, *Discrete Comput. Geom.* **19** (1998), 485–519.
9. Ch. Chojnacki (A. Hanani): Über wesentlich unplättbare Kurven im dreidimensionalen Raume, *Fund. Math.* **23** (1934), 135–142.
10. K. Clarkson, H. Edelsbrunner, L. Guibas, M. Sharir, E. Welzl: Combinatorial complexity bounds for arrangements of curves and surfaces, *Discrete and Computational Geometry* **5** (1990), 99-160.
11. H. Edelsbrunner, L. Guibas, J. Hershberger, J. Pach, R. Pollack, R. Seidel, M. Sharir, and J. Snoeyink: On arrangements of Jordan arcs with three intersections per pair, *Discrete Comput. Geom.* **4** (1989), 523–539.
12. A. Efrat: The complexity of the union of $(\alpha, \beta)$-covered objects, *Proceedings of the 15th Annual Symposium on Computational Geometry*, ACM Press, 1999, 134–142.
13. A. Efrat and M. J. Katz: On the union of $\kappa$-curved objects. *Comput. Geom. Theory Appl.* **14** (1999), 241–254.
14. A. Efrat, G. Rote, and M. Sharir: On the union of fat wedges and separating a collection of segments by a line, *Comput. Geom. Theory Appl.* **3** (1993), 277–288.
15. A. Efrat and M. Sharir: On the complexity of the union of fat convex objects in the plane, *Discrete Comput. Geom.* **23** (2000), 171–189.

16. P. Erdös: On extremal problems of graphs and hypergraphs, *Israel J. Math.* **2** (1964), 183–190.

17. L. Guibas and M. Sharir: Combinatorics and algorithms of arrangements, in: *New Trends in Discrete and Computational Geometry* (J. Pach, ed.), Springer-Verlag, Berlin, 1993, 9–36.

18. P. Gupta, R. Janardan, and M. Smid: A technique for adding range restrictions to generalized searching problems, *Inform. Process. Lett.* **64** (1997), 263–269.

19. D. Halperin and M. Sharir: New bounds for lower envelopes in three dimensions, with applications to visibility in terrains, *Discrete Comput. Geom.* **12** (1994), 313–326.

20. S. Hart and M. Sharir: Nonlinearity of Davenport-Schinzel sequences and of generalized path compression schemes, *Combinatorica* **6** (1986), 151–177.

21. G. O. H. Katona: On a problem of L. Fejes Toth, *Studia Sci. Math. Hungar.* **12** (1977), 77–80.

22. M. J. Katz: 3-D vertical ray shooting and 2-D point enclosure, range searching, and arc shooting amidst convex fat objects, *Comput. Geom. Theory Appl.* **8** (1997), 299–316.

23. M. D. Kovalev: A property of convex sets and its application (Russian), *Mat. Zametki* **44** (1988), 89–99. English translation: *Math. Notes* 44 (1988), 537–543.

24. K. Kedem, R. Livne, J. Pach and M. Sharir: On the union of Jordan regions and collision-free translational motion amidst polygonal obstacles, *Discrete Comput. Geom.* **1** (1986), 59–71.

25. K. Kedem and M. Sharir: An efficient motion-planning algorithm for a convex polygonal object in two-dimensional polygonal space, *Discrete Comput. Geom.* **5** (1990), 43–75.

26. M. van Kreveld: On fat partitioning, fat covering and the union size of polygons, *Computational Geometry: Theory and Applications* **9** (1998), 197–210.

27. D. Leven and M. Sharir: Planning a purely translational motion for a convex object in two-dimensional space using generalized Voronoi diagrams, *Discrete Comput. Geom.* **2** (1987), 9–31.

28. L. Lovász, J. Pach, and M. Szegedy: On Conway's thrackle conjecture, *Discrete Comput. Geom.* **18** (1997), 369–376.

29. T. Lozano-Pérez and M. A. Wesley: An algorithm for planning collision-free paths among polyhedral obstacles, *Commun. ACM* **22** (1979), 560–570.

30. J. Matoušek, J. Pach, M. Sharir, S. Sifrony, and E. Welzl: Fat triangles determine linearly many holes, *SIAM Journal of Computing* **23** (1994), 154–169.

31. P. McMullen: On the upper-bound conjecture for convex polytopes, *J. Combinatorial Theory, Ser. B* **10** (1971), 187–200.

32. J. Pach and P.K. Agarwal: *Combinatorial Geometry*, J. Wiley and Sons, New York, 1995.

33. J. Pach, I. Safruti, and M. Sharir, The union of congruent cubes in three dimensions, 17th ACM Symposium on Computational Geometry, 2001, accepted.

34. J. Pach and M. Sharir: The upper envelope of piecewise linear functions and the boundary of a region enclosed by convex plates: combinatorial analysis, *Discrete Comput. Geom.* **4** (1989), 291–309.

35. J. Pach and M. Sharir: On the boundary of the union of planar convex sets, *Discrete Comput. Geom.* **21** (1999), 321–328.

36. J. Pach and G. Tardos: On the boundary complexity of the union of fat triangles, *Proceedings of 41st Annual Symposium on Foundations of Computer Science, Los Angeles, 2000.*

37. F. van der Stappen: *Motion Planning amidst Fat Obstacles (Ph. D. Thesis,* Faculteit Wiskunde & Informatica, Universiteit Utrecht, 1994.

38. F. van der Stappen, D. Halperin, and M. Overmars: The complexity of the free space for a robot moving amidst fat obstacles, *Computational Geometry: Theory and Applications* **3** (1993), 353–373.

39. J. T. Schwartz and M. Sharir: On the "piano movers" problem I,II, *Comm. Pure Applied Math.* **36** (1983), 345–398 and *Adv. Applied Math.* **4** (1983), 298–351.

40. J. T. Schwartz and M. Sharir: A survey of motion planning and related geometric algorithms, in: *Geometric Reasoning* (D. Kapur and J. Mundy, eds.), MIT Press, Cambridge, MA, 1989, 157–169.

41. J. T. Schwartz and M. Sharir: Algorithmic motion planning in robotics, in: *Handbook of Theoretical Computer Science* (J. van Leeuwen, ed.), Elsevier, Amsterdam, 1990, 391–430.

42. M. Sharir: Almost tight upper bounds for lower envelopes in higher dimensions, *Discrete Comput. Geom.* **12** (1994), 327–345.

43. M. Sharir and P.K. Agarwal: *Davenport-Schinzel Sequences and Their Geometric Applications,* Cambridge University Press, Cambridge, 1995.

44. W. T. Tutte: Toward a theory of crossing numbers, *Journal of Combinatorial Theory* **8** (1970), 45–53.

45. S. Whitesides and R. Zhao: $K$-admissible collections of Jordan curves and offsets of circular arc figures, *Technical Report SOCS 90.08,* McGill University, Montreal, 1990.

46. A. Wiernik and M. Sharir: Planar realizations of nonlinear Davenport-Schinzel sequences by segments, *Discrete Comput. Geom.* **3** (1988), 15–47.

# Structure Theorems for Systems of Segments

János Pach[1*] and József Solymosi[2**]

[1] Rényi Institute of the Hungarian Academy of Sciences
and New York University, Courant Institute
pach@renyi.hu

[2] Institute for Theoretical Computer Science, ETH Zürich
solymosi@inf.ethz.ch

**Abstract.** We study intersection properties of systems of segments in the plane. In particular, we show that there exists a constant $c > 0$ such that every system $\mathcal{S}$ of $n$ straight-line segments in the plane has two at least $cn$-element subsystems $\mathcal{S}_1, \mathcal{S}_2 \subset \mathcal{S}$ such that either every segment in $\mathcal{S}_1$ intersects all elements of $\mathcal{S}_2$, or no segment in $\mathcal{S}_1$ intersects any element of $\mathcal{S}_2$. We also propose a fast approximate solution for reporting *most* intersections among $n$ segments in the plane.

## 1 Introduction

The problem of detecting and reporting intersections among straight-line segments in the plane is one of the oldest and most extensively studied topics in computational geometry. It is a basic ingredient of many hidden surface removal algorithms, and has numerous other applications in computer graphics, motion planning, geographic information systems, etc. The first efficient techniques were developed by Shamos and Hoey [SH76] and Bentley and Ottmann [BO79] more than twenty years ago. The running times of the best known algorithms, due to Balaban [B95] and Chazelle and Edelsbrunner [CE92], are $O(n \log n + I)$, where $n$ and $I$ denote the number of segments and the number of intersections, resp. (See also [PS91].)

In the present paper, we discuss some structural properties of *intersection graphs* of segments, i.e., graphs that can be obtained by assigning a vertex to every element of a system of segments $\mathcal{S}$ in the plane, and connecting two of them by an edge if and only if their intersection is non-empty. Throughout this paper, we assume that the elements of $\mathcal{S}$ are in *general position*, i.e., no two segments are parallel and no three of their endpoints are collinear. In particular, if two elements of $\mathcal{S}$ intersect, then they determine a proper crossing.

We prove the following Ramsey-type result.

* Supported by NSF grant CR-97-32101, PSC-CUNY Research Award 61392-0030, and OTKA-T-020914.
** Supported by the joint Berlin/Zürich graduate program Combinatorics, Geometry, Computation, financed by German Science Foundation (DFG) and ETH Zürich. On leave from MTA SZTAKI.

J. Akiyama, M. Kano, and M. Urabe (Eds.): JCDCG 2000, LNCS 2098, pp. 308–317, 2001.

**Theorem 1.** *There exists a constant $C > 0$ such that every system $\mathcal{S}$ of $n$ segments in the plane has two disjoint subsystems $\mathcal{S}_1, \mathcal{S}_2 \subset \mathcal{S}$ such that $|\mathcal{S}_1|, |\mathcal{S}_2| \geq Cn$ and*

*(i) either every segment in $\mathcal{S}_1$ crosses all segments in $\mathcal{S}_2$,*

*(ii) or no segment in $\mathcal{S}_1$ crosses any segment in $\mathcal{S}_2$.*

In the sequel, $A$ stands for an absolute constant smaller than $10^6$. Theorem 1 is a direct corollary of the following two complementary results.

**Theorem 2.** *Any system $\mathcal{S}$ of $n$ segments in the plane with at least $cn^2$ crossings $(c > 0)$ has two disjoint subsystems, $\mathcal{S}_1, \mathcal{S}_2 \subset \mathcal{S}$, such that $|\mathcal{S}_1|, |\mathcal{S}_2| \geq \frac{(2c)^A}{660}n$ and every segment in $\mathcal{S}_1$ crosses all segments in $\mathcal{S}_2$.*

**Theorem 3.** *Any system $\mathcal{S}$ of $n$ segments in the plane with at least $cn^2$ non-crossing pairs $(c > 0)$ has two disjoint subsystems, $\mathcal{S}_1, \mathcal{S}_2 \subset \mathcal{S}$, such that $|\mathcal{S}_1|, |\mathcal{S}_2| \geq \frac{(c/5)^A}{330}n$ and no segment in $\mathcal{S}_1$ crosses any segment in $\mathcal{S}_2$.*

The above results, combined with Szemerédi's Regularity Lemma [S78], can be used to establish a fairly strong structure theorem for intersection graphs of segments. We say that two sets have *almost the same* number of elements if their sizes differ by at most a factor of 2.

**Theorem 4.** *For any $\varepsilon > 0$, there exists an integer $K = K(\varepsilon)$ with the property that any system $\mathcal{S}$ of segments in the plane can be partitioned into $K + 1$ sub-families, $\mathcal{S}_0, \mathcal{S}_1, \ldots, \mathcal{S}_K$ such that $|\mathcal{S}_0| < \varepsilon|\mathcal{S}|$, all other subfamilies have almost the same size, and for all but at most $\varepsilon K^2$ pairs $1 \leq i, j \leq K$,*

*(i) either every segment in $\mathcal{S}_i$ crosses all segments in $\mathcal{S}_j$,*

*(ii) or no segment in $\mathcal{S}_i$ crosses any segment in $\mathcal{S}_j$.*

Fix an element $s_i$ in each $\mathcal{S}_i$. For any $s \in \mathcal{S}$, let $f(s) := s_i$ if and only if $s$ belongs $\mathcal{S}_i$ ($0 \leq i \leq K$). We can think of $f(s)$ as the segment *representing* $s$. According to Theorem 4, with a very small error, two randomly selected elements $s, t \in \mathcal{S}$ cross each other if and only if $f(s) \cap f(t)$ is non-empty.

A *geometric graph* is a graph whose vertices are points in general position in the plane (i.e., no three points are on a line) and whose edges are straight-line segments connecting these points. Our last two results are easy corollaries to Theorems 2 and 3, respectively.

**Theorem 5.** *Any geometric graph $G$ with $n$ vertices and at least $cn^2$ edges $(c > 0)$ has two disjoint sets of edges $E_1, E_2 \subset E(G)$ such that $|E_1|, |E_2| \geq (c/32)^{A+3}\binom{n}{2}$ and every edge in $E_1$ crosses all edges in $E_2$.*

**Theorem 6.** *Any geometric graph $G$ with $n$ vertices and at least $cn^2$ edges $(c > 0)$ has two disjoint sets of edges $E_1, E_2 \subset E(G)$ such that $|E_1|, |E_2| \geq (c/34)^{A+3}\binom{n}{2}$ and no edge in $E_1$ crosses any edge in $E_2$.*

The rest of the paper is organized as follows. In Section 2, we establish Theorems 2 and 3. Theorem 4 is proved in Section 3. The last section contains the proofs of Theorems 5 and 6, as well as some concluding remarks.

## 2    Proofs of Theorems 2 and 3

Three sets of points in the plane are said to be *separable* if each of them can be separated from the other two by a straight line. Given three separable sets, there is no straight line which intersects the convex hull of all of them.

**Lemma 1.** *Every set of $n$ points in general position in the plane has three separable subsets of size $\lfloor n/6 \rfloor$.*

*Proof.* Assume without loss of generality that $n$ is divisible by 6, and let $P$ be an $n$-element point set. Choose two lines that divide the plane into 4 regions, containing $n, 2n, n$, and $2n$ points of $P$ in their interiors, in this cyclic order. Let $P_1, P_2, P_3$, and $P_4$ denote the corresponding subsets of $P$. By the *ham-sandwich theorem*, there is a line $\ell$ which simultaneously cuts $P_2$ and $P_4$ into two halves of equal size (see Fig. 1). Then $\ell$ avoids either the convex hull of $P_1$ or that of $P_3$. Assume, by symmetry, that $P_1$ is 'above' $\ell$. Then $P_1$ and the parts of $P_2$ and $P_4$ 'below' $\ell$ are three separable sets. $\square$

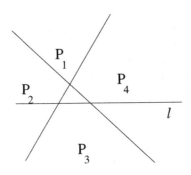

**Fig. 1.**

**Lemma 2.** *Let $\mathcal{S}$ and $\mathcal{T}$ be two systems of segments in general position in the plane. Then there are two subsystems $\mathcal{S}^* \subseteq \mathcal{S}$, $\mathcal{T}^* \subseteq \mathcal{T}$ such that $|\mathcal{S}^*| \geq \lfloor |\mathcal{S}|/330 \rfloor$, $|\mathcal{T}^*| \geq \lfloor |\mathcal{T}|/330 \rfloor$, and*
   *(i) either every segment in $\mathcal{S}^*$ crosses all segments in $\mathcal{T}^*$,*
   *(ii) or no segment in $\mathcal{S}^*$ crosses any segment in $\mathcal{T}^*$.*

*Proof.* Let $|\mathcal{S}| = m, |\mathcal{T}| = n$, and suppose, for simplicity, that both $m$ and $n$ are multiples of 330. Let $P$ be the set of endpoints of all segments in $\mathcal{S}$. By Lemma 1, there are three separable $m/3$-element subsets, $P_1, P_2, P_3 \subseteq P$. Color a segment $t \in \mathcal{T}$ with color $i$ if its supporting line does not intersect the convex hull of $P_i$ ($i = 1, 2, 3$). Let $\mathcal{T}_i$ denote the segments of color $i$. At least one third of the elements of $T$ get the same color, so we can assume with no loss of generality that $|T_1| \geq n/3$.

If there are at least $m/330$ segments in $S$, both of whose endpoints belong to $P_1$, then we are done, because these segments are disjoint from all elements of $T_1$.

Hence, we can assume that at least $(1/3 - 2/330)m = 18m/55$ elements of $S$ have precisely one of their endpoints in $P_1$. Let $Q$ denote the set of other endpoints of these segments. Let us choose three separable subsets $Q_1, Q_2, Q_3 \subseteq Q$, each of size at least $|Q|/6 = 3m/55$. Just as before, color a segment $t \in T_1$ with color $i$ if its supporting line does not intersect the convex hull of $Q_i$ ($i = 1, 2, 3$). Again, at least $|T_1|/3 \geq n/9$ elements of $T_1$ get the same color, say color 1; they form a subsystem $T_{11} \subseteq T_1$.

Let $S_{11}$ denote set of all elements of $S$ with one endpoint in $P_1$ and the other in $Q_1$. Clearly, we have $|S_{11}| = |Q_1| \geq 3m/55$.

Let us repeat now the whole procedure with $T_{11}$ in the place of $S$ and $S_{11}$ in the place of $T$. We obtain two subsets, $T' \subseteq T_{11}$ and $S' \subseteq S_{11}$, satisfying

$$|T'| \geq \frac{3|T_{11}|}{55} \geq \frac{n}{165}, \quad |S'| \geq \frac{|S_{11}|}{9} \geq \frac{m}{165}.$$

We can assume that at least half of the supporting lines of the elements of $T'$ cross the convex hull of $S'$, for otherwise we would obtain two non-crossing systems of at least $|T'|/2$ and $|S'|$ segments. The set of all elements of $T'$, whose supporting lines cross the convex hull of $S'$ is denoted by $T^*$. Similarly, we can assume that the supporting lines of at least half of the elements of $S'$ cross the convex hull of $T^*$; otherwise, we could find two non-crossing systems of at least $|T^*|$ and $|S'|/2$ segments. Let $S^*$ denote the set of all elements of $S'$, whose supporting lines cross the convex hull of $T^*$. It follows from the definitions that every element of $S^*$ crosses all elements of $T^*$ and that

$$|S^*| \geq \frac{|S|'}{2} \geq \frac{m}{330}, \quad |T^*| \geq \frac{|T|'}{2} \geq \frac{n}{330}. \quad \square$$

Given any system of segments, $S$ and $T$, in general position in the plane, define their *crossing density*, $\delta(S, T)$, as the number of crossing pairs $(s, t)$, $s \in S$, $t \in T$ divided by $|S| \cdot |T|$. Clearly, we have $0 \leq \delta(S, T) \leq 1$.

Theorems 2 and 3 readily follow from the next result.

**Theorem 7.** *There exists a constant $A < 10^6$ satisfying the following condition. Let $S$ and $T$ be any sets of segments in general position in the plane, and suppose that their crossing density is at least $c > 0$. Then there are two disjoint subsystems $S' \subseteq S$, $T' \subseteq T$ such that*

$$|S'| \geq \frac{c^A}{330}|S|, \quad |T'| \geq \frac{c^A}{330}|T|,$$

*and every segment in $S'$ crosses all segments in $T'$.*

*Proof.* Let $|\mathcal{S}| = m, |\mathcal{T}| = n$, and suppose first that both $m$ and $n$ are powers of 330. According to our assumption, $\delta(\mathcal{S}, \mathcal{T}) \geq c$.

Applying Lemma 2, we obtain two subsystems, $\mathcal{S}^* \subset \mathcal{S}, \mathcal{T}^* \subset \mathcal{T}$, such that $|\mathcal{S}^*| = m/330, |\mathcal{T}^*| = n/330$, and $\delta(\mathcal{S}^*, \mathcal{T}^*)$ is either 1 or 0. In the first case we are done, so assume $\delta(\mathcal{S}^*, \mathcal{T}^*) = 0$. Then we have

$$c \leq \delta(\mathcal{S}, \mathcal{T}) = \frac{329}{330^2}\delta(\mathcal{S}, \mathcal{T} - \mathcal{T}^*) + \frac{329}{330^2}\delta(\mathcal{S} - \mathcal{S}^*, \mathcal{T}) + \frac{329^2}{330^2}\delta(\mathcal{S} - \mathcal{S}^*, \mathcal{T} - \mathcal{T}^*).$$

Therefore, at least one of the crossing densities $\delta(\mathcal{S}, \mathcal{T} - \mathcal{T}^*)$, $\delta(\mathcal{S} - \mathcal{S}^*, \mathcal{T})$, $\delta(\mathcal{S} - \mathcal{S}^*, \mathcal{T} - \mathcal{T}^*)$ exceeds

$$c_1 := c\frac{330^2}{330^2 - 1}.$$

In other words, there exist two subsystems, $\mathcal{S}_1 \subset \mathcal{S}, \mathcal{T}_1 \subset \mathcal{T}$, with $|\mathcal{S}_1| \geq m/330$, $|\mathcal{T}_1| \geq n/330$ such that $\delta(\mathcal{S}_1, \mathcal{T}_1) \geq c_1$.

Applying Lemma 2 to $\mathcal{S}_1$ and $\mathcal{T}_1$, we obtain two subsystems $\mathcal{S}^{**} \subset \mathcal{S}_1, \mathcal{T}^{**} \subset \mathcal{T}_1$, such that $|\mathcal{S}^{**}| \geq m/330^2, |\mathcal{T}^{**}| \geq n/330^2$, and $\delta(\mathcal{S}^{**}, \mathcal{T}^{**})$ is either 1 or 0. Again, we can assume that $\delta(\mathcal{S}^{**}, \mathcal{T}^{**}) = 0$, otherwise we are done. As before, we can find two subsystems, $\mathcal{S}_2 \subset \mathcal{S}_1, \mathcal{T}_2 \subset \mathcal{T}_1$, with $|\mathcal{S}_2| \geq m/330^2, |\mathcal{T}_2| \geq n/330^2$ such that

$$\delta(\mathcal{S}_2, \mathcal{T}_2) \geq c_2 := c\left(\frac{330^2}{330^2 - 1}\right)^2.$$

Since the crossing density between any two sets is at most 1, after some

$$k \leq \frac{\log \frac{1}{c}}{\log \frac{330^2}{330^2 - 1}}$$

steps, this procedure will terminate. That is, when we apply Lemma 2 for the $k$-th time, we obtain two subsystems $\mathcal{S}' \subseteq \mathcal{S}, \mathcal{T}' \subseteq \mathcal{T}$ such that $|\mathcal{S}'| \geq m/330^k$, $|\mathcal{T}'| \geq n/330^k$, and $\delta(\mathcal{S}', \mathcal{T}') = 1$. Thus, every element of $\mathcal{S}'$ crosses all elements of $\mathcal{T}'$, and $|\mathcal{S}'| \geq c^A m$, $|\mathcal{T}'| \geq c^A n$, where

$$A \leq \frac{\log 330}{\log \frac{330^2}{330^2 - 1}} < 10^6.$$

This completes the proof of Theorem 7 in the case when $m$ and $n$ are powers of 330. Otherwise, using an easy averaging argument, we can find $\mathcal{S}_0 \subseteq \mathcal{S}, \mathcal{T}_0 \subseteq \mathcal{T}$, whose sizes are powers of 330, $|\mathcal{S}_0| \geq m/330, |\mathcal{T}_0| \geq n/330$, and $\delta(\mathcal{S}_0, \mathcal{T}_0) \geq c$. Applying the above argument to $\mathcal{S}_0$ and $\mathcal{T}_0$, the result follows. $\square$

*Proof of Theorem 2.* Assume, for simplicity, that $n$ is even. Given a system of $n$ segments in general position in the plane, which determine at least $cn^2$ crossings, one can partition it into two equal parts so that the crossing density between them is at least $2c$ (see e.g. [PA95]). Applying Theorem 7 to these parts, the result follows. $\square$

Theorem 3 can be established analogously, by repeated application of Lemma 2. However, here we deduce it from Theorems 2 and 3.

*Proof of Theorem 3.* Let $\mathcal{S}$ be a set of $n$ segments in general position in the plane with at least $cn^2$ non-crossing pairs. For any $s \in \mathcal{S}$, let $\ell(s)$ denote the supporting line of $s$. The set $\ell(s) \setminus s$ consists of two half-lines; denote them by $h_1(s)$ and $h_2(s)$. Let $\mathcal{H}_1 := \{h_1(s) : s \in \mathcal{S}\}$, $\mathcal{H}_2 := \{h_2(s) : s \in \mathcal{S}\}$, $\mathcal{T} := \mathcal{S} \cup \mathcal{H}_1 \cup \mathcal{H}_2$. Further, for any $h \in \mathcal{H}_1 \cup \mathcal{H}_2$, let $s(h)$ be the unique segment $s \in \mathcal{S}$, for which $h_1(s)$ or $h_2(s)$ is equal to $s$.

Note that if two segments $s, t \in \mathcal{S}$ do not cross each other, then the crossing between their supporting lines, $\ell(s)$ and $\ell(t)$, gives rise to a crossing between a pair of elements of $\mathcal{T}$, involving at least one half-line. Therefore, the number of crossing pairs in $\mathcal{T}$ involving at least one half-line is at least $cn^2$. There are three possibilities:

1. for some $i = 1, 2$, the number of crossing pairs in $\mathcal{H}_i$ is at least $cn^2/5$;
2. the number of crossing pairs between $\mathcal{H}_1$ and $\mathcal{H}_2$ is at least $cn^2/5$;
3. for some $i = 1, 2$, the number of crossing pairs between $\mathcal{H}_i$ and $\mathcal{S}$ is at least $cn^2/5$.

In Case 1, applying Theorem 2 to $\mathcal{H}_i$, we obtain two subsystems, $\mathcal{H}_{i1}, \mathcal{H}_{i2} \subset \mathcal{H}$, whose sizes are at least $\frac{(2c/5)^A}{660}n > \frac{(c/5)^A}{330}$, and every half-line in $\mathcal{H}_{i1}$ crosses all half-lines in $\mathcal{H}_{i2}$. Then $\mathcal{S}_1 := \{s(h) : h \in \mathcal{H}_{i1}\}$ and $\mathcal{S}_2 := \{s(h) : h \in \mathcal{H}_{i2}\}$ meet the requirements in Theorem 3.

In Case 2, apply Theorem 7 to obtain $\mathcal{H}_1' \subseteq \mathcal{H}_1$, $\mathcal{H}_2' \subseteq \mathcal{H}_2$, whose sizes are at least $\frac{(c/5)^A}{330}n$, and every element of $\mathcal{H}_1'$ crosses all elements of $\mathcal{H}_2'$. Setting $\mathcal{S}_1 := \{s(h) : h \in \mathcal{H}_1'\}$, and $\mathcal{S}_2 := \{s(h) : h \in \mathcal{H}_2'\}$, the result follows. Case 3 can be treated similarly. $\square$

# 3   Proof of Theorem 4

The proof is based on a variant of Szemerédi's Regularity Lemma, which was discovered by Komlós (see [KS96]) and can be established by an elegant argument.

For any graph $G$ and for any disjoint subsets $X, Y \subset V(G)$, let $E(X, Y) \subseteq E(G)$ denote the set of edges of $G$ running between $X$ and $Y$. Clearly, we have $|E(X, Y)| \le |X||Y|$. For any $\gamma, \delta > 0$, we call the pair $(X, Y)$ $(\gamma, \delta)$-*superregular* if for every $X' \subseteq X$ and $Y' \subseteq Y$ satisfying

$$|X'| \ge \gamma|X|, \quad |Y'| \ge \gamma|Y|,$$

we have

$$|E(X', Y')| \ge \delta|X||Y|.$$

**Lemma 3.** ([KS96]) *Let $\gamma > 0$ be a sufficiently small constant, and let $\delta > 0$.*

*Then any graph with n vertices and at least $\delta n^2$ edges has a $(\gamma, \delta)$-superregular pair $(X, Y)$ with*

$$|X| = |Y| \geq \delta^{1/\gamma^2} n.$$

First we establish

**Theorem 8.** *For every $\delta > 0$, there exists an integer $k = k(\delta) > 0$ with the following property. The intersection graph $G$ of any system of $n$ segments in the plane has $k$ bipartite subgraphs, which altogether cover all but at most $\delta n^2$ edges of $G$.*

*Proof.* Set $G_0 := G$, and let $\beta$ be a small positive constant to be specified later. Suppose that for some $i \geq 1$ we have already defined $G_{i-1}$. If $G$ does not have a complete bipartite subgraph $H_i$, which contains at least $\beta n^2$ edges of $G_{i-1}$, then stop. Otherwise, pick such a subgraph $H_i$, and let $G_i$ denote the graph obtained from $G_{i-1}$ by the deletion of all edges belonging to $H_i$. Obviously, this procedure will terminate in

$$j \leq \frac{|E(G)|}{\beta n^2} \leq \frac{1}{2\beta}$$

steps, with a graph $G_j$.

We claim that $G_j$ has fewer than $\delta n^2$ edges, provided that $\beta$ is sufficiently small. Suppose that this is not true. Then, according to Lemma 3, $G_j$ has a $(\gamma, \delta)$-superregular pair $(X, Y)$ with

$$|X| = |Y| \geq \delta^{1/\gamma^2} n,$$

where $\gamma := \frac{\delta^A}{330}$. Let $\mathcal{S}_X$ and $\mathcal{S}_Y$ denote the corresponding families of segments. In view of Theorem 7, there are two disjoint subsystems $\mathcal{T}_X \subseteq \mathcal{S}_X$ and $\mathcal{T}_Y \subseteq \mathcal{S}_Y$ with

$$|\mathcal{T}_X| = |\mathcal{T}_Y| \geq \frac{\delta^A}{330}|X| = \gamma|X| = \gamma|Y|$$

such that every segment in $\mathcal{T}_X$ crosses all elements of $\mathcal{T}_Y$. Let $X'$ and $Y'$ denote the subsets of $X$ and $Y$, corresponding to $\mathcal{T}_X$ and $\mathcal{T}_Y$, resp. Then $X'$ and $Y'$ induce a complete bipartite subgraph in $G$. Furthermore, using the fact that $(X, Y)$ is a $(\gamma, \delta)$-superregular pair in $G_j$, we obtain that at least

$$\delta|X'||Y'| \geq \delta\gamma^2\delta^{2/\gamma^2} n^2$$

edges between $X'$ and $Y'$ belong to $G_j$. Therefore, if we choose $\beta$ so small that this last quantity exceeds $\beta n^2$, then we could continue our procedure and define the next graph $G_{j+1}$. This contradiction completes the proof. $\square$

Obviously, a similar result holds for $\overline{G}$, the *complement* of a segment intersection graph $G$.

**Theorem 9.** *For every $\delta > 0$, there exists an integer $k = k(\delta) > 0$ with the following property. The complement $\overline{G}$ of the intersection graph $G$ of any system of $n$ segments in in the plane has $k$ bipartite subgraphs, which altogether cover all but at most $\delta n^2$ edges of $\overline{G}$.*

Now we are in a position to prove Theorem 4. Let $G$ denote the intersection graph of $\mathcal{S}$. Let $\delta$ be a small positive constant which will be specified later. By Theorems 8 and 9, there is a family

$$\mathcal{F} = \{A_1, B_1, A_2, B_2, \ldots, A_{2k}, B_{2k}\}$$

of subsets of $V(G)$ such that

1. $A_i$ and $B_i$ are disjoint $(1 \leq i \leq 2k)$;
2. $A_i \times B_i$ is contained either in $E(G)$ or in $E(\overline{G})$ $(1 \leq i \leq 2k)$;
3. all but at most $2\delta n^2$ pairs $\{u, v\} \subset V(G)$ are covered by $\cup_{i=1}^{2k} A_i \times B_i$.

We say that two vertices of $V(G)$ are of the *same type*, if every member of $\mathcal{F}$ contains both or neither of them. The number of different types is at most $3^{2k}$. A given type is *negligible*, if fewer than

$$s := \frac{\varepsilon n}{3^{2k}}$$

vertices have it. Letting $V_0$ denote the set of all vertices with negligible types, we have $|V_0| < \varepsilon n$.

Divide the elements of $V(G) - V_0$ into groups $V_1, V_2, \ldots, V_K$ of almost the same size: for every $1 \leq i \leq K$, let $s \leq |V_i| \leq 2s$. Clearly, we have

$$\frac{(1 - \varepsilon)n}{2s} \leq K \leq \frac{n}{s}.$$

A pair $(i, j)$, $1 \leq i \neq j \leq K$ is called *exceptional*, if $V_i \times V_j$ is not contained in $\cup_{i=1}^{2k} A_i \times B_i$. For every non-exceptional pair $(i, j)$, $V_i$ and $V_j$ induce a complete bipartite subgraph either in $G$ or in $\overline{G}$.

Let $m$ denote the number of exceptional pairs. The total number of pairs $\{u, v\} \subset V(G)$ for which $u \in V_i$, $v \in V_j$ for some exceptional pair $(i, j)$ is at least $ms^2$. On the other hand, by condition 3 above, this number cannot exceed $2\delta n^2$. Thus, we obtain that

$$\frac{m}{K^2} \leq \frac{2\delta n^2}{s^2 K^2} \leq \frac{2\delta n^2}{s^2} \frac{4s^2}{(1 - \varepsilon)^2 n^2} = \frac{8\delta}{(1 - \varepsilon)^2}.$$

This is smaller than $\varepsilon$, if $\delta$ is sufficiently small, so the partition of $\mathcal{S}$ corresponding to $V_0 \cup V_1 \cup \ldots \cup V_K$ meets all the requirements of Theorem 4.

# 4    Concluding Remarks

First we show how Theorems 5 and 6 follow from the previous results.

*Proof of Theorem 5.* Let $G$ be a geometric graph with $n$ vertices and at least $cn^2$ edges. The next result of Ajtai, Chvátal, Newborn, Szemerédi [ACNS82] and, independently, Leighton [L83] (see also [PA95], [PT97]) implies that there are at least $\frac{c}{64}e^2$ crossings pairs of edges.

**Lemma 4.** *Let $G$ be a geometric graph with $n$ vertices and $e > 4n$ edges, for some $c > 0$. Then $G$ has at least $\frac{e^3}{64n^2}$ crossing pairs of edges.*

Thus, we can apply Theorem 2 to the system $\mathcal{S} = E(G)$. We obtain two subsets $E_1, E_2 \in E(G)$ such that every edge in $E_1$ crosses all edges in $E_2$, and $|E_1| = |E_2| \geq \frac{(c/32)^A}{336}cn^2 > (c/32)^{A+2}\binom{n}{2}$. □

Theorem 6 can be proved similarly. The only difference is that, instead of Theorem 2 and Lemma 4, we have to use Theorem 3 and

**Lemma 5.** ([P91]) *Let $G$ be a geometric graph with $n$ vertices and $e \geq 3n/2$ edges, for some $c > 0$. Then $G$ has at least $\frac{4e^3}{27n^2}$ pairs of edges that do not cross and do not share an endpoint.*

The above theorems can also be established using Szemerédi's Regularity Lemma [S78]. However, then the dependence on $c$ of the sizes of the homogeneous subsystems whose existence is guaranteed by our results gets much worse.

According to an old theorem of Kővári, Sós, and Turán [KST54], every graph with $n$ vertices and at least $cn^2$ edges has a complete bipartite subgraph with $c' \log n$ vertices in its classes, where $c' > 0$ is a suitable constant depending on $c$. This immediately implies that Theorem 2 holds with the much weaker bound $c' \log n$ instead of $c'n$.

For some computational aspects of recognizing intersection graphs of segments, see [KM94].

# References

ACNS82. M. Ajtai, V. Chvátal, M. Newborn, and E. Szemerédi, Crossing-free subgraphs, *Annals of Discrete Mathematics* **12** (1982), 9–12.

B95.    I. J. Balaban, An optimal algorithm for finding segment intersections, in: *Proc. 11th Annual ACM Symposium on Computational Geometry*, ACM, 1995, 211–219.

BO79.   J. L. Bentley and T. A. Ottmann, Algorithms for reporting and counting geometric intersections, *IEEE Trans. Comput.* **C-28** (1979), 643–647.

CE92.   B. Chazelle and H. Edelsbrunner, An optimal algorithm for intersecting line segments in the plane, *J. ACM* **39** (1992), 1–54.

CS89.   K. Clarkson and P. Shor, Applications of random sampling in computational geometry II, *Discrete Comput. Geom.* **4** (1989), 387–421.

KS96.   J. Komlós and M. Simonovits, Szemerédi's regularity lemma and its applications in graph theory, in: *Bolyai Society Mathematical Studies* **2**, em Combinatorics, Paul Erdős is Eighty (Volume 2), Budapest, 1996, 295–352.

KM94.    J. Kratochvíl and J. Matoušek, Intersection graphs of segments, *Journal of Combinatorial Theory, Series B* **62** (1994), 289–315.

KST54.   T. Kővári, V. T. Sós, and P. Turán, On a problem of K. Zarankiewicz, *Colloq. Math.* **3** (1954), 50–57.

L83.      T. Leighton, *Complexity Issues in VLSI, Foundations of Computing Series*, MIT Press, Cambridge, MA, 1983.

P91.      J. Pach, Notes on geometric graph theory, in *Discrete and Computational Geometry: Papers from the DIMACS Special Year (J. E. Goodman, R. Pollack, and W. Steiger, eds.) DIMACS Series in Discrete Mathematics and Theoretical Computer Science* **6**, American Mathematical Society, Providence, RI, 1991, 273–285.

PA95.    J. Pach and P.K. Agarwal, *Combinatorial Geometry*, J. Wiley & Sons, New York, 1995.

PS91.    J. Pach and M. Sharir, On vertical visibility in arrangements of segments and the queue size in the Bentley-Ottmann line sweeping algorithm, *SIAM J. Comput.* **20** (1991), 460–470.

PT97.    J. Pach and G. Tóth, Graphs drawn with few crossings per edges, *Combinatorica* **17** (1997), 427–439.

SH76.    M. I. Shamos and D. Hoey, Geometric intersection problems, in: *Proc. 17th Annual IEEE Symp. on Found. Comput. Sci.* (1975), 151–162.

S78.     E. Szemerédi, Regular partitions of graphs, in: *Problèmes combinatoires et théorie des graphes (Colloq. Internat. CNRS, Univ. Orsay, Orsay, 1976), Colloq. Internat. CNRS,* **260**, CNRS, Paris, 1978, 399–401.

# 3–Dimensional Single Active Layer Routing

András Recski and Dávid Szeszlér

Department of Computer Science and Information Theory
Budapest University of Technology and Economics, H-1521 Budapest, Hungary
{recski,szeszler}@cs.bme.hu

**Abstract.** Suppose that the terminals to be interconnected are situated in a rectangular area of length $n$ and width $w$ and the routing should be realized in a box of size $w' \times n' \times h$ over this rectangle (single active layer routing) where $w' = cw$ and $n \leq n' \leq n + 1$. We prove that it is always possible with height $h = O(n)$ and in time $t = O(n)$ for a fixed $w$ and both estimates are best possible (as far as the order of magnitude of $n$ is concerned). The more theoretical case when the terminals are situated in two opposite parallel planes of the box (the 3–dimensional analogue of channel routing) is also studied.

## 1 Introduction

Traditionally, the detailed routing phase of the design of VLSI (Very Large Scale Integrated) circuits was considered as **a 2-dimensional problem,** gradually extended to $2, 3, \ldots$ layers. Even within this problem single row routing and channel routing (where the terminals to be interconnected are on one side, or on two opposite sides, respectively, of a rectangle) are the better understood subproblems, where the inputs are essentially one-dimensional (one or two lists of terminals, of length $n$, also called the *length* of the channel). Here the main aim is to realize the routing, and since its 'horizontal' size is given, its 'vertical' size, or *width*, $w$ should be minimized. An important quantity is the *density* of the problem: the maximum number $d$ so that there exists a vertical straight line cutting $d$ nets into two.

In case of the 2-layer Manhattan model

- (1) single row routing is always possible in $O(n)$ time with $w = d$, see [12, 17],
- (2) channel routing is **NP**-hard [18, 26],
- (3) but this latter becomes always possible in $O(nd)$ time with $w = O(d)$ if we may extend the length of the channel by introducing additional columns [3, 6, 14]. In fact, the number of these additional columns may be as large as $O(\sqrt{n})$, see [3, pp. 212–213].

In spite of (2), there are plenty of practically effective algorithms available, which can solve 'difficult' problems (of length 150...200) with width around 20 [17, 10]. In case of switchbox routing (where the terminals are on all the four sides of a rectangle – a real 2-dimensional problem) instances with length 23 and width 15...16 are already 'difficult' problems on 2 layers [5, 8, 15, 22].

J. Akiyama, M. Kano, and M. Urabe (Eds.): JCDCG 2000, LNCS 2098, pp. 318–329, 2001.
© Springer-Verlag Berlin Heidelberg 2001

If more than two layers are permitted, both channel and switchbox routing become easier, see for example [7] and [4, 25], respectively.

As technology permits more and more layers, a 'real' **3-dimensional** approach becomes reasonable. There are plenty of deep results in this area, see [1, 2, 9, 11, 13, 19, 20, 21, 23, 24], for example. Most of them embed certain 'universal-purpose' graphs (like $n$-permuters, $n$-rearrangeable permutation networks, shuffle-exchange graphs) into the 3-dimensional grid, ensuring that *pairs* of terminals can be connected, moreover, in some papers along *edge-disjoint* paths. Our result below is of much simpler structure but it allows *multiterminal nets* as well, and ensures *vertex disjoint* paths (or Steiner-trees) for the interconnections of the terminals within each net.

Throughout, except in the last section, we restrict ourselves to the *single active layer case* (all the terminals are on a single plane and the third dimension (above this plane, with *height* $h$) is for interconnections only). The terminals occupy certain gridpoints of an $n \times w$ rectangle. Henceforth we will use 'vertical direction' to refer to the direction of $h$ (that is, to the direction perpendicular to this $n \times w$ rectangle) and not for the direction of $w$.

One can easily see even in small instances like $4 \times 1$ that a routing is usually impossible unless either the length $n$ or the width $w$ may be extended by introducing extra rows or columns between rows and columns of the original grid (compare with (3) above). If it is allowed to introduce both extra rows and extra columns then there is a trivial upper bound of $h = O(wn)$ for the height of the routing; see Lemma 1 below.

Our main result is that if $w$ is fixed and $n$ becomes large, a routing of height $O(n)$ can be attained not only in the aforementioned trivial way, but even if the new length $n'$ satisfies $n \leq n' \leq n + 1$ and only $w$ is extended to $w' = cw$, where $c$ is a suitably chosen constant (we will show that $c \geq 8$ suffices). In view of Lemma 2 this linear bound is best possible. Moreover, our algorithm realizes this in $O(n)$ time, which is also essentially best possible.

Throughout this paper we are going to think of $w$ as fixed and try to obtain bounds for the height as a function of $n$ only.

## 2   Definitions and Main Results

The vertices of a given (planar) grid of size $n \times w$ are called *terminals*. A *net* $N$ is a set of terminals. A *single active layer routing problem* is a set $\mathcal{N} = \{N_1, N_2, \ldots, N_t\}$ of pairwise disjoint nets. $n$ and $w$ are the *length* and the *width* of the routing problem, respectively.

By a *spacing of $s_w$ in direction $w$* we are going to mean that we introduce $s_w - 1$ pieces of extra columns between every two consecutive columns (and also to the right hand side of the rightmost column) of the original grid. This way the width of the grid is extended to $w' = s_w \cdot w$. A *spacing of $s_n$ in direction $n$* is defined analogously.

A *solution with a given spacing $s_w$ and $s_n$* of a routing problem $\mathcal{N} = \{N_1, N_2, \ldots, N_t\}$ is a set $\mathcal{T} = \{T_1, T_2, \ldots, T_t\}$ of pairwise vertex–disjoint Steiner–

trees in the cubic grid of size $(w \cdot s_w) \times (n \cdot s_n) \times h$ (above the original planar grid containing the terminals) such that the terminal set of $T_i$ is $N_i$ for every $1 \leq i \leq t$. $h$ is called the *height* of the routing.

Since $w$ is fixed, one can also imagine the input as $w$ *rows* of terminals (each of length $n$) or as a set of $\binom{w}{2}$ channel routing problems, each with length $n$ and with a given density. Let $D$ be the maximum of these densities. Clearly, $D \leq n$.

**Theorem 1.** *If $s_w \geq 8$ then for any fixed value of $w$ and for any $n$ a single active layer routing problem can always be solved in time $t = O(n)$ and with height $h = O(n)$ such that the length $n$ is preserved or increased by at most one. Both linear bounds are best possible.*

Our algorithm gives $t = O(w^3 n)$ and $h = O(wD)$.

# 3    Straightforward Bounds

**Lemma 1.** *If $s_w \geq 2$ and $s_n \geq 2$ then every routing problem can be solved with height $h \leq \frac{w}{2}n$.*

*Proof.* We assign a separate layer to each net. For every terminal we introduce a vertical wire segment to connect the terminal with the layer of its net. The interconnection of the terminals of each net can now be performed trivially on its layer using the extra rows and columns guaranteed by the spacing in both directions.

Since 1–terminal nets can be disregarded, the number of nets is at most $\frac{1}{2}nw$ thus $h \leq \frac{w}{2}n$ follows immediately.

**Lemma 2.** *For any given $n$ there exists a routing problem that cannot be solved with height $h$ smaller than $\frac{n}{2s_w}$.*

*Proof.* Let, for simplicity, the width and the length be even, let $w = 2a$ and $n = 2b$. Consider the following example (the idea is very similar to those in [4, 16]). Suppose that each net consists of two terminals in central-symmetric position as shown in Figure 1.

The number of nets is $an$. Since each net is cut into two by the central vertical line $e$, any routing with width $w' = s_w w$ and height $h$ must satisfy $w'h \geq an$. Therefore $h \geq (w/2w')n$, hence $h \geq \frac{n}{2s_w}$.

Since in the above example $D = n$, this also proves the lower bound $h = \Omega(D)$.

The straightforward lower bound for the time is the length of the input, that is, $t = \Omega(wn)$.

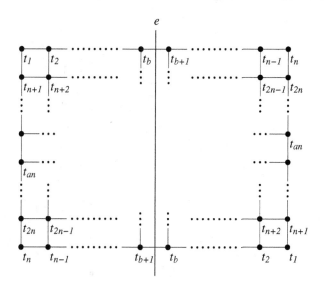

**Fig. 1.**

## 4    Two Simple Steps

**1.** Suppose at first that $w = 1$. Then what we have is essentially a single row routing problem with density $d$. Each net determines an interval of length at most $n$ and these intervals can be packed in a vertex-disjoint way into $d$ parallel lines, usually called *tracks*, using the 'left-edge algorithm' [12, 17]. Using the classical 2-layer Manhattan model, we can arrange the tracks in a horizontal plane, as shown in the top of Figure 2, thus realizing a routing with $w' = d$ and $h = 2$. However, alternatively these tracks can occupy either a vertical plane, leading to $w' = 2$ and $h = d$, or two vertical planes, leading to $w' = 3$ and $h = \lceil d/2 \rceil$, see the middle and the bottom drawing of Figure 2, respectively. (Theoretically one can pack the tracks to more vertical planes and thus ensure $h = \lceil 3d/(2w') \rceil$ for larger values of $w'$ as well but it does not seem to be interesting.) Throughout in Figures 2, 3 and 6 continuous lines denote wires while dotted lines are for the indication of coplanarity only.

Similarly, if $w = 2$ then we have a channel routing problem with density $d$ and using the same linear time algorithm we can always realize a routing with $w' = d + 1$ and $h = 3$ or with $w' = 3$ and $h = d + 1$, see Figure 3. This method is also well-known, it dates back at least to [7].

The right hand side of Figure 3 shows the essential idea of our algorithm: immediately at the level of the terminals (when leaving the 'single active layer') we increase the width from $w$ to $w' = s_w w$ in order to make enough space 'between the rows of the terminals' for the vertical plane(-s) containing the interconnecting wires. The same process is illustrated for $w = 6$ in Figure 5 below.

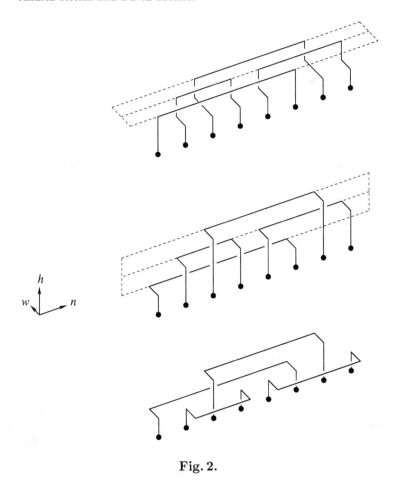

**Fig. 2.**

Later we shall need the following observation: if a larger horizontal area is available, we can decrease the height to $h = 2\lceil d/(w' - 2)\rceil + 2$. For this we can arrange the tracks in $\lceil d/(w' - 2)\rceil$ parallel horizontal planes – however, we need an 'empty' plane between two consecutive planes of tracks to ensure that the endpoints of the intervals can always reach the terminals, even if they are 'in the wrong side', as terminals $t_2$ and $t_3$ in Figure 3. Therefore the basic quantity $hw'$ can be upper bounded by essentially $3d/2$ for $w = 1$ but only by $2d$ for $w = 2$.

**2.** Let us turn now to the general problem with width $w$. Since the terminals occupy certain gridpoints of an $n \times w$ rectangle, we consider them as a collection of $w$ parallel *rows*, each of length $n$. We wish to solve $\binom{w}{2}$ channel routing problems one after the other. Figure 4 illustrates this for $w = 6$. At first (going from the bottom of the figure to the top) we solve those $w - 1$ channel routings where the rows are adjacent (first 'floor'), then those $w - 2$ ones where the distance of two rows is two (second 'floor') etc.

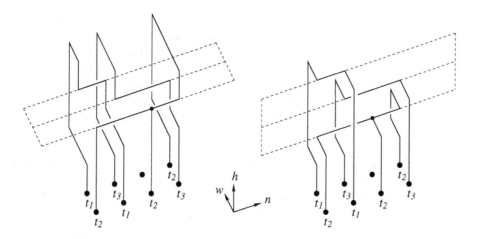

**Fig. 3.**

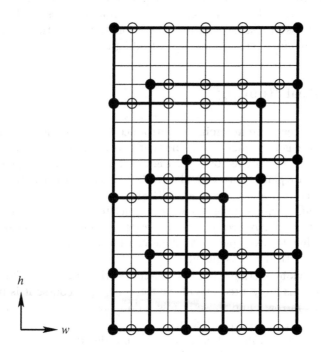

**Fig. 4.**

The $w - 1$ channel routings at the first floor (actually the $w - 1$ vertical planes containing the wire segments of these channel routings) do not interfere with one another (these are illustrated by the $w - 1$ empty dots in the bottom horizontal line of the figure).

On the other hand, the $w - 2$ channel routings at the second floor do not have this property hence this floor will have two levels, one for the row pairs 1 and 3, 3 and 5, etc and one for the row pairs 2 and 4, 4 and 6 etc. In general, floor $f$ contains $l_f$ levels where $l_f = f$ if $1 \leq f \leq \lfloor w/2 \rfloor$ and $l_f = w - f$ if $\lfloor w/2 \rfloor < f \leq w - 1$. Solid dots illustrate the rows, the vertical lines in the figure show that the terminals within a row may appear at different floors. Empty dots indicate the areas where these two rows can be interconnected. Hence such an empty dot may indicate a contribution of at most $D$ to the final height (compare with the right hand side of Figure 3).

Observe that there are two empty dots between two solid ones in the second floor, three empty dots between two solid ones in the third floor etc. Hence the total height requirement is *not* $D \times \sum_{f=1}^{w-1} l_f = O(w^2 D)$ but only $2D \times \sum_{f=1}^{w-1} \frac{1}{f} l_f = O(wD)$. The extra constant 2 is due to the necessary empty planes between the consecutive planes of tracks, as explained in the last remark in Step 1 above, concerning the empty plane between the consecutive planes of tracks.

In the introduction we mentioned that the width of the input must be extended to $w' = s_w w$. Figure 4 might give the wrong impression that $s_w = 2$ suffices. However, as we shall see in the next section, the realization of the 'crossings' in the figure requires much more space, leading to $s_w = 8$.

## 5     The Real Routing

For future reference we are going to introduce the following terminology. By a *w–plane* we are going to mean a plane that is perpendicular to the width of the routing, that is, to the 'vector' $w$ of Figures 2, 3 and 6. Analogously, *h–planes* and *n–planes* are planes perpendicular to the height and the length of the routing (or to the vectors $h$ and $n$), respectively. Similarly, by a *w–wire segment* we are going to mean a wire segment that is parallel to the width of the routing or to the vector $w$. *h–wire segments* and *n–wire segments* are defined in the same way. (Note that for example an *h–plane* is a horizontal plane, while an *h–wire* segment is a vertical wire segment.)

In the previous section Figure 4 illustrated the order how the $\binom{w}{2}$ channel routing problems are routed one above the other. (Of course it is possible that the terminals in row $i$ and those in row $k$ do not share any net and therefore a whole level within a floor is missing.)

However, Figure 4 may alternatively be considered as a 'cross-section' of the routing by an $n$–plane of size $w' \times h$ (and then there are $n$ copies of these cross-sections, one behind the other). In this sense an empty dot in a particular level indicates a whole $w$–plane of length $n$ and of height at most $2\lceil D/f \rceil$ containing several wires one above the other for that channel routing. Hence a horizontal line in Figure 4 between a solid and an empty dot indicates a wire segment going towards this plane – but it may go for a wire segment running in this plane or it may wish to avoid it and go for a wire segment in one of the other parallel planes (that is, towards one of the further empty dots).

It is very important to realize, therefore, that there are two types of 'crossings' which have to be avoided if we wish to realize the final 3-dimensional routing along the $n$–planes one after the other:

- Type 1 — A vertical line, connecting two solid dots, and a horizontal line, connecting two empty dots, may cross each other in Figure 4.
- Type 2 — A horizontal line which passes through an empty dot in Figure 4 may, in fact, not use that particular $n$–wire (which is perpendicular to the actual $n$–plane).

The basic point is that crossings of Type 2 can be avoided by a detour within the actual $n$–plane (increasing the height by one and the width by two) but crossings of Type 1 can be avoided outside the $n$–plane only. Therefore we must use the adjacent $n$–plane as well. But what happens if in this latter $n$–plane there is another vertical line interconnecting two solid dots (or in the real routing: a $h$–wire segment coming from a terminal) blocking the detour?

We avoid this problem in the following way: Since $w' > w$, we must, in any case, start the routing in each $n$–plane by 'expanding' the $w$ terminals into larger distances. This increases the height by $w/2$ and can be performed like in either of the ways shown in Figure 5. Now, if we use the two kinds of expansions alternatively then the terminals within a single row will form a zigzag pattern and hence two $h$–wires that 'should' block each other in two consecutive $n$–planes, will actually be shifted by two units away.

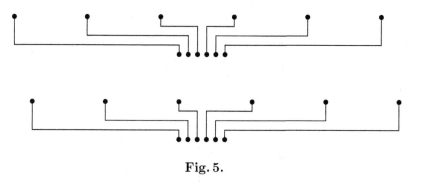

**Fig. 5.**

This way we have ensured that a Type 1 detour will not be blocked by an $h$–wire segment coming from a terminal. However, one difficulty remains to be solved: A Type 1 detour can still cross a $w$–wire segment that goes in the same $h$–plane one unit behind (connecting two $w$–planes within a level). In order to handle this, let us number the $n$–planes from 'front' to 'back' with the numbers $1, 2, \ldots, n$ and the $h$–planes form 'bottom' to 'top' with the numbers $1, 2, \ldots, h$. Now let us declare the following rule: If a $w$–wire segment goes in an $n$–plane numbered with an even number, then it must go in an $h$–plane also numbered with an even number; similarly, if a $w$–wire segment goes in an $n$–plane numbered with an odd number, then it must go in an $h$–plane also numbered with an odd number. Since the height of a level is always the double of the number of tracks in it, the above rule can obviously be fulfilled. (This way some of the Type 2 detours will become unnecessary: If a $w$–wire segment reaches a $w$–plane (containing some of the tracks within a level) between two consecutive tracks, then the $w$–wire segment can cross the $w$–plane without meeting the tracks, there is no need for a Type 2 detour.)

Figure 6 illustrates most of these situations in a single drawing. Recall that continuous lines are wires, dotted lines are for the indication of coplanarity only. Observe that $t_1, t_2, \ldots$ are terminals within a single row (illustrating the aforementioned zigzag pattern), while the terminals $t_1', t_2', \ldots$ form the next row. It might be instructive to recall that the detour between $A$ and $B$ is of Type 1 while that between $C$ and $D$ is of Type 2.

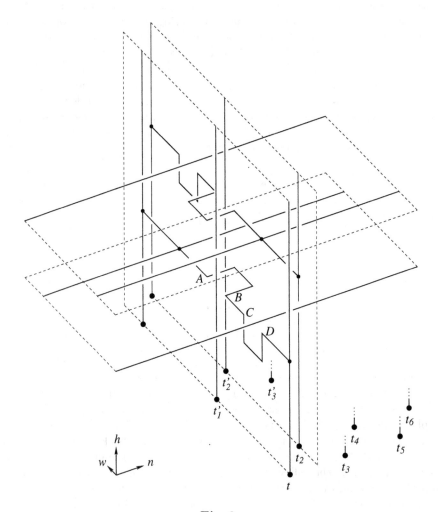

**Fig. 6.**

Figure 7 shows a part of Figure 6 again in order to explain why $s_w$ needs to be as large as 8.

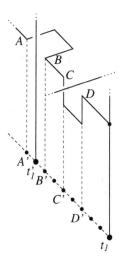

**Fig. 7.**

# 6   3–Dimensional Channel Routing

Throughout this paper we have dealt with the single active layer routing problem, which is, in a sense, the 3–dimensional analogue of single row routing. However, the 3–dimensional analogue of channel routing may also be of interest not only from a technical point of view (see [11] for example), but also in a theoretical sense: in contrast to the essential difference in complexity between single row routing and channel routing in the 2–layer Manhattan model (see (1) and (2) in the Introduction), there does not seem to be such a difference between their 3–dimensional analogues. The *3–dimensional channel routing problem* is defined as two parallel rectangular planes of size $n \times w$ containing all the terminals to be interconnected (by vertex disjoint Steiner trees) in a box of size $n' \times w' \times h$ between the two parallel planes. As before, we suppose that $w$ is fixed and $n$ can be very large hence we allow $w' = s_w w$ but $n \le n' \le n+1$ only.

**Theorem 2.** *For any fixed value of $w$ and for any $n$ such a 3-dimensional channel can always be routed in $t = O(n)$ time such that the length $n$ is preserved or increased by at most one, the width is extended to $w' = s_w w$ and the required height is $h = O(n)$. Both linear bounds are best possible.*

The lower bound is the same as in Lemma 2. The routing can be performed basically along the same line as explained in Sections 4 and 5 for the single active layer case. Just like the two 'expansions' of Figure 6 are alternating along the direction $n$, the expansions opposite to each other on the two parallel planes should also be shifted by two units away.

## Acknowledgments

Grants No. OTKA 29772 and 30122 of the Hungarian National Science Foundation, grant No. FKFP 409/1997 of the Hungarian Ministry of Education and the NATO Collaborative Linkage Grant PST.CLG.976383 are gratefully acknowledged.

## References

[1] Aggarwal, A., M. Klawe, D. Lichtenstein, N. Linial, and A. Wigderson, 1991. A lower bound on the area of permutation layouts, *Algorithmica* 6, 241-255.

[2] Aggarwal, A., J. Kleinberg, and D. P. Williamson, 2000. Node-disjoint paths on the mesh and a new trade-off in VLSI layout, *SIAM J. Computing* 29, 1321-1333.

[3] Baker S. B., S. N. Bhatt, and F. T. Leighton, 1984. An approximation algorithm for Manhattan routing, *Advances in Computing Research* 2, 205–229.

[4] Boros, E., A. Recski and F. Wettl, 1995. Unconstrained multilayer switchbox routing, *Ann. Oper. Res.* 58, 481-491.

[5] Burstein, M. and R. Pelavin, 1983. Hierarchical wire routing, *IEEE Trans. Computer-Aided Design* CAD-2, 223-234.

[6] Burstein, M. and R. Pelavin, 1983. Hierarchical channel router, *Integration, the VLSI Journal* 1, 21-38.

[7] Chan, Wan S., 1983. A new channel routing algorithm, *in:* R. Bryant, ed. *3rd Caltech Conference on VLSI*, Comp. Sci. Press, Rockville, 117-139.

[8] Cohoon, J.P. and P. L. Heck, 1988. BEAVER: A computational-geometry-based tool for switchbox routing, *IEEE Trans. Computer-Aided Design* CAD-7, 684-697.

[9] Cutler, M. and Y. Shiloach, 1978. Permutation layout, *Networks* 8, 253-278.

[10] Deutsch, D, 1976. A dogleg channel router, *Proc. 13rd Design Automation Conf.* 425-433.

[11] Enbody, R.J., G. Lynn, and K. H. Tan, 1991. Routing the 3–D chip, *Proc. 28th ACM/IEEE Design Automation Conf.* 132–137.

[12] Gallai T. His unpublished results have been announced in A. Hajnal and J. Surányi, 1958. Über die Auflösung von Graphen in vollständige Teilgraphen, *Ann. Univ. Sci. Budapest Eötvös Sect. Math.* 1, 115-123.

[13] Games, R. A., 1986. Optimal book embeddings of the FFT, Benes, and barrel shifter networks, *Algorithmica* 1, 233-250.

[14] Gao, S. and M. Kaufmann, 1994. Channel routing of multiterminal nets, *J. of the ACM* 41, 791-818.

[15] Grötschel, M., A. Martin, and R. Weismantel, 1993. Routing in grid graphs by cutting planes, *in:* G. Rinaldi and L. Wolsey, eds. *Integer programming and combinatorial optimization,* 447-461.

[16] Hambrusch, S. E., 1985. Channel routing in overlap models, *IEEE Trans. Computer-Aided Design of Integrated Circ. Syst.* CAD-4, 23-30.

[17] Hashimoto, A. and J. Stevens, 1971. Wire routing by optimizing channel assignment, *Proc. 8th Design Automation Conf.* 214-224.

[18] LaPaugh A. S., 1980. A polynomial time algorithm for optimal routing around a rectangle, *Proc. 21st FOCS Symp.* 282-293.

[19] Leighton, T., 1983. *Complexity issues in VLSI: Optimal layouts for the shuffle-exchange graph and other networks,* The MIT Press, Cambridge, MA.

[20] Leighton, T. and A. L. Rosenberg, 1986. Three-dimensional circuit layouts, *SIAM J. Computing* 15, 793-813.

[21] Leighton, T., S. Rao and A. Srinivasan, 1999. New algorithmic aspects of the Local Lemma with applications to routing and partitioning, *Proc. Tenth Annual ACM-SIAM Symp. on Discrete Algorithms,* ACM/SIAM, New York and Philadelphia, 643-652.

[22] Luk, W.K., 1985. A greedy switch-box router, *Integration, the VLSI Journal* 3, 129-149.

[23] Rosenberg, A.L., 1983. Three-dimensional VLSI: A case study, *J. ACM.* 30, 397-416.

[24] Shirakawa, I., 1980. Some comments on permutation layout, *Networks* 10, 179-182.

[25] Szeszlér D., 1997. Switchbox routing in the multilayer Manhattan model, *Ann. Univ. Sci. Budapest Eötvös Sect. Math,* 40, 155–164.

[26] Szymanski, T. G., 1985. Dogleg channel routing is **NP**-complete, *IEEE Trans. Computer-Aided Design of Integrated Circ. Syst.* CAD-4, 31-41.

# Nonregular triangulations, view graphs of triangulations, and linear programming duality

Fumihiko Takeuchi

fumi@is.s.u-tokyo.ac.jp

Dept. of Information Science, Univ. of Tokyo, 113-0033, Japan

**Abstract.** For a triangulation and a point, we define a directed graph representing the order of the maximal dimensional simplices in the triangulation viewed from the point. We prove that triangulations having a cycle the reverse of which is not a cycle in this graph viewed from some point are forming a (proper) subclass of nonregular triangulations. We use linear programming duality to investigate further properties of nonregular triangulations in connection with this graph.

## 1 Introduction

Let $\mathcal{A} = \{\boldsymbol{p}_1, \ldots, \boldsymbol{p}_n\} \subset \mathbb{R}^d$ be a point configuration with its convex hull $\text{conv}(\mathcal{A})$ being a $d$-dimensional polytope. A *triangulation* $\Delta$ *of* $\mathcal{A}$ is a geometric simplicial complex with its vertices among $\mathcal{A}$ and the union of its faces equal to $\text{conv}(\mathcal{A})$. A triangulation is *regular* (or *coherent*) if it can appear as the projection of the lower faces of the boundary complex of a $(d+1)$-dimensional polytope in $\mathbb{R}^{d+1}$. If not, the triangulation is *nonregular*. (See, for example, [6] [11].)

Starting from the study of generalized hypergeometric functions, Gel'fand, Kapranov & Zelevinskiĭ showed that regular triangulations of a point configuration are forming a polytopal structure described by the secondary polytope [4] [5]. In connection with Gröbner bases, Sturmfels showed that initial ideals for the affine toric ideal determined by a point configuration correspond to the regular triangulations of the point configuration [9] [10]. Regular triangulations are a generalization of the Delaunay triangulation well known in computational geometry, and have also been used extensively in this field [2].

Though nonregular triangulations are known to be behaving differently from regular triangulations, they are not well understood yet. Santos showed a nonregular triangulation with no flips indicating that a flip graph can be disconnected, which never happens when restricted to regular triangulations [8]. Ohsugi & Hibi showed the existence of a point configuration with no unimodular regular triangulations, but with a unimodular nonregular triangulation [7]. Also, de Loera, Hoşten, Santos & Sturmfels showed that cyclic polytopes can have exponential number of nonregular triangulations compared to polynomial number of regular ones [1]. The aim of this paper is to put some insight into nonregular triangulations.

J. Akiyama et al. (Eds.): JCDCG 2000, LNCS 2098, pp. 330-338, 2001.

In the sequel, we fix a triangulation $\Delta$. For the triangulation $\Delta$ and a point $v$ in $\mathbb{R}^d$, we define the *graph* $G_v$ *of* $\Delta$ *viewed from* $v$ as the directed graph with its vertices corresponding to the $d$-simplices in $\Delta$ and a directed edge $\overrightarrow{\sigma\tau}$ existing when $\sigma$, $\tau$ are adjacent and $v$ belongs to the closed halfspace having the affine hull $\mathrm{aff}(\sigma \cap \tau)$ as its boundary and including $\sigma$. When $v \in \mathrm{aff}(\sigma \cap \tau)$, both edges $\overrightarrow{\sigma\tau}$, $\overrightarrow{\tau\sigma}$ appear in $G_v$. The graph $G_v$ is a directed graph whose underlying undirected graph is the adjacency graph of the $d$-simplices in $\Delta$. Of course, $G_v$ might differ for different choices of $v$. Though there are infinitely many choices of viewpoints $v$, there are only finitely many view graphs $G_v$.

A sequence of vertices $\sigma_1, \sigma_2, \ldots, \sigma_i, \sigma_1$ in $G_v$ forms a *cycle* when $\overrightarrow{\sigma_1\sigma_2}, \ldots,$ $\overrightarrow{\sigma_{i-1}\sigma_i}, \overrightarrow{\sigma_i\sigma_1}$ are edges of $G_v$ and $\sigma_i \neq \sigma_j$ for $i \neq j$. We define a cycle $\sigma_1, \sigma_2, \ldots, \sigma_i,$ $\sigma_1$ to be *contradicting* when the reverse sequence $\sigma_1, \sigma_i, \ldots, \sigma_2, \sigma_1$ is not a cycle in $G_v$. For vertices $\sigma_1, \ldots, \sigma_i$ in $G_v$, the edges $\overrightarrow{\sigma_1\sigma_2}, \ldots, \overrightarrow{\sigma_{i-1}\sigma_i}, \overrightarrow{\sigma_2\sigma_1}, \ldots, \overrightarrow{\sigma_i\sigma_{i-1}}$ exist if and only if $v \in \mathrm{aff}(\sigma_1 \cap \cdots \cap \sigma_i)$.

The regularity of a triangulation can be stated as a linear programming problem, so regularity and linear programming obviously have a connection. An interesting point in our argument is that we use linear programming duality to analyze in further detail some properties of nonregular triangulations.

For any triangulation, the condition of regularity can be written as a linear programming problem as follows. Let the variable $w = (w_1, \ldots, w_n)$ represent the $(d+1)$-coordinates of the lifting (or weight) of the vertices $p_1, \ldots, p_n$, such that the triangulation is lifted to a piecewise linear function $f_w$ from $\mathrm{conv}(\mathcal{A})$ to $\mathbb{R}$. For each $(d-1)$-simplex in $\Delta$ not in the boundary of the convex hull $\mathrm{conv}(\mathcal{A})$, the local convexity of $f_w$ can be expressed by a linear inequality involving only the vertices of the two adjacent $d$-simplices in $\Delta$ (see Section 2.1). Gather the inequality constraints that correspond to each such interior $(d-1)$-simplex. Altogether, we get a system of inequalities $Aw > 0$ ($0$ is the zero vector), and the triangulation is regular if and only if this has a solution. Easily, this is equivalent to $Aw \geq 1$ ($1$ is the vector with all entries one) having a solution. Thus, by linear programming duality (or Farkas' lemma), the triangulation is nonregular if and only if the *dual problem* $uA = 0$, $u \geq 0$ has a nonzero solution.

Our main theorem constructs a nonzero solution of the dual problem combinatorially and explicitly from a contradicting cycle in a graph of the triangulation viewed from some point.

**Theorem.** *For a triangulation $\Delta$, if a graph $G_v$ viewed from some point $v$ contains a contradicting cycle, in correspondence with this cycle, we can make a nonzero solution of the dual problem stated above. Thus, $\Delta$ is nonregular. The support set (i.e. collection of nonzero elements) of this solution becomes a subset of the edges forming the cycle. On the other hand, the reverse of this claim is not true. There exists a nonregular triangulation with none of its view graphs $G_v$ containing a contradicting cycle. (See Example 5)*

The theorem says that triangulations containing a contradicting cycle in its graph $G_v$ viewed from some point $v$ are forming a (proper) subclass of nonregular triangulations. This subclass of triangulations is interesting in that their nonregularity are described more combinatorially using graphs. On the other hand, regularity or nonregularity, defined by linear inequalities, are of continuous nature. This is the first attempt to give a (combinatorial) subclass of nonregular triangulations. Even if we consider contradicting closed paths instead of contradicting cycles, allowing to pass the same vertex more than once, the class of the triangulations having such contradicting thing in its view graph does not change, because any contradicting closed path includes a contradicting cycle.

Checking that Example 5 is a counterexample for the reverse of the implication in the theorem (i.e. the view graph from any viewpoint does not contain a contradicting cycle), can be done by extensive enumeration of view graphs. However, by describing nonregularity as a linear programming problem, and using linear programming duality, we prove the counterexample in a more elegant way.

A similar but different directed graph of a triangulation viewed from a point has been studied by Edelsbrunner [3]. This was in the context of computer vision, and his graph represents the in_front/behind relation among simplices of any dimension, even not adjacent to each other. When our graph and the restriction of Edelsbrunner's graph to $d$-simplices are compared, neither includes the other in general. However, if we take the transitive closure of our graph, it includes his graph as a (possibly proper) subgraph. Our graph might be more appropriate in describing combinatorial structures of triangulations, because their underlying undirected graphs are the adjacency graphs of $d$-simplices. Edelsbrunner proves that if a triangulation is regular, his graph viewed from any point is "acyclic". The line shelling argument in a note there gives a proof for the contrapositive of our theorem, but without explicit construction of a solution of the dual problem.

We first prepare basic results, and then prove our main theorem (Section 2). Finally, we give illustrative examples and a counterexample for the reverse of the main theorem (Section 3).

## 2    Regularity, linear programming, and duality

### 2.1    Inequalities for regularity

A triangulation $\Delta$ of the point configuration $p_1, \ldots, p_n$ is regular if there exists a lifting (or weight) $w_1, \ldots, w_n \in \mathbb{R}$ such that the projection with respect to the $x_{d+1}$ axis of the lower faces of the boundary complex of the $(d + 1)$-dimensional polytope $\mathrm{conv}(\binom{p_1}{w_1}, \ldots, \binom{p_n}{w_n})$ becomes $\Delta$. This condition on the lifting is equivalent to the condition that the function from $\mathrm{conv}(\mathcal{A})$ to $\mathbb{R}$ obtained by interpolating the lifting according to the triangulation in a piecewise linear fashion is convex. This implies (in fact, is equivalent to) the local convexity of this function in the neighborhood of every $(d - 1)$-simplex in $\Delta$ which is not on the boundary of $\Delta$. These conditions can be described by inequalities with $w_1, \ldots, w_n$ the variables.

A global criterion for convexity is therefore as follows:

– For each $d$-simplex $\text{conv}(\boldsymbol{p}_{i_0}, \ldots, \boldsymbol{p}_{i_d})$ in $\Delta$, and any point $\boldsymbol{p}_j \notin \{\boldsymbol{p}_{i_0}, \ldots, \boldsymbol{p}_{i_d}\}$, the lifted point $\binom{\boldsymbol{p}_j}{w_j}$ is above the hyperplane $\text{aff}(\binom{\boldsymbol{p}_{i_0}}{w_{i_0}}, \ldots, \binom{\boldsymbol{p}_{i_d}}{w_{i_d}})$ in $\mathbb{R}^{d+1}$:

$$\begin{vmatrix} 1 & \cdots & 1 \\ \boldsymbol{p}_{i_0} & \cdots & \boldsymbol{p}_{i_d} \end{vmatrix} \begin{vmatrix} 1 & \cdots & 1 & 1 \\ \boldsymbol{p}_{i_0} & \cdots & \boldsymbol{p}_{i_d} & \boldsymbol{p}_j \\ w_{i_0} & \cdots & w_{i_d} & w_j \end{vmatrix} > 0.$$

A local criterion for convexity can be expressed with much fewer inequalities as follows:

– For each interior $(d-1)$-simplex $\text{conv}(\boldsymbol{p}_{i_1}, \ldots, \boldsymbol{p}_{i_d})$ in $\Delta$, where the two $d$-simplices $\text{conv}(\boldsymbol{p}_{i_0}, \boldsymbol{p}_{i_1}, \ldots, \boldsymbol{p}_{i_d})$ and $\text{conv}(\boldsymbol{p}_{i_1}, \ldots, \boldsymbol{p}_{i_d}, \boldsymbol{p}_{i_{d+1}})$ are intersecting, the lifted point $\binom{\boldsymbol{p}_{i_{d+1}}}{w_{i_{d+1}}}$ is above the hyperplane $\text{aff}(\binom{\boldsymbol{p}_{i_0}}{w_{i_0}}, \ldots, \binom{\boldsymbol{p}_{i_d}}{w_{i_d}})$ in $\mathbb{R}^{d+1}$:

$$\begin{vmatrix} 1 & \cdots & 1 \\ \boldsymbol{p}_{i_0} & \cdots & \boldsymbol{p}_{i_d} \end{vmatrix} \begin{vmatrix} 1 & \cdots & 1 & 1 \\ \boldsymbol{p}_{i_0} & \cdots & \boldsymbol{p}_{i_d} & \boldsymbol{p}_{i_{d+1}} \\ w_{i_0} & \cdots & w_{i_d} & w_{i_{d+1}} \end{vmatrix} > 0. \qquad (*)$$

The equivalence of these two convexity conditions is proved easily by reducing to the one dimensional case.

The collection of inequalities $(*)$ for all interior $(d-1)$-simplices in $\Delta$ form a linear program which we denote by

$$A\boldsymbol{w} > 0.$$

We say the matrix $A$ of this linear program *represents the regularity of* $\Delta$. Note that this program has solutions if and only if the program $A\boldsymbol{w} \geq \mathbf{1}$ has solutions. Let $m$ be the number of interior $(d-1)$-simplices in $\Delta$. The matrix $A$ is an $m \times n$ matrix.

**Lemma 1** *For a triangulation* $\Delta$, *and the matrix $A$ representing its regularity,* $\Delta$ *is regular if and only if there exists* $\boldsymbol{w} \in \mathbb{R}^n$ *such that* $A\boldsymbol{w} \geq \mathbf{1}$. *By linear programming duality (or Farkas' lemma),* $\Delta$ *is nonregular if and only if there exists* $\boldsymbol{u} \geq \mathbf{0}$, $\boldsymbol{u} \neq \mathbf{0}$ *such that* $\boldsymbol{u}A = \mathbf{0}$.

Thus, the (non)regularity of $\Delta$ can be judged by the existence of a nonzero point in the polyhedron $\{\boldsymbol{u} \geq \mathbf{0} : \boldsymbol{u}A = \mathbf{0}\} \subset \mathbb{R}^m$ of the set of solutions of the dual problem.

## 2.2  A nonzero solution of the dual problem from a contradicting cycle

Here, we give an explicit construction of a nonzero solution of the dual problem $\boldsymbol{u}A = \mathbf{0}, \boldsymbol{u} \geq \mathbf{0}$, from a contradicting cycle in the graph $G_v$ viewed from some point $\boldsymbol{v}$. For $\boldsymbol{v} \in \mathbb{R}^d$, a $d$-simplex $\sigma$ in $\Delta$, and $\boldsymbol{w} \in \mathbb{R}^n$, define $x_{d+1}(\boldsymbol{v}, \sigma, \boldsymbol{w})$ as follows: consider the projection along the $(d+1)$-coordinate of the point $\boldsymbol{v}$ to the affine hull of $f_{\boldsymbol{w}}(\sigma)$, the lifting of the $d$-simplex $\sigma$ by $\boldsymbol{w}$, in $\mathbb{R}^{d+1}$, and let $x_{d+1}(\boldsymbol{v}, \sigma, \boldsymbol{w})$ be the $x_{d+1}$ coordinate of this point.

**Lemma 2** *Let $\Delta$ be a triangulation, $A$ the matrix representing its regularity, and $v \in \mathbb{R}^d$. For an edge $\overrightarrow{\sigma\tau}$ in the graph $G_v$ viewed from $v$, there exists a constant $\alpha_{\sigma\cap\tau} \geq 0$ such that*

$$x_{d+1}(v, \sigma, w) - x_{d+1}(v, \tau, w) = \alpha_{\sigma\cap\tau} A_{\sigma\cap\tau,*} w \qquad \text{(for any } w \in \mathbb{R}^n\text{)},$$

*where $A_{\sigma\cap\tau,*}$ is the row of $A$ corresponding to the interior $(d-1)$-simplex $\sigma\cap\tau$ in $\Delta$. Furthermore, $v \in \mathrm{aff}(\sigma\cap\tau)$ if and only if $\alpha_{\sigma\cap\tau} = 0$.*

*Proof.* Straightforward.

We now construct a nonzero solution of the dual problem from a contradicting cycle. This will prove the forward implication in our main theorem.

*Proof.* (**main theorem**) Suppose we have a contradicting cycle $\sigma_1, \sigma_2, \ldots, \sigma_i, \sigma_1$ in $G_v$. By Lemma 2, we can find $\alpha_{\sigma_1\cap\sigma_2}, \ldots, \alpha_{\sigma_i\cap\sigma_1} \geq 0$, or their collection as a vector $\alpha \geq 0$, satisfying for any $w \in \mathbb{R}^n$,

$$x_{d+1}(v, \sigma_1, w) - x_{d+1}(v, \sigma_2, w)$$
$$\cdots$$
$$+ x_{d+1}(v, \sigma_i, w) - x_{d+1}(v, \sigma_1, w)$$
$$= \alpha_{\sigma_1\cap\sigma_2} A_{\sigma_1\cap\sigma_2,*} w$$
$$\cdots$$
$$+ \alpha_{\sigma_i\cap\sigma_1} A_{\sigma_i\cap\sigma_1,*} w$$
$$= \alpha A w$$
$$= 0.$$

(The elements corresponding to the edges not in the cycle have 0 as their value in vector $\alpha$.) Thus, $\alpha A = 0$. Since we took a contradicting cycle, by Lemma 2, $\alpha \neq 0$. Hence, we obtain a nonzero solution of the dual problem $uA = 0, u \geq 0$. This together with Lemma 1 proves the forward implication in the main theorem.

## 2.3    Recognizing nonregularity or finding contradicting cycles

Judging whether the given triangulation $\Delta$ is (non)regular reduces to judging whether the system of inequalities $Aw \geq 1$ has a solution $w$, where the matrix $A$ represents the regularity of $\Delta$ in the sense described above. This is a linear programming problem, and can be computed in polynomial time for fixed dimension $d$, for example, using interior point method.

One way to judge if a triangulation $\Delta$ has a contradicting cycle in some view graph $G_v$ is to enumerate all possible view graphs and enumerate the cycles there. The generation of view graphs can be done, for example, by generating all graphs viewed from the minimal cells in the hyperplane arrangement made by the affine hulls of the interior $(d-1)$-simplices in $\Delta$.

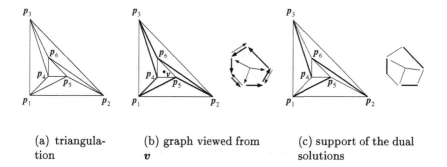

(a) triangula-
tion

(b) graph viewed from
$v$

(c) support of the dual
solutions

**Fig. 1.** Example 3.

## 3   Examples

**Example 3 (A nonregular triangulation with 6 vertices)** For the point configuration

$$p_1 = (0\,0), \qquad p_2 = (4\,0), \qquad p_3 = (0\,4),$$
$$p_4 = (1\,1), \qquad p_5 = (2\,1), \qquad p_6 = (1\,2),$$

we consider the triangulation $\Delta$ indicated in Fig. 1(a) below. The graph $G_v$
viewed from $v = (\frac{4}{3}\,\frac{4}{3})$ is in Fig. 1(b), since $v$ lies on $p_1p_4$, $p_2p_5$, and $p_3p_6$. It has
one contradicting cycle $p_1p_4p_5, p_1p_2p_5, p_2p_5p_6, p_2p_3p_6, p_3p_4p_6, p_1p_3p_4, p_1p_4p_5$
denoted by bold edges. The matrix representing the regularity of $\Delta$ is

$$A = \begin{array}{c|cccccc} & w_1 & w_2 & w_3 & w_4 & w_5 & w_6 \\ \hline p_1p_4 & 3 & & 1 & -8 & 4 & \\ p_1p_5 & -1 & 1 & & 4 & -4 & \\ p_2p_5 & 1 & 3 & & & -8 & 4 \\ p_2p_6 & & -1 & 1 & & 4 & -4 \\ p_3p_4 & 1 & & -1 & -4 & & 4 \\ p_3p_6 & & 1 & 3 & 4 & & -8 \\ p_4p_5 & 1 & & & -3 & 1 & 1 \\ p_4p_6 & & & 1 & 1 & 1 & -3 \\ p_5p_6 & & 1 & & 1 & -3 & 1 \end{array}.$$

The polyhedron of the solutions of the dual problem is a single ray

$$\{u \geq 0 : Au = 0\} = \mathbb{R}_{\geq 0}(0\,1\,0\,1\,1\,0\,0\,0\,0),$$

where interior 1-simplices are indexed lexicographically. The support of the
nonzero solutions is denoted by bold edges in Fig. 1(c). Remark that they are
included in the (underlying undirected) edges of the contradicting cycle.

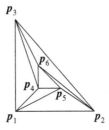

**Fig. 2.** Triangulation of Example 4.

**Example 4 (Another nonregular triangulation with 6 vertices)** The vertex $p_2$ in Example 3 is perturbed. The point configuration becomes

$$p_1 = (0\,0), \qquad p_2 = (\tfrac{7}{2}\,0), \qquad p_3 = (0\,4),$$
$$p_4 = (1\,1), \qquad p_5 = (2\,1), \qquad p_6 = (1\,2).$$

The triangulation $\Delta$ is indicated in Fig. 2 below. Each of the graph viewed from $v_1 = (\tfrac{5}{4}\,\tfrac{3}{2})$, $v_2 = (\tfrac{4}{3}\,\tfrac{4}{3})$, or $v_3 = (\tfrac{7}{5}\,\tfrac{7}{5})$ has a unique contradicting cycle. The matrix representing the regularity of $\Delta$ is

$$A = \begin{array}{c|cccccc} & w_1 & w_2 & w_3 & w_4 & w_5 & w_6 \\ \hline p_1p_4 & 3 & & 1 & -8 & 4 & \\ p_1p_5 & -1 & 1 & & \tfrac{7}{2} & -\tfrac{7}{2} & \\ p_2p_5 & \tfrac{1}{2} & 3 & & & -7 & \tfrac{7}{2} \\ p_2p_6 & & -1 & \tfrac{1}{2} & & 3 & -\tfrac{5}{2} \\ p_3p_4 & 1 & & -1 & -4 & & 4 \\ p_3p_6 & & 1 & \tfrac{5}{2} & 3 & & -\tfrac{13}{2} \\ p_4p_5 & 1 & & & -3 & 1 & 1 \\ p_4p_6 & & & 1 & 1 & 1 & -3 \\ p_5p_6 & & 1 & & \tfrac{1}{2} & -\tfrac{5}{2} & 1 \end{array} .$$

The polyhedron of the solutions of the dual problem is a cone

$$\{u \geq 0 : Au = 0\}$$
$$= \mathbb{R}_{\geq 0}(1\,8\,0\,8\,5\,0\,0\,0\,0)$$
$$+ \mathbb{R}_{\geq 0}(0\,8\,2\,14\,7\,0\,0\,0\,0)$$
$$+ \mathbb{R}_{\geq 0}(0\,6\,0\,7\,6\,1\,0\,0\,0)$$
$$+ \mathbb{R}_{\geq 0}(0\,2\,0\,2\,1\,0\,1\,0\,0)$$
$$+ \mathbb{R}_{\geq 0}(0\,2\,0\,2\,2\,0\,0\,1\,0)$$
$$+ \mathbb{R}_{\geq 0}(0\,1\,0\,2\,1\,0\,0\,0\,1),$$

where interior 1-simplices are indexed lexicographically. The first three rays correspond to the solutions made by the contradicting cycles in view graphs

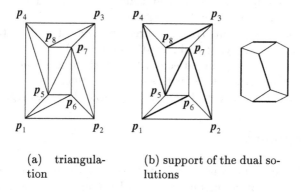

(a)  triangula-
tion

(b) support of the dual solutions

**Fig. 3.** Example 5.

$G_{v_1}, G_{v_2}, G_{v_3}$, as in Subsection 2.2. The latter three rays have no such correspondence.

**Example 5 (Counterexample for the reverse of the main theorem)** With
the point configuration

$$p_1 = (0\ 0), \qquad p_2 = (3\ 0), \qquad p_3 = (3\ 4), \qquad p_4 = (0\ 4),$$
$$p_5 = (1\ 1), \qquad p_6 = (2\ 1), \qquad p_7 = (2\ 3), \qquad p_8 = (1\ 3),$$

the triangulation $\Delta$ indicated in Fig. 3(a) below is a nonregular triangulation
with none of its view graphs $G_v$ containing a contradicting cycle. The matrix
representing the regularity of $\Delta$ is

$$A = \begin{array}{c|cccc|cccc}
 & w_1 & w_2 & w_3 & w_4 & w_5 & w_6 & w_7 & w_8 \\
\hline
p_1 p_5 & 3 & & & 1 & -8 & 4 & & \\
p_1 p_6 & -1 & 1 & & & 3 & -3 & & \\
p_2 p_6 & 2 & 4 & & & & -9 & 3 & \\
p_2 p_7 & & -2 & 2 & & & 4 & -4 & \\
p_3 p_7 & & 1 & 3 & & & & -8 & 4 \\
p_3 p_8 & & & -1 & 1 & & & 3 & -3 \\
p_4 p_8 & & & 2 & 4 & 3 & & & -9 \\
p_4 p_5 & 2 & & & -2 & -4 & & & 4 \\
\hline
p_5 p_6 & 2 & & & & -4 & 1 & 1 & \\
p_6 p_7 & & 2 & & & 2 & -5 & 1 & \\
p_7 p_8 & & & 2 & & 1 & & -4 & 1 \\
p_5 p_8 & & & & 2 & 1 & & 2 & -5 \\
p_5 p_7 & & & & & -2 & 2 & -2 & 2 \\
\end{array}.$$

The polyhedron of the solutions of the dual problem is a single ray

$$\{u \ge 0 : Au = 0\} = \mathbb{R}_{\ge 0}(0\ 2\ 0\ 1\ 0\ 2\ 0\ 1\ 0\ 0\ 0\ 0\ 1),$$

where interior 1-simplices are indexed lexicographically. The support of the nonzero solutions is denoted by bold edges in Fig. 3(b). If a contradicting cycle existed for some view graph $G_v$, this (directed) cycle should contain all of the bold edges (in its underlying undirected counterpart). However, there are no cycles containing all of these bold edges. Hence, there exists no view graph $G_v$ containing a contradicting cycle for this example. (Remark: If we take the edge $p_6 p_8$ instead of $p_5 p_7$, this new flipped triangulation becomes regular.)

## Acknowledgments

The author thanks Kenji Kashiwabara for bringing the problem to the author's interest, and Hervé Brönnimann, Masahiro Hachimori, Hiroshi Imai, Mary Inaba, Francisco Santos, Akihisa Tamura, and an anonymous referee for comments and encouragements.

## References

1. JESÚS A. DE LOERA, SERKAN HOŞTEN, FRANCISCO SANTOS, AND BERND STURM-FELS, The polytope of all triangulations of a point configuration, *Doc. Math.*, **1** (1996) 103–119.
2. HERBERT EDELSBRUNNER, *Algorithms in combinatorial geometry*, Springer-Verlag, Berlin, 1987.
3. HERBERT EDELSBRUNNER, An acyclicity theorem for cell complexes in $d$ dimension, *Combinatorica*, **10** (1990) 251–260.
4. ISRAEL M. GELFAND, MIKHAIL M. KAPRANOV, AND ANDREI V. ZELEVINSKY, *Discriminants, resultants and multidimensional determinants*, Birkhäuser, Boston, 1994.
5. ISRAEL M. GEL'FAND, ANDREI V. ZELEVINSKIĬ, AND MIKHAIL M. KAPRANOV, Newton polyhedra of principal $A$-determinants, *Soviet Math. Dokl.*, **40** (1990) 278–281.
6. CARL W. LEE, Regular triangulations of convex polytopes, DIMACS Series in Discrete Mathematics and Theoretical Computer Science 4, Amer. Math. Soc., 1991, 443–456.
7. HIDEFUMI OHSUGI AND TAKAYUKI HIBI, A normal $(0, 1)$-polytope none of whose regular triangulations is unimodular, *Discrete Comput. Geom.*, **21** (1999) 201–204.
8. FRANCISCO SANTOS, A point configuration whose space of triangulations is disconnected, *J. Amer. Math. Soc.*, **13** (2000) 611–637.
9. BERND STURMFELS, Gröbner bases of toric varieties, *Tôhoku Math. J.*, **43** (1991) 249–261.
10. BERND STURMFELS, *Gröbner bases and convex polytopes*, American Mathematical Society, Providence, RI, 1996.
11. GÜNTER M. ZIEGLER, *Lectures on polytopes*, Springer-Verlag, New York, 1995.

# Efficient Algorithm for Searching a Polygonal Room with a Door

Xuehou Tan

Tokai University, Numazu 410-0395, Japan
tan@fc.u-tokai.ac.jp

**Abstract.** We study the problem of searching for a mobile intruder in a polygonal room $P$ with a door $d$ by a mobile searcher. This is to decide whether there exists a search schedule to detect the intruder with allowing him to exit the door $d$, no matter how he moves, and if so, generate a search schedule. A search is called the $k$-searcher if he holds $k$ flashlights and can see along only $k$ rays emanating from his position. This problem had first illuminated by a $k$-searcher. For a 1-searcher we present an optimal $O(n \lg + m)$ time and $O(n)$ space algorithm for solving a search, if it exists, where $n$ is the number of vertices of $P$ and $m$ $(< n^2)$ is the minimum number of search instructions required to clear $P$. This improves upon the previous $O(n^2)$ time and space. Optimality of our algorithm is assured by identifying critical visibility events occurring in $P$ and processing them exactly... Furthermore our method can easily be extended to solve the problem of searching a room by a 2-searcher. The extension is based on an application of the notion of visibility to weak link 2-visibility.

## 1 Introduction

Recently, much attention has been devoted to the problem of searching a room for a mobile intruder in a polygonal region $P$ by a searcher [3,6,11]. The objective is to decide whether there exists a search schedule to detect the intruder who moves continuously with an arbitrary speed, and if so, generate a search schedule. This problem had been first introduced by Suzuki and Yamashita [9]. Both the searcher and the intruder are modeled as points that can move continuously in $P$. A searcher is called the $k$-searcher if he holds $k$ flashlights, where $k$ is a positive integer, and he can see only along the rays emanating from his flashlights. The searcher can change the flashlights continuously with a bounded speed, and the ray directions. A polygonal region $P$ is said to be searchable if a search schedule exists for the $k$-searcher. The problem of searching a polygonal room by a $k$-searcher has first been introduced in [11].

Several restricted variants of the search problem have been studied in situations where re-entrance is allowed. In the study of searchability, which we consider two guards and the problem of moving around a room, essays on the boundary of the polygon, as an application, can be formed by a search groups...

J. Akiyama, M. Kano, and M. Urabe (Eds.): JCDCG 2000, LNCS 2098, pp. 339-350, 2001.

of e sahe r [15 4,0]. Wh d          (e ntac st )oh poly gn on day ig ntrduc shr m    s a hp b emN ntrude r sha b t nt he dag d thugh          Le b. ga van $O(n^2$ ad pa es t t other m  sa rb hd hf k≥a goom h b y          k-se ar hiah b by (k−1)-ahe [ 7 ] H the h ateabae mh bnd ad ni fie (de t hehate ah   -se arh 8]diffe f m hr 2s ahe rs 7 ].

I p ig u ni    fl ad   fl ints h tint o the m  se ah prb m 1 -ah pre s at h     $O(og\ n+m)$ bad $O(n)$ s g ith m  fge ratingh  sd u b fis ts whe re ah e numb r of re s fl $Pm (\le n^2$) th ninu m  numbe fs ah - stru t seu irdo be a P. The t y gith m    b b y d t g rt h b t ur rd   n P d d g th se ah shd bh em    Fh m our th xt ndo sb t he p h ng m     by a 2-a h f The ta d on ge ne b af t in fl h b h a h k          -2-vis b ly

## 2  P elim  a y

Wl fis g ade  fit n s t he oo m  sa p rb h he w b kto   -y ad pr b n [1, 4, ])

### 2.1  Basic definitions

A ri Rs mb p g     (ige 1  fit rse h sh b )wt h do or d h b oun d ay  B n iv as u m t h ah e r P   a ge nebo t b h e he. Th a n h re ev r t s f P   ae h d no t he dg t nis h a pi T w pont s   $x, y$ P ae l o bu tu k vi b e f t h h s t $\overline{xg}$ one g t he nie n t y ont ad P. F t i n   $R\ Q$ P w s ay h t  R is we k a b viy e f Q f ry p t h R vis b m  s pt h  Q.

L s(t)b e t h pe n  k-s a he r ad $f_1(t)$,.., $f_k(t)$ n of n d t fl  flh g tah bu n d P  a t m t pe v y A p t x P l be d e  te te d or l un te d  at t h r t d e of he he g ts  $s(t)f_1(t)$.., $s(t)f_k(t)$ An y gh hagt n t a h e n t ru d m  (whe pt b n k nowt ho h e s e ar h a e p b e of mg bi tr ar y  fl t be n  tm n a te d o h e r wis it be c l A s a h h d ul e of k t h b f P  a u p S =< $s, f_1, \cdots, f_k$ > of k+1 n h u os fn t c n s  $s, f_1, .., f_k: [0 1] \to P$ su b h tat r u de r s b a d t at h t n fl  $s(\ )f_1(\ )$,.., $s(\ )f_k(\ )$ wt h a t h w g h t o t h a g P P  do be  k-a h h e xt s a se ah d u bof h ks ahe r 6r P.

### 2.2  Two-guard walkability of corridors

A ri d P  is a b y h t rac e   u a de t  h b oun d - ar Th v  -y ad pr b m ak h e s wh   P su b h t aw (be  -

at $\bar{v}$ )gard $l$ a d ow h g t poly gal h a s          hd   Ro a t fr m
t o  p e ise ad he o he  m e r is in  wh          h d
r a e  w n u t u  w s b  Br w n t s     $p, p' \in$ ws a h      p e e d e s
$p'$ ad  $p'$ s e e d p  y n t e         pb e   $p'$ whe hre s ng $L$ f m s to
t W w r t e $p < p'$.The d fih oh   R s  n r c.      l $l(t$ d $r(t)$ de no e
h g n sf w g a s          hd  r n L a d $R$ e s pe y.  A w kn
y R t h e be 6 m u h s f h u of n t o s           $l : [0,1] \to L$
ad  $r : [0,1] \to R$ he    $l(0) = r(0) = u, l(1) = r(1) = $ d   $l(x$ d  $r(x)$ a e
mutu y  h   $x$. An  g n t   $\overline{l(x)r(x)}$ s c  a l  w k  s  g e t n
Aoa  w k n P f m a w h  g e   t $\overline{p_0 q_0}$ to th o th r $\overline{p_1 q_1}$ wh e  $p_0 < p_1$
ad  $q_0 < q$ h a f l  f l t h  d o n   $l(0) = p_0$ , $r(0) = q_0$  $l(1) = p_1$ d
$r(1) = q_1$ . Th w k s t r  g t b h       hd  s t o f n t o s

B t x   $x$  h h a , h        $Succ(x)$ n t e t h e  r t 6 t h e
h y s u e d n g     $x$ a P   $red(x$ h t n d a t p r e d n g
$x$. A v r t f e  f e f s n t e o a k g a r h        80° h e w s e
t  e x Ap o t a n d   f h o r e  f l e x s t s h a 6 r g h  t s : t h e
b u d g h t f m   a s t x   r h n  r R o b y   $Backw(r)$,
s h e f i r s p n 6 P  h b y a b t h  h h e d r t o f m   $Succ(r)$
b r h e 6 d s h o   f w(r) h e  f i r p n h y t h b u t h o
at r h e t o f m  $\underline{Pred(r)}$ t o  Se F g 1. W e  f h t a t o of
h e g t   $\overline{rBackw(r)}$ or $\overline{rForw(r)}$ as f m  t o $Backw(r)$ $Forw(r)$ .

A p r f o r e  f l t e  $p \in L$ q $\in R$ m       a d e d k  6 a , o
an u d d h 6 h  $q < Backw(p) \in R$ ad  $p < Backw(q) \in L$ h h
(F g. 1 a) , o n  d e d k  6 a  v d d h o f   $q > Forw(p) \in R$
ad  $p > Forw(q) \in L$ h h  (F g. 1 b) A p of e  f l  r t s $p, p' \in L$ ( p.
$q, q' \in R$ s s h o  6 m  a w e d  p $\propto p'$ fd  $Forw(p') < Backw(p)$ h R ( p.
$q < q'$ a d $Forw(q') < Backw(q)$ h L) h h. Se F g c  -d.

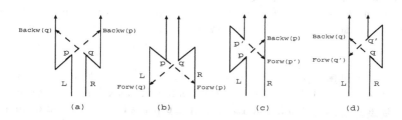

**Fig. 1** .Dea h ks a m h

**Lemma 1** [4,5] A o r h o B w h  e if d h y if th  h s  L d  R a e
m tu h l a y a k  y s b e  d o d e d l k o s   . F u th r m o, if th d  it a
th t th r e  a e a n d  g s h  R s d d , e o d r h o P s t r  g t w h h  e. It
tak $\Theta(sn)$ tim e to t e h th two-w  h l it y of  o r  h d o, $O(n)$ tim e to o t p t a
str  g t w h h, d  $O(n \lg n + m)$ tim e to o t p t a  w k, w r e m $(\leq n^2)$ s th
m n in h n m h r o f s a h s  t r o t s  .

# 3   Polygonal rooms searchable by a 1-searcher

A searcher is a 1-searcher... 

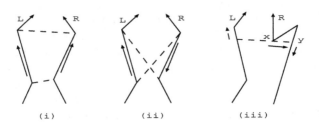

(i)          (ii)          (iii)

**Fig. 2** Search instructions of a searcher.

                                                                    7

Let us denote the points on the boundary clockwise of the boundary of $P$, starting at the door $d$, and proceeding, we consider two points $d_l$ and $d_r$ such that $d_l \le p \le d_r$ for a point $p$ on the boundary of $P$. Since finding an edge that hides the Search in 2.2. But in $P$, $Succ(x)$ denotes the vertex succeeding $x$ and $Pred(x)$ the vertex that is preceding. Once a reflex vertex $r$ is behind and forward rays to $Backw(r)$ and $Forw(r)$ which are the first points of $P$ hit by the bus shot in the directions from $Succ(r)$ to $r$ and from $Pred(r)$ to $r$ respectively. Any skew edge $d$-$d_l$ (e.g, Fig. 3 a) and e) find (So that the boundary of $P$ are two edges from $d_l$ to $d_r$, then in the quadratic edges that should be non-redundant.) Finally, we can see an $a$-$d_l$-$b_l$ any point on the boundary of $P$ that should be the ones which indeed are the last hidden door $d$.

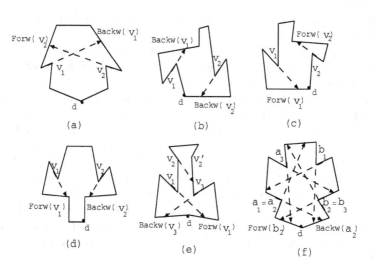

(a) (b) (c)

(d) (e) (f)

Fig. 3. The conditions A2 and A3.

**Theorem 1** A polygonal room $P$ is the achievable if one of the following conditions is true.

(A1) There are two reflex vertices $v_1$ and $v_2$ such that $v_1 < Backw(v_1), Forw(v_2) < v_2$ (Fig. 3a), $v_1 < Backw(v_1) < v_2 < Backw(v_2)$ (Fig. 3b), $Forw(v_1) < v_1 < Forw(v_2) < v_2$ (Fig. 3c), or $Forw(v_1) < v_1 < v_2 < Forw(v_2)$ (Fig. 3d).

(A2) There are three reflex vertices $v_1$, $v_2$ and $v_3$ such that $v_1 Forw(v_1)$ intersects with $v_3 Backw(v_3)$ and $v_1 < v_2 < Backw(v_2) < v_3$ or $v_1 < Forw(v_2) < v_2 < v_3$ (Fig. 3e).

(A3) There are two chains $<a_1, a_2, a_3>$ and $<b_1, b_2, b_3>$ (some of $a_i$ and $b_j$, $1 \le j \le 3$, may be identical etc.) such that (A3-a) $a_1$ and $a_3$ may see $Succ(a_2)$-directed edge also $a_2 < Backw(a_2)$, (A3-b) $b_1$ and $b_3$ may see $Pred(b_2)$-directed edge also $Forw(b_2) < b_2$, and (A3-c) all vertices $v$ between $a_3$ and $b_1$ are viewed and looked to (Fig. 3f).

**Proof.** ... first ... the ... A1 (Fig. 3) ... the ... and ... whether the ... $Succ(v_1)$ ... $red(v_2)$ ... . ( ... ... [5, 11]) ... the ... $Succ(v_1)$ ... $red(v_2)$ ... be ... such ... the ... starting ... ... The ... is shown in Fig. 3 below ...

... the ... A2, ... find three ... $Pred(v_1)$, $Succ(v_2)$ and $Succ(v_3)$ in Fig. 3 e) ... the ... the ... two ... is ... both the third ... It ... the ... the ... of ... and ... P ... 1 ... are ... .

... the ... A3 $Succ(a_3)$ ... to be ... P ... $red(a_1)$, $Succ(a_2)$ and $Succ(a_3)$ and $Pred(b_1)$ ... are ... P ... $red(b_1)$, $Pred(b_2)$ and $Succ(b_3)$; ... . Furthermore ... $v$ ... $a$ and $b_1$ ... $v$-de ... accessible ... should exist (Fig. 3 f).

Finally ... that the ... construction ... are ... ... □

**Remark.** The ... A1, A2 and A3 ... the ... N1, N2 and N3 in Theorem 5 [ ... ]. Note that ... given ... been ... redundancy.

... difficult ... to ... here ... Using the ... adapt the ... of P ... ... denote ... reflect ... by ... P ... by ... ... to ... the boundary. A ... visibility ... produces ... ... P ... $r < Backw(r)$ or $r > Forw(r)$ ... But ... both ... simultaneously. ) ... C ... P (C) is ... the ... P ... (C') where C' is the ... ... where ... ... [1, 2, 13].

**Theorem 2** *A ... P is 1-searchable if one of the conditions of Theorem ...*

**Proof.** Assume that the ... of P is visible from ... P ... simple ... . Let $r_1, ..., r$ be the sequence of the ... boundary of P, ... Let $Ray(r_i)$ be the ... the ... $r_i$. Without loss of ... $\overline{r_1Ray(r_1)}$ ... inside ... a from $r_i$ to $Ray(r_i)$. (Otherwise ... the ... $\overline{r_lRay(r_l)}$, and ... be ... counterclockwise. )

In ... by ... $L(x,y)$ ... $R(x,y)$ the ... from $x$ to $y$ with $x < y$. Usually ... $x = r_{i-1}$ and $y = r_i$ ... $L(x,y)$ . If a ... ... a ... from ... by ... the ... ... ... ...

*Case 1. The condition A3-b is true.*

*Ca 1.1 Th  d  is  to  th  ig  b  f  l  e  th  c   s $\overline{r_iRay(r_i)}$, 1 $\leq l$.*
If he  r  t  wh  by           (  be  s  se    se   fine  d t  th
  b  he  b  s  by  d  e  by             r  the first  h  te (  o
F g. 5  d  F g  6 ) . E  sc  r  t   $r_l$  pre  d  *( e  $r_l < r^*$) .
O  s  h  u  H        d  $r^*$. A,  w   s  h  p  h  e
1  he  r  s  g  s  he  b  of  her  e  h  m       th  b  t
(  v  d  f  m  d) .

*Ca 1.1.1 i  .First  h  s*  $L(d_l, r_1)$  d  $R(Ray(r_1), d_r$  em  u  t  u
  y  b,  he  r  wh  he  r  he  t  a  t  b  e          $r_1$ (F g4a -b) ,
  o  d  **A1**  **A2**  we  t  r  u (F g. 4  -d) . N  d  e  u  b  n
$L(d_l, r_1)$  d  $R(Ray(r_1), d_r$  b  s  t  h  e  s  t  c  e  $r_1$  h  c  w  e ,  the  id-
**A1** or **A2**  w  r  e t  r  (F g4  -f)  H  n  e ,  t  her  g  $P$ $(r_1Ray(r_1))$  b  ab
b  y  d  s  or  1  -s  e  ar  b  m   e  t  e  d  b  $\overline{r_1Ray(r_1)}$ .

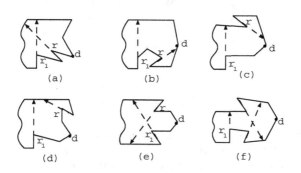

(a)      (b)      (c)

(d)      (e)      (f)

**Fig. 4 .Case 1.1 1**

*Ca 1.1.2  $< i \leq l$. I  f  h  e  d  b  n*         **A1**  t  h  u  t  $\overline{r_{i-1}Ray(r_{i-1})}$
n  t  e  s  wh  $r_iRay(r_i)$  C  s  d  e  f  s  he  s  t  u  a  wh  e  t  he  h  $L(r_{i-1}, r_i)$
  s  b  b  o              $R(Ray(r_{i-1}), Ray(r_i))$  If  o  n  t  s  $L(r_{i-1}, r_i)$  ar  b
  he  n  t  e  p  n  f  $r_{i-1}Ray(r_{i-1})$  d  $r_iRay(r_i)$  d  h  e  e  fl  e  x
  t  r  n  $R(Ray(r_{i-1}), Ray(r_i))$  h  h  $\underline{Backw(r) < r_i < r^* < r < Ray(r_i)}$
h  b  a  he          f  b  t  f  m  $r_{i-1}Ray(r_{i-1})$  $r_iRay(r_i)$  If  h  e
are  s  m  r  s  $R(Ray(r_{i-1}), Ray(r_i))$  u  h  h  $Backw(r) < r_i < r^* <$
$r < Ray(r_i)$  b  a  n         s  t  p  d  wh  he  first  s  r  s  a-
  n  t  d (F g 5 )  d  t  t  o  p  b  n        1  he  r  W
h  h  t  he  r  t  r  t  m  $\overline{sr}$  o  $r^*$. S  t  he  fi  t  t  a  -
g  $Backw(r) < r_i < r^* < r < Ray(r_i)$  h  n  e  h  h
$R(r^*, r)$  o  t  he  r  w  d  g  n  $L(s, r^*)$  w  e  $r_i$  w  n  c  h,
  he  w  he  t  h  t  s  b  n  $r_{i-1}$  d  $r_i$. The  a  e  n  d  -
  s  b  h  s  $L(s, r^*)$  d  $R(r^*, r)$  t  h  w  d  b  n  **A1**  o
**A3-b**  w  r  u  h  h  $L(s, r^*)$  b  o  $R(r^*, r)$  he  r  w  $r_i$
  a  b  l,  t  he  e  he  r  t  b  t  w  $r_{i-1}$  d  $r_i$,  he
b  g  t  n  $L(s, r^*)$  $r^*$  d  s  f  he  d  n  **A3-b**. A,  he
h  $R(r^*, r)$  s  b  $L(s, r^*)$  he  r  he  r  s  e  h
  t  h  h  b  he  f  h  r   e  n  b  first  t  s  g

$Backw(r) < r_i < r^* < r < Ray(r_i)$. Hence $r$ is a proof, and with $r$ ... for $r^*$.

In the situation where points in $\overline{L(r_{i-1}, r_i)}$ are not visible ... step 6, ... $\overline{r_{i-1}Ray(r_{i-1})}$ and $\overline{r_iRay(r_i)}$ ... be the first ... in $\overline{L(r_{i-1}, r_i)}$ such that $P$ ... $red(r)$ is visible to ... (It is possible ... 6 $Succ(r)$ ... is visible to these two points ... otherwise the vertex $r$ first ... $\overline{r_{i-1}Ray(r_{i-1})}$ ... $\overline{rForw(r)}$ ... so on (Fig. 5) ... from which $P$ is ... such that $r'$ satisfying $Backw(r') < r_i < r^* < r' < Ray(r_i)$ is encountered.

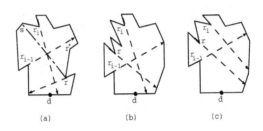

(a)          (b)          (c)

**Fig. 5** .Case 1.1.2

Consider the situation where $\overline{L(r_{i-1}, r_i)}$ is wholly in $R(Ray(r_{i-1}), Ray(r_i))$. Let $r$ be the vertex in $\overline{L(r_{i-1}, r_i)}$ whose forward ray to $Forw(r)$ is the minimum among the one with ... (Fig. 5). Since $Forw(r) > Ray(r_i)$, these ... given $Forw(r)$ intersects with $\overline{r_{i-1}Ray(r_{i-1})}$, but does not with $\overline{r_iRay(r_i)}$. Because 6 ... in ... $Forw(r)$, ... in $\overline{L(r_{i-1}, r)}$ is wholly in $R(Ray(r_{i-1}), Forw(r))$. As done ... flashlight be at ... from $\overline{r_{i-1}Ray(r_{i-1})}$ to $rForw(r)$ (The situation is shown Fig. 5 c and our ... otherwise, the condition A2 ... retried). The ... $\overline{L(r, r_i)}$ is now wholly in $R(Ray(r_i), Forw(r))$. Then there is to ... otherwise that ... **A1** or **A2** are true. Because the ... $r$ and $r_i$, and ... or between ... point ... the ... in **A1** or **A2** are true. The flashlight is then from $\overline{rForw(r)}$ to $\sigma_iRay(r_i)$ using a ...

*Case 1.1.3* The region $P(\overline{r_lRay(r_l)})$ is encountered. If ... $P - P(\overline{r_lRay(r_l)})$ is ... between the point ... $r$ ... $Ray(r_l)$ ... and thus ... from $P$ ... by moving ... to the boundary ... $P - P(\overline{r_lRay(r_l)})$. Otherwise, ... be the first ... from $r_l$ to $Ray(r_l)$ such that ... $P$ ... $red(r)$ from being visible to $Ray(r_l)$ and is forward ray to $Forw(r)$ ... among the ... the ... (Fig. 6). Obviously, $Forw(r) > Ray(r_l)$. Since $R(Ray(r_l), Forw(r))$ is wholly ... is visible to $\overline{L(r_l, r)}$, we can rotate the flashlight from $\overline{r_lRay(r_l)}$ to $rForw(r)$. If ... $r^*$ is ... then from $\overline{rForw(r)}$ with ... $r^*$, ... as the forward ray from $\overline{rForw(r)}$ to $\sigma r^*$. Because 6 minimum of $Forw(r)$ and ... the vertex $r^*$, then ... in $\overline{L(r, r^*)}$ is wholly in $R(r^*, Forw(r))$. As the ...

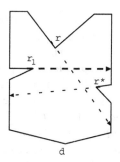

**Fig. 6** .Case 1.1.3 .

b $R(r^*, Forw(r))$ is w ak y is b t o $L(r, r^*)$ b w s t h e b n g r t x in $R(r^*, Forw(r))$ e t h c b, ot og bw t h a d $r^*$ s a f **A3-b**, Th a n o d s bt w h a s b w s t h e c od t n **A1** or **A3-b** w r t r u Th e a e h h e b a $L(r, r^*)$ p $R(r^*, Forw(r))$; ot b w h e e w r e o t h b t c s a e $r$ Th u b n p r o d, a b w b r o o Rd . If $r^*$ s a b h h f om t o $Forw(r)$ b b r o t a g r o d u r e s p b 6 m d . In h s , a s a b s u 5 c b ar n g o om $P$ a u t u b t p u t

*Ca 1.2. Th d o d s to th r g t e $\overline{r_i Ray(r_i)}$ to th l t p me $\overline{r_j Ray(r_j)}$ (i < j).* As a m t h a r s h e fis c r t t s u b d s t o t h t f $\overline{r_j Ray(r_j)}$ W t h a h e r s a m d o $r_j$. Th e e ar n d-d e d b or $r_j$-d b p h e w s b n l b n **A1** o **A3-b** w e s b f l. Th b a $R(r_j, d_r)$ b b to $L(d_l, r_j)$ pt b w s $r_j$ a d h e b g t s $R(d_l, r_j)$ b l s a f h e d b **A1**. Th n $L(d_l, r_j)$ s w ak y b $R(r_j, d_r)$ p h e r w s $r_j$ w r n b c r t b p h d t n **A1** w t r u H n e, t h b n s p r o g a d h w h b o om $P$ s t h b e d u s n y m d t o $r_j$.

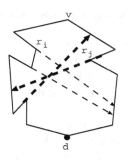

**Fig. 7** .Case 1.2

We show that if $m$ ... $d$ ... $r_i$ Let $r_i$ be the ... $\overline{r_i Ray(r_i)}$ ... between $r_i$ and $r_j$. The ... from $d$ ... $r_i$ and $r_j$ ... any ... $d$ to $r_i$ or from $r_j$ to $d$ ... $d$ ... A1 we true (Fig. 7), ... for $r_j$ or $r_i$; the ... (resp. $r_j$) ... first ... $\overline{r_i Ray(r_i)}$ (resp. $\overline{r_i Ray(r_j)}$).

*Case* 2. The ... A3-a is ... It is ... Case ...

*Case* 3. Both A3-a and A3-b are true. Since A3-c is true in this case, ... vertex $d'$ ... $<a_1, a_2, a_3>$ satisfying A3-a ... $<b_1, b_2, b_3>$ satisfying A3-b ... $d'$ and $d'$-... (Fig. 7).

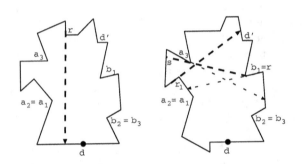

**Fig. 8** .Case 3 .

Assume first ... $L(d_l, d')$ and $R(d', d_r)$ are mutually ... Since the $d$-... or $d'$-... the ... from $d$ to $d'$. Now we show that ... from $d$ ... $d'$ ... If ... the ... of ... the ... the ... as in ... Case 1.2, we ... show that ... from $d$ ... $d'$ is ... If ... to ... the ... $r_i < d' < Ray(r_i)$ ... $\leq i \leq l$ ... the ... $L(d_l, d')$ and $R(d', d_r)$ are not mutually ... As in Case 1 and Case 1.2, we have the ... $\overline{r_i Ray(r_i)}$ ... $i = 1, \ldots, l$. It is ... shown in Fig. 5 ... no ... the ... a straight ... from $s$ to $d'$. Remember ... $r_i < d' < Ray(r_i)$, ... $d'$ ... $<b_1, b_2, b_3>$ satisfying A3-b before $d'$ ... If the ... shown in Fig. 5 ... ... the ... from $\overline{r_i Ray(r_i)}$ to $d'$.

If the ... $L(d_l, d')$ is ... visible from ... $R(d', d_r)$, then $R(d', d_r)$ ... visible from ... $L(d_l, d')$; ... A1 are true. (The ... that ... when $R(d', d_r)$ ... visible to ... $L(d_l, d')$ ... be ... analogously.) With ... r ... that ... $r$ ... first ... r-... in $L(d_l, d')$ ... the $Pred(r)$ ... visible to ... point ... $R(d', d_r)$ (Fig 8a). Then, $L(d_l, Pred(r))$ and $R(Pred(r), d_r)$ are mutually weakly ... No ... $Pred(r)$-... ; otherwise ... triple ... $<b_1, b_2, b_3>$ satisfying the and in **A3-b** be $d'$ ... Here is ... $d$ ... $red(r)$. Since $red(r)$ has the property ... $d'$, as shown by ... $d$ to $Pred(r)$ is optimal.

... the situation where the ... in $L(d_l, d')$ such that $Succ(r)$ is visible from a point in $R(d', d_r)$. In this case ... the ... $d$ is to the first essential cut ... Assume that there are $Succ(r_l)$ ... ; otherwise ... $d$ to $Succ(r_l)$. As ... and ... where ... $\overline{(r_l Ray(r_l))}$. If the situation is shown in Fig. 5 ... is counterclockwise ... $\overline{(r_l Ray(r_l))}$, ... $r_i < d' < Ray(r_i)$ ... otherwise the ... in **A2** we ... the ... straight ... from $\overline{sr}$ to $d'$. Suppose that the ... $\overline{(r_l Ray(r_l))}$ is the ... d. If $d'$ is to the ... $\overline{r_l Ray(r_l)}$, ... the ... is a straight ... from $r_l Ray(r_l)$ to $d'$. If $d'$ is to the ... of $r_l Ray(r_l)$ ... in ... 3 ... rotate the ... with the ... at ... is stopped when ... there ... or ... $P$ ... reflex ... where $d' < r$ ... $Backw(r) < r$ ... s.t. ... the opposite ... the ... (Fig. 8b). The ... the ... a straight ... from $\overline{sr}$ to $d'$. Since $P$ ... $\overline{(r_l Ray(r_l))}$ ... the ... the duo ... presented is optimal.

In the ... when ... $P$ ... d. ... other situation ... the new rule ... $\Box$

**Theorem 3** *For a polygonal room $P$, it ... be ... a ... time search ... schedule ... in $O(\log n + m)$ time and $O(n)$ space, if it exists, where $n$ is the number of ... of $P$, $m (\le n^2)$ is the minimum number of search instructions required to search $P$.*

**Proof.** ... takes $O(n \log n)$ ... into ... the 1-search ... of polygon ... After ... is ... A1, A2 and A3 ... rules the construction algorithm ... Theorem 1 ... output a ... the duo. Now ... that the ... should be ... It ... ... $P(\overline{r_1 Ray(r_1)})$ ... and the search ... should be ... in ... 1 ... 6 ... Theorem 2 ... If the situation shown in Fig. ... is counterclockwise ... 1, 2, ... the ... the duo for ... $P(\overline{(r_i Ray(r_i))})$ is optimal to ... Thus ... these ... the algorithm ... is ... used. Once the ... that shown in Fig. 5 ... in ... 2, as straight ... is a ... optimal search schedule. Similarly ... the search ... in ... 1 13 ... is optimal. The optimal search ... should in ... 2 as in ... 8 is argued ... discussed ... proof of Theorem 2. It ... the proof. $\Box$

## 4 Extension

... room of ... a polygon ... by ... be extended to solve the search problem for 2-searchers. As shown in [1, 10, 11], ... that the ... is based on ... the notion of ... k-visibility or the notion of k-2-visibility. The important issue ... need easy ... k-2-rights [1]. Since ... argue only do ... the ... k-2-rays ... $O(n^2)$

in [1 0]. By noting that a solution ... (v) he dribonider ... those a problem writing ... l

**Theorem 4** *Fo ... ab y g ... r ... it ... b l ... tg en ... rtea as ah b d u ... of a 2s n h r in $O(n^2)$ tine ... d $O(n)$ ... p ... , if ite xi ts.*

**Proof.** One d n h a x t e n d e d b t r a c t □

# Refe nc es

1. D.Crs I Suki and M Yma sh a, Seah ngoa n b lei n trdi n a ori d *Int. J. Comput. Geom. & Appl.* **5** ( 19 3 19

2. L .JG bl C.L atm b e,S.M L a adl e,D.Li m and R M ani, F ndi ng a nupre il ab e ar g ma woks pae wi th bl ce, i n *P oc. IEEE int. Conf. Robotics and Automation,* 19 7

3. .L .JG bl C.L atm b e,S.M badl e, D.L n and R M ani, N si bli ty bs d pursui tasi oi n ap by ga n a vi runn, *Int. J. Comp. Geom. & Appl.* **9** ( 19 4 174 .

4. P. H. ffm m ptin al algi th n fo h tw grd rb m, *Int. J. Comput. Geom. & Appl.* **6** ( 19 1 l

5. C. Ick i m nd R Kl e n, Th tw grd pr b m, *Int. J. Comput. Geom. & Appl.* **2** ( 19 3 7 28

6. S.M L all e, B Si m and G.Slutki, A alg i th fo search n gap by gri n wi th fushi gt n *P oc. 16th Annu. ACM Symp. Comput. Geom.* ( 2 0 00 9- 29

7. H L, S.Y Sh n and K.Y Chw a, N si bli ty bs d pursui tasi oi n ap by ga ro m w i th a d *P oc. 15th Annu. ACM Symp. Comput. Geom.* ( 19 28- 9.

8. H L e, S.M rk and K.Y Chw a, Search ng p by gri on w i th bl by a lsear h *Int. J. Comp. Geom. & Appl.* **10** ( 2 00 21- 2.

9. I Suki and M Yma sh a, Search ng fo n b lei n truds i m ap by ga rg n, *SIAM J. Comp.* **21** ( 19 86 888

10. X Tam, ffi n t b ut n b hoi r lear c h pr b m, N bs Cup t Sc i . **1763** ( r d e. JCDCG'98) ( 2 03 3 3 1

11. X Tan, Seah ng a i m ple p by g by ksear h Prpri nt( als an x trd abst t n L etc Notes Comput. Sci. **1969** ( 2 00 ) 53 - 54 .

12. X Tan, T.H m and .Y ngki, A i m en tal algi th fo r truti ng sh ts w a th n rt s *Int. J. Comp. Geom. & Appl.* **3** ( 19 ) 3 6 5- 3

13. .X Tan, T. H m and .Y ngki, Cori gh to "A i m en al algi th fo ctruti ng sh t w at h rr t s ," *Int. J. Comp. Geom. & Appl.* **3** ( 19 3 19 3 3 .

14. M nash a, H U n o, I Suki and K am, "Search ng fo b le n trud es i n ap by gl rg n by ag o nb le sear hs In *P o. 13th Annu. ACM Symp. Comput. Geom.* ( 19 4 9.

# A New Structure of Cylinder Packing

Yoshinori Teshima[1], Yoshinori Watanabe[2], and Tohru Ogawa[3]

[1] Accelerator Laboratory, High Energy Accelerator Research Organization (KEK)
Tsukuba-shi, Ibaraki-ken, 305-0801 JAPAN[†]
yteshima@post.kek.jp
http://kafka.bk.tsukuba.ac.jp
[2] Laboratory for Entrepreneurship, University of Tsukuba, Tsukuba-shi, Ibaraki-ke
305-8573 JAPAN
watanabe@kafka.bk.tsukuba.ac.jp
[3] Institute of Applied Physics, University of Tsukuba, Tsukuba-shi, Ibaraki-ken
305-8573 JAPAN
ogawa-t@koalanet.ne.jp

**Abstract.** We report a new periodic structure of the cylinder packing. All the cylinders are congruent and the length of the cylinders are infinite and their directions are restricted to only six directions of $\langle 110 \rangle$. Each cylinder is fixed by cylinders of other directions, so that the whole structure sustains mechanical stability. The packing density equals to $\frac{(351\sqrt{2}-108\sqrt{6})\pi}{1936}$ ($\sim 0.376219$), which lies between two values ever known: 0.494 or 0.247. The arrangement of parallel cylinders forms a certain 2D rhombic lattice common to all of six $\langle 110 \rangle$ directions. Nevertheless the way of fixing cylinders is different in all of six directions: the cylinders of two directions are supported with the rhombus-type, and the cylinders of other four directions are supported with the equilateral-triangle-type. The structure containing the equilateral-triangle-type has never been known.

## 1 Introduction

The history of the research on the cylinder packing is much shorter than that on the sphere packing. Few mathematicians have treated the problem[1]. Some simple structures of the cylinder packing appear in the books about solid puzzles(e.g. Holden[2], Coffin[3] ). O'Keeffe and Andersson applied the cylinder packing to the science[4][5][6][7]. They are crystal chemists and explained the garnet structure famous for its complexity by using a periodic cylinder packing restricted to only four directions of $\langle 111 \rangle$.

In engineering, some structures of the cylinder packing are used for the composite materials. Such structures are light and tough against the stress from various directions. Some periodic structures were designed (Hatta[8], Hijikata

---

[†] Teshima's present address: Advanced Engineering Center, RIKEN(The Institute of Physical and Chemical Research), Hirosawa 2-1, Wakoh-shi, Saitama-ken, 351-0198 JAPAN. E-mail: kippoh@riken.go.jp

J. Akiyama et al. (Eds.): JCDCG 2000, LNCS 2098, pp. 351-361, 2001.

and Fukuta[9]). Stimulated by their researches, the authors investigated other periodic structures with six directions [10][11][12][13][14], and paid much effort exhaustively to find all the possible structures with high symmetry and density. More precisely, all the cylinders are supported by contacts of the cylinders in more than four directions among other five. Watanabe found the remarkable fact that each structure could continuously modify itself without changing its directions of cylinders, its packing density and its stability at the same time when the cylinders are kept tangent one another. An animation can be demonstrated on his web-site[14].

Quasiperiodic packings of cylinders were also reported. An architect Hizume extended Coffin's structure and Ogawa cooperated the research[15][16][13]. The structures have the icosahedral symmetry, and are composed of six directions indicated by $\langle 1\tau 0 \rangle$ ($\tau = (1 + \sqrt{5})/2$). Other quasiperiodic packings were discussed in recent papers[17][18][19].

The structures of the cylinder packing have to be researched more systematically. The present paper suggests further possibility of $\langle 110 \rangle$ six-axes structures. The theory of $\langle 110 \rangle$ structure will change to general one.

## 2    The New $\langle 110 \rangle$-Structure

The structure is a six-axes periodic structure made of cylinders with one diameter $d$. Some similar structures whose packing density are 0.494 or 0.247 have been known and are listed on Table 1 in Appendix. For the sake of convenience, we call the structures respectively Type-I(density 0.494), Type-II(density 0.247), and Type-III(the present structure).

The six directions are $A(1,1,0)$, $B(1,-1,0)$, $C(1,0,1)$, $D(-1,0,1)$, $E(0,1,1)$, $F(0,1,-1)$. If we classify these directions according to perpendicularity, they separate into three groups: $A$–$B$, $C$–$D$, and $E$–$F$. Another classification is made by the relations of sixty degrees:

- $B$, $D$ and $F$ are sixty degrees each other. A common normal vector is $(1,1,1)$
- $A$, $C$ and $F$ are sixty degrees each other. Normal vector $(-1,1,1)$
- $A$, $D$ and $E$ are sixty degrees each other. Normal vector $(1,-1,1)$
- $B$, $C$ and $E$ are sixty degrees each other. Normal vector $(1,1,-1)$

When we care about cylinders parallel to $A$-direction in the structure of Type-III, they form a rhombic lattice on a plane perpendicular to $A$-direction. Each of other five directions also forms the rhombic lattice on a plane perpendicular to the direction. Moreover, the six rhombic lattices are congruent. A rhombus of the rhombic lattice has two diagonals, and we express that the length of the longer diagonal is $2a$ and that the length of the shorter diagonal is $\sqrt{2}a$. The relation between $a$ and $d$ is

$$a = (2/3)(1 + 2\sqrt{3})d. \qquad (1)$$

The following expression is written with $a$ and $d$ to make formulae simple.

## 2.1  Equations describing the structure

For the description of the structure of cylinders, we use the line equation as the center line of the cylinder. Whether two cylinders contact each other or not is confirmed by comparing their distance with the diameter $d$. For Type-III, the equations of one cylinder about each direction are shown:

$$A_1 = \begin{pmatrix} 1 \\ 1 \\ 0 \end{pmatrix} t + \begin{pmatrix} 0 \\ 0 \\ -\frac{a}{4} \end{pmatrix}, \tag{2}$$

$$B_1 = \begin{pmatrix} 1 \\ -1 \\ 0 \end{pmatrix} t + \begin{pmatrix} 0 \\ 0 \\ \frac{a}{4} \end{pmatrix}, \tag{3}$$

$$C_1 = \begin{pmatrix} 1 \\ 0 \\ 1 \end{pmatrix} t + \begin{pmatrix} 0 \\ -\frac{2a-d}{2} \\ 0 \end{pmatrix}, \tag{4}$$

$$D_1 = \begin{pmatrix} -1 \\ 0 \\ 1 \end{pmatrix} t + \begin{pmatrix} 0 \\ \frac{2a-d}{2} \\ 0 \end{pmatrix}, \tag{5}$$

$$E_1 = \begin{pmatrix} 0 \\ 1 \\ 1 \end{pmatrix} t + \begin{pmatrix} \frac{a-d}{2} \\ 0 \\ 0 \end{pmatrix}, \tag{6}$$

$$F_1 = \begin{pmatrix} 0 \\ 1 \\ -1 \end{pmatrix} t + \begin{pmatrix} -\frac{a-d}{2} \\ 0 \\ 0 \end{pmatrix}, \tag{7}$$

where, $t$ is a parameter and $-\infty < t < \infty$ for cylinders with infinite length.

All other cylinders $A_{mn}$ of A-direction are obtained by parallel movements of $A_1$:

$$A_{mn} = A_1 + m a_1 + n a_2 \tag{8}$$

where, $a_1$ and $a_2$ are vectors causing the parallel movements, and $m$ and $n$ are integers. In the same way, all the cylinders can be generated by parallel movements of one cylinder in other directions. All the vectors of the parallel movements are shown as follows:

$$A: a_1 = (\tfrac{a}{2}, -\tfrac{a}{2}, a), \quad a_2 = (-\tfrac{a}{2}, \tfrac{a}{2}, a),$$
$$B: b_1 = (-\tfrac{a}{2}, -\tfrac{a}{2}, a), \quad b_2 = (\tfrac{a}{2}, \tfrac{a}{2}, a),$$
$$C: c_1 = (-\tfrac{a}{2}, a, \tfrac{a}{2}), \quad c_2 = (\tfrac{a}{2}, a, -\tfrac{a}{2}),$$
$$D: d_1 = (-\tfrac{a}{2}, a, -\tfrac{a}{2}), \quad d_2 = (\tfrac{a}{2}, a, \tfrac{a}{2}),$$
$$E: e_1 = (a, \tfrac{a}{2}, -\tfrac{a}{2}), \quad e_2 = (a, -\tfrac{a}{2}, \tfrac{a}{2}),$$
$$F: f_1 = (a, -\tfrac{a}{2}, -\tfrac{a}{2}), \quad f_2 = (a, \tfrac{a}{2}, \tfrac{a}{2}).$$

An overview of this structure is shown in Fig. 1 and the projections to the planes perpendicular to $\langle 110 \rangle$-directions are in Fig. 2.

## 2.2  Packing Density

It is easy to calculate the packing density for the present periodic cylinder system. Because the cylinders don't overlap, whole packing density is six times as high as the density of one direction. The density of one direction is determined by the ratio of the area of a circle and of a rhombus. Hence, the whole packing density $D$ is

$$D = \frac{\pi(d/2)^2}{(\sqrt{2}a)a} \cdot 6 = \frac{(351\sqrt{2} - 108\sqrt{6})\pi}{1936} \simeq 0.376219\cdots. \tag{9}$$

The density lies between two values ever known: $\sqrt{2}\pi/9 \simeq 0.494$(Type-I) and $\sqrt{2}\pi/18 \simeq 0.247$(Type-II).

## 2.3  The Supported States

If we restrict the directions of cylinders to $\langle 110 \rangle$, there can be two typesof support, that is, the rhombic-type and the equilateral-triangle-type. In the rhombic-type, a cylinder is fixed by four directions which are sixty degrees against the cylinder. In the equilateral-triangle-type, a cylinder is fixed by two directions of sixty degrees and one perpendicular direction against the direction. Concerning the present structure, the cylinders of $A$- and $B$-directions are fixed with the rhombus-type and other four directions are fixed with the equilateral-triangle-type. The more detail is shown as follows:

- $A$-direction is fixed by $C$-, $D$-, $E$-, and $F$-directions
- $B$-direction is fixed by $C$-, $D$-, $E$-, and $F$-directions
- $C$-direction is fixed by $A$-, $B$-, and $D$-directions
- $D$-direction is fixed by $A$-, $B$-, and $C$-directions
- $E$-direction is fixed by $A$-, $B$-, and $F$-directions
- $F$-direction is fixed by $A$-, $B$-, and $E$-directions.

These are confirmed on Fig. 2. On the other hand, the supported state for Type-I and Type-II is shown in Fig. 3 and Fig. 4. Only the rhombus-type supporting is contained in Type-I and Type-II.

It is effective to investigate the structure about its arrangement along $\langle 111 \rangle$-directions. $B$-, $D$-, $F$-directions are perpendicular to $(1,1,1)$-direction, the fact mentioned at the beginning of Sec. 2. Therefore the cylinders of the $B$-, $D$-, $F$-directions are stacked up along $(1,1,1)$-direction. In Type-III, the order of appearance along $(1,1,1)$-direction is $\cdots FBDFBDFBD\cdots$ and there is a gap $(= 0.43647d)$ between $D$ and $F$. A list about all of four $\langle 111 \rangle$-directions is shown below, where the mark $\sqcup$ means the gap whose size is common to four directions.

- Along $(1,1,1)$: $\cdots FBD\sqcup FBD\sqcup FBD\sqcup FBD\cdots$
- Along $(-1,1,1)$: $\cdots CAF\sqcup CAF\sqcup CAF\sqcup CAF\cdots$
- Along $(1,-1,1)$: $\cdots DAE\sqcup DAE\sqcup DAE\sqcup DAE\cdots$
- Along $(1,1,-1)$: $\cdots CBE\sqcup CBE\sqcup CBE\sqcup CBE\cdots$

Concerning Type-I and Type-II, there is no gap along $\langle 111 \rangle$-directions. The fact brings Type-I and Type-II only the rhombus-type support.

# 3   Conclusion

An unknown ⟨110⟩ six-axes structure was reported. The parallel cylinders of the structure form a certain 2D rhombic lattice common to all of six directions. The six directions, $A$, $B$, $\cdots$, $F$, are classified into two groups $\alpha$ and $\beta$: $A$ and $B$ belong to $\alpha$, and all the others to $\beta$. All the cylinders of group $\alpha$ are supported by contact with the cylinders of four directions of group $\beta$, so that the contacting four cylinders construct a rhombus. All the cylinders of group $\beta$ are supported by cylinders of three directions: $A$, $B$ and the perpendicular pair of itself, so that the three cylinders construct a equilateral triangle.

Researching of the derivative structures from the present structure and distinguishing the space group are further problems.

The authors are grateful to Y. Nakashima, T. Hirata, D. Nagy, W. Sasaki and H. Nakagawa for their beneficial comments and their encouragement.

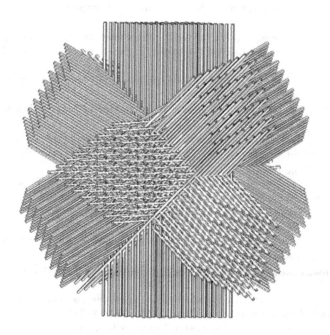

**Fig. 1.** An overview of the structure Type-III. Not only the present structure but also all ⟨110⟩-structures show the exterior of the rhombic dodecahedron if the models are made of many cylinders with the same length.

# References

1. Bezdek, A., Kurperberg, W.: Maximum Density Space Packing with Congruent Circular Cylinders of Infinite Length. Mathematika **37** (1990) 74–80

2. Holden A.: Shapes space and symmetry. Columbia University Press, New York (1971)

3. Coffin, S. T.: The Puzzling World of Polyhedral Dissections. Oxford University Press (1990)

4. O'Keeffe, M., Andersson, S.: Rod Packings and Crystal Chemistry. Acta Cryst. **A33** (1977) 914–923

5. O'Keeffe, M.: Cubic Cylinder Packings. Acta Cryst. **A48** (1992) 879–884

6. O'Keeffe, M., Hyde, B. G.: Crystal Structures. BookCrafters, Chelsea Michigan (1996)

7. O'Keeffe, M., Plévert, J., Teshima, Y., Watanabe, Y., Ogawa, T.: The Invariant Cubic Rod(Cylinder) Packings: Symmetries and Coordinates. Acta Cryst. **A57** (2001) 110–111

8. Hatta, H.: Elastic Moduli and Thermal Expansion Coefficients of Three Dimensional Fabric Composites. Nihon Fukugoh Zairyou Gakkaishi (J. Japan Soc. Composite Materials) 14 (1988) 73–80

9. Hijikata, A., Fukuta, K.: Structure of Three-Dimensional Multi-Axis Composites. Nihon Fukugoh Zairyou Gakkaishi (J. Japan Soc. Composite Materials) 18 (1992) 231–238

10. Ogawa, T., Teshima, Y., Watanabe, Y.: Geometry and Crystallography of Rod-system. Nihon Fukugoh Zairyou Gakkaishi (J. Japan Soc. Composite Materials) 21 (1995) 165–173

11. Teshima, Y.: A Study for Possible Structures of Congruent Cylinders. Bussei Kenkyu (Kyoto) **65** (1995) 405–439

12. Watanabe, Y., Teshima, Y., Ogawa, T.: Geometry of Rod Packings. Symmetry, Culture and Science 7 (1996) 81–84

13. Ogawa, T., Teshima, Y., Watanabe, Y.: Geometry and Crystallography of Self-Supporting Rod Structures. In: Ogawa, T. et al. (eds.): Katachi ∪ Symmetry. Springer-Verlag, Tokyo (1996) 239–246

14. Watanabe, Y.: Geometry of Rod Packings. A dissertation of Tsukuba University (1999), http://kafka.bk.tsukuba.ac.jp/~watanabe/rod/rod.html

15. Hizume A.: Hoshi-Kago. In: Katachi-no-Bunkakai (eds.): Asia no Katachi wo Yomu. Kousakusha, Tokyo (1993) 208–225

16. Ogawa, T.: Rittai-Tajiku-Ori. In: Katachi-no-Bunkakai (eds.): Asia no Katachi wo Yomu. Kousakusha, Tokyo (1993) 226–235

17. Parkhouse, J. G., Kelly, A.: The Regular Packing of Fibres in Three Dimensions. Proc. R. Soc. Lond. A **454** (1998) 1889-1909

18. Dunneau, M., Audier, M.: Icosahedral quasiperiodic packing of fibers parallel to fivefold and threefold axes. Acta Cryst. **A55** (1999) 746–754

19. Audier, M., Dunneau, M.: Icosahedral quasiperiodic packing of fibers parallel to fivefold and threefold axes. Acta Cryst. **A56** (2000) 49–61

# Appendix: $N$-axes structures

Table 1. $N$-axes structures

| $N$ | Direction | Contact-Index[†] | Density | Figure |
|---|---|---|---|---|
| 1 | [001] | $6^1$ | $\sqrt{3}\pi/6 \simeq 0.907$ | Fig. 5 |
|   |   | $4^1$ | $\pi/4 \simeq 0.785$ | Fig. 6 |
| 2 | [100], [010] | $4^2$ | $\pi/4 \simeq 0.785$ | Fig. 7 |
| ( $k$ |   | $4^k$ | $\pi/4 \simeq 0.785$ | — ) |
| 3 | ⟨100⟩ | $4^3$ | $3\pi/16 \simeq 0.589$ | Fig. 8 |
| 4 | ⟨111⟩ | $6^4$ | $\sqrt{3}\pi/8 \simeq 0.680$ | Fig. 9 |
|   |   | $3^4$ | $\sqrt{3}\pi/18 \simeq 0.302$ | Fig. 10 |
| 6 | ⟨110⟩ | $4^6$ | $\sqrt{2}\pi/9 \simeq 0.494$ | Fig. 3 |
|   |   | $5^6$ | $\sqrt{2}\pi/9 \simeq 0.494$ | — |
|   |   | $4^6$ | $\sqrt{2}\pi/18 \simeq 0.247$ | Fig. 4 |
|   |   | $4^4 5^2$ | $\sqrt{2}\pi/18 \simeq 0.247$ | — |
|   |   | $4^2 5^4$ | $\sqrt{2}\pi/18 \simeq 0.247$ | — |
|   |   | $5^6$ | $\sqrt{2}\pi/18 \simeq 0.247$ | — |
|   |   | $3^4 4^2$ | $\frac{(351\sqrt{2}-158\sqrt{6})\pi}{1936} \simeq 0.376$ | Fig. 1, Fig. 2 |

⟨100⟩ ≡ [100], [010], [001].
⟨111⟩ ≡ [111], [1$\bar{1}$1], [$\bar{1}$1$\bar{1}$], [$\bar{1}$11].
⟨110⟩ ≡ [110], [101], [011], [1$\bar{1}$0], [$\bar{1}$01], [01$\bar{1}$].
†"Contact Index" shows the contact number with other cylinders in the each projection.
A good example to understand the index is shown in Fig. 2

$A(1,1,0)$                          $B(1,-1,0)$

$C(1,0,1)$                          $D(-1,0,1)$

$E(0,1,1)$                          $F(0,1,-1)$

Fig. 2. Projection of the structure Type-III to each plane perpendicular to $\langle 110\rangle$. $A$- and $B$-directions are fixed with the rhombic-type and other four directions are fixed with the equilateral-triangle-type. Thus, four directions are fixed by tri-gon, and two directions are fixed by tetra-gon, so we express the fixing state of the structure by $3^4 4^2$. (This is "Contact Index" in Table 1.) On these figures, upper direction is $+z$ for the projections of $A$ and $B$, $+y$ for $C$ and $D$, and $+x$ for $E$ and $F$.

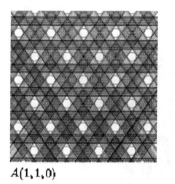

$A(1,1,0)$

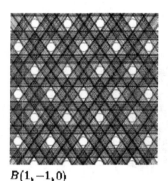

$B(1,-1,0)$

Fig. 3. Projection of the structure Type-I, whose packing density is $\sqrt{2}\pi/9 \simeq 0.494$. The figure is projections to $A$- and $B$-directions. The same projections as them are acquired for other four directions, therefore they are omitted. A cylinder is fixed with the rhombic-type in all directions, so the contact index is $4^6$. Notice the existence of the structure with another contact index: $5^6$ and with the same packing density. The fixing states described as another index contain the rhombic-type supporting.

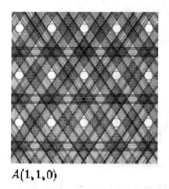

$A(1,1,0)$

$B(1,-1,0)$

Fig. 4. Projection of the structure Type-II, whose packing density is $\sqrt{2}\pi/18 \simeq 0.247$. The figure is projections to $A$- and $B$-directions this time too, and other four are also omitted. Though the arrangement of cylinders in one direction is rectangular now, a cylinder is still fixed with the rhombic-type in all directions, so the contact index is $4^6$. Notice the existence of the structures with other contact indexes: $4^4 5^2, 4^2 5^4$ or $5^6$ and with the same packing density. The fixing states described as the other indexes always contain the rhombic-type supporting.

**Fig. 5.** An overview of one-axis structure. Parallel cylinders form the honeycomb lattice. The packing density is $\sqrt{3}\pi/6 \simeq 0.907$. The contact index is $6^1$. This structure is the densest of all cylinder packings

**Fig. 6.** An overview of another typical one-axis structure. Parallel cylinders form the square lattice. This is mechanically unstable. The packing density is $\pi/4 \simeq 0.785$, and the contact index is $4^1$

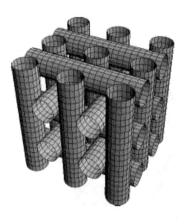

**Fig. 7.** Two-axes structure. The directions are [100] and [010] as typical ones. Arbitrary N-axes structures are possible if we extend this two-layer structure. Such N-axes structure have the same density and the same contact index as the two-axes structure. This structure is also unstable between different layer. The contact index is $4^2$

**Fig. 8.** Three-axes structure. The directions are ⟨100⟩. The packing density is $3\pi/16 \simeq 0.589$. The contact index is $4^3$ which is the square-type supporting

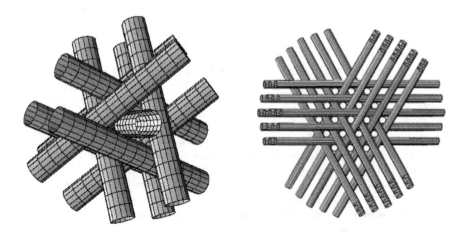

**Fig. 9.** Four-axes structure I(An overview and a projection). The directions are (111). The packing density is $\sqrt{3}\pi/8 \simeq 0.680$. The contact index is $6^4$ which is the hexagonal-type supporting

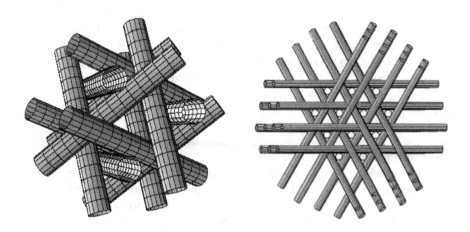

**Fig. 10.** Four-axes structure II(An overview and a projection). The directions are still (111) however the packing density is $\sqrt{3}\pi/18 \simeq 0.302$. and the contact index is $3^4$. Two four-axes structures cannot come and go each other by any removal of cylinders

# Efficient algorithms for the minimum diameter bridge problem

Takeshi Tokuyama

Graduate School of Information Sciences, Tohoku University
Sendai 980-8577 Japan, tokuyama@dais.is.tohoku.ac.jp

**Abstract.** We give efficient algorithms for constructing a bridge between two convex regions in a fixed dimensional space so that the diameter of the bridge graph is minimized. If both the set of vertices and the set of half-spaces defining the faces of the convex regions are given, we have an optimal linear time randomized algorithm. If only the vertices are given, we give a subquadratic time algorithm, and if only the half-spaces are given, we give a nearly quadratic time algorithm.

## 1 Introduction

Consider two convex regions $R_1$ and $R_2$ in the $d$-dimensional Euclidean space. The minimum diameter bridge problem is to find these two convex regions (called $R_1$ and $R_2$), together with $p_1 \in R_1$ and $p_2 \in R_2$ such that the diameter

$$d(p_1, p_2) + \max_{v \in V}(d(v, p_1)) + \max_{w \in W}(d(p_2, w))$$

of the chain $R_1$ and $R_2$ connected by the bridge is minimized. Here $d(x, y)$ is the Euclidean distance, and $V$ and $W$ are vertex sets of $R_1$ and $R_2$ respectively. The problem has been considered for $d = 3$ in the plane. A linear algorithm is given in [2] by using a lower envelope technique and a duality matching giving in [5].
Some approximation algorithms for the higher dimensional case are considered by Chon [8]. The algorithm considered is too difficult to be quite reduced to that of [3].

The minimum diameter bridge problem is related to the following minimum separation point problem. Given two convex regions $R_1$ and $R_2$ in the Euclidean space, determine the minimum distance between $R_1$ and $R_2$. This is more closely related to an $LP$-type problem considered in computational geometry [4], and it can be solved in $O(n)$ time by using a dual information of discrete algorithms and mathematical programming methods, where the moving vertices. Although the minimum diameter bridge problem is more complicated than the minimum separation point problem, we can reduce it to a minimum parametric optimization problem and apply both $LP$-type approaches and the parametric searching techniques proposed in a minimax problem [13].

J. Akiyama, M. Kano, and M. Urabe (Eds.): JCDCG 2000, LNCS 2098, pp. 362-369, 2001.

The earch reffer the two ways are prese ... the problem: the ... type sta a toin we are give ... the s of every ... Ves of $R_i$ for $i = ...,2$. ... the dual repsa a toin $R$ are ... mains intersections of close dhalf spaes. ... the minip sta a toin of the ve te xepe se ... and repese ration are given.

If we are give n an ... dre prese ... ration we can de sig $O(nM)$ to p tal expected time diame ... dia ... gorithm, $n$ with the ... sol ... con ve xre ... Misch ... r 6 defrig haf spae ... similar represe n ... tion has, $M$ is al e ashe ... rof ace s of ... ee gons) . This is a improve men re ve r ... which ... deter ministic time al g ... be alredy known, sin ... ethe kno w ... n e ... gach ... s tha e con ve xhll st ... rtch ... ing 6 he ve ric e sh ... the bir ay of one x e gons ... give n apa ... input. h o he rw ... swere cho st orte tic im ur m th prve ... that hese 6 lines defrig ge sis know n.

If we a ... earlyg iv em a ve ex re ... prese ... ration and $M$ ca n be mn le rthm. Furt he r mre , we do n o kno w ... te m th $O(an M^{4/2})$ ... tial g o ri th ... mx he ime dre prese ... ration e $M$ is es in Ho r s ha ... ca e , we pre ... $\tilde{O}(e n^{2\lfloor(d+1)/2\rfloor}/\lfloor(d+3)/2\rfloor)$ ... tial g o ri h m, with the big ... O ... ma ig n o imp dy l o g ... th m 6 rs . The te co mpl e x is sq d a ic 6 ny c o nsart numb r d ... i m s os , ... s $\tilde{O}(n^{4/3})$ ad $\tilde{O}(n^{3/2})$ 6 r d ... 4 ad 5 $\le \le$ 6 e spe c If we al ... egy ... i ve m al e p re se ... ratio n we giv $O(M^2)$ ... al go ri th m

## 2 Preliminary

### 2.1 Parametric minmax problem

Consider an optimaproble m $Q$ that has a ... ve c to $x = (x_1, x, ..., x_\nu)$ a maxim in p b ... jc tv efur to $g(x)$ ... nd e ... ge ... m c o n s ... ... on

If we ... pl ... an ... go lve f ... ction $(g ... x)$ w ... ih a In y $g_\lambda(x)$ ot im o s ly de pe ... men ... na ... dime nsio nl re ... ap ... ... re ... to $\lambda = (\lambda_1, \lambda_2, .., \lambda_d)$ , we h ... ve p a a ... me tric fa ... il $Q(\lambda)$ of ... eo ... rig ... opt im ... to p ro ble ... wh e e ... e sd uto ns c o ... mal ... yd e ... pe ... nd o ... th ... ee $z(\lambda)$ e ts he optim al v a e ... the ob jc tve f t o $g_\lambda(x)$ 6 Q ... $\lambda$) Le $\mathcal{F}$ tb a (pos s bly ... nde ... d one x p d y go m al e g ... hed imensio n s p ... asubh at $\lambda$ sh ... d ec o me d . W ll $\mathcal{F}$ th ... d sb l e g ... h e p ... rs ... h at it is d ff e nt fn he ... f asib ere ... ga ... of

The p ... b ef fi n ... ng the v al $\lambda \in \mathcal{F}$ min ... ing $z(\lambda)$ ... al e ... min max pa am ... mini a ... t op b ... ge e m al, this ... hat pro be m. We v er he ob jc c tve fr t oisac ... ve x fin c toin the pa a me te s, we h ... e 6 ll w i ... g e s h [1]

**Theorem 1.** *Co s ... ra n op i mi a t ... po b ... t ha a t ns bs ... , $T_D$ se - q ... ul t ... me d s b h s a ... ge p ... ra l l ... p ... t th ... w at h ... Q) a a l l t me o r ... N ... e s os r by u si y l ... e s ... pra t s u ... tr hsp ct t o p a - am t se . ... t ... pp e s is g wth n $O(T_{pre})$ i m e mp t y t o d ...*

$T_D$, $\tau_B$ and $N$. He also ...

We only give some intuition of the idea here ...

## 3  Solutions of the minimum diameter bridge problem

### 3.1  If a mixed representation is given

The minimum diameter bridge problem ...

### 3.2  Solution using multidimensional parametric searching

Suppose that the points $p_1$ and $p_2$ are given. Then, ...

a dimension ... parameter ... the problem is basically ... sion of ... the ... neighbor problem. ... $i = 1, 2$, the ... the ... nearest neighbor distance $\varepsilon_i = v_i(dv, p)$ is the ... representation ... of the Euclidean distance ... if we consider it as ... when we compute the coordinate values $\delta p_i$. The Euclidean distance ... is a convex ... function ... Hence ... point which is a sphere ... a convex ... function is a convex function.

Thus, we consider ... dimensional ... searching. We would have ... have $O(\log n)$ ... an $O(n)$ processors ... algorithm ... running time ... the ... maximum finding ... algorithm. The ... is ... the ... algorithm ... is the problem of finding the Voronoi cell of ... We have ... the parallel algorithm using the ... data structure. We give ... description of the data structure, ... although its ... (although highly technical) ... techniques ... geometry:

Consider a set of $U = \{u_1, u_2 .., u_n\}$ in the $d$-dimensional space, where $u_i = (eu(1), u(2;).., u(d,i))$. For point $(x(1), ..., x(d))$ the square of the distance between $u_i$ and $x$ is $(dx, u)^2 = \sum_{j=1}^d (x_j + u_j)^2$. If we consider $(de, u_i)^2 - \sum_{i=1}^d (x_j^2)$, $(f, u_i)$ is a linear function $m(j)$ $(j = 1, 2,.., n)$. A point $u_k$ is the ... the ... point ... from $x$ ... if $f(p, u_k)$ is the largest.

Let us consider ... dual problem. Let $\ell_i$ be the hyperplane in the $(d+1)$-dimensional space defined by $D(\ell) = f(\ell) , ..., (x_1,...,x_d, u)$. Let the points $D(\ell) = (2u_{1,i}, 2u_{2,i},..., 2u(d,i), \sum_{j=1}^d(u_{j}^2))$. Let $\tilde{q} = (q(1), q(2,).., q(d), 1)$ ... point $q$ in the ... $d$-dimensional space. We ... the point $u_i$ is ... the ... from $q$ and only if the inner product of $\tilde{q}$ and $D(\ell)$ is the largest.

Let $S(U) = \{D(\ell_i) : i = 1, 2,.., n\}$. Consider a parameter $r \le n^{1-\delta}$ ... constant $c$.

We partition ... more ... use ... (Theorem 3.1 of [5]). The ... points is divided into ... many subsets $S_i$ ($i = 1, 2,... c$) such that $|S_i| \le n/cr$. Each $S_i$ is contained in ... simplices ... and any hyperplane ... those simplices ... it is at most $c'r^{1-1/(d+1)}$ ... constant. Moreover, if the hyperplane has ... points ... in ... of the ... passes ... it, the number of simplices ... is at most $c'r^{1-1/\lfloor(d+1)/2\rfloor}$. The data structure is constructed in $O(n)$ ... time.

When a point $D(\ell_j)$ ... in the ... reproduces ... with a given ... We ... to have vertices of simplices ... in descending order with respect to the inner product with $\tilde{q}$. Thus, we have ... dealing with each simplex, consider the first vertex in the list and ... solve ... vertices of a simplex). Thus, we have these $c'r^{1-1/\lfloor(d+1)/2\rfloor}$ simplices in the list. Consider the points $Y$ that obtain the data ... of $S_i$ corresponding the ... simplices $D(\ell_j)$. The ... maximizing the inner product with $\tilde{q}$ must be contained in $Y$. By definition, $Y$ contains at most $nr^{-1/\lfloor(d+1)/2\rfloor}$ points. Hence, we define $D(\ell)$ in $O(\log n)$ ... pa-

**Theorem 2.** *The minimum diameter problem can be solved* $O(M + \log p)$ *in e.*

### 3.3 Solution as an LP-type problem

If ... half plane ... the define of $R_1$ is ... we ... whe the ... points ... If ... in $h$, $(B \setminus W \cup \{h\}) = (B \setminus W)$. Othe ... wise ... we ... the ... clusive ... the ... go ith m ... we ... iz e ... their ... ratio ... the ... type plane ... $(B \setminus W \cup \{h\})$.

... using the ... analysis ... the problem can ... solve d$O(iM + )$ ... in expe c te d time ... the ... in $V_i$ ($i = 1, 2, .., k$) and $M = \sum_i M_i$.

**Theorem 3.** *The minimum diameter ... edge problem ... solved in ... $O(M + )$ ... and ... expec ted time.*

## 3.4  If a vertex representation is given

... would like ... aside ... the case w he re h ... edge ... in $R$ as ... the vertex ... representation $\{ \sum_{e \in V_i} (\theta) | 0 \leq (\theta) \leq , \sum_{v \in V} \theta = 1\}$ ... compute the independe ... set ... to apply ... them ... that the property ... se c ... the ... have the set $\theta$ ... $\leq 3$ and $\theta$ ... $n^{\lfloor d/2 \rfloor}$ ... if $\geq 4$

... can fo m ... at the pro ble m ... it LP pro ble m ... here we ... only in ... re ... the ... feasib es ... path the pa ... t rs pa ... set $V_1$ and $V_2$. Le ... $W$ ... are se to... ver tic es o $R$ and $R_2$ ... the ... in $h$ eo righ pro ble m $W_m = V_i$ fo r $i = , 2$ we o ... de ... the case w he re the ... en ... re laio ... we $W_i$ and $V_i$.

... sat $m$ $Q_1 = Q_2 = \emptyset$ and ... case the fe a ble re g ... in se ... ver tic $W_1 \cup W_2$ ... ys ... h a ... $Q_i$ is the c o me ... $i = 1, 2$ ... te xp l ic it yo ... $Q_i$ ... S ... pose t ha t the c ne ... so l utio n is ... an e ... da $p_1 \in Q_1$ and $p_2 \in Q_2$ and we kn w set $X_i \subset W_i$ ... tic es ... defin ge fi e (w thout loss ge ne al ... we ... ass ... that the fac e ... d dim ns ion s impl e $F_i$ x) ... ping $p_i$ ... $i = 1, 2$

If a we tw (wx ih ... loss o f ... re h ... ... $\in W$ in se te d, we c on s ide r the ... in ... ons plane ... $F_1$ and, a me ... whe the r here is to te $p_1 \in$ to $F_2$ ... c uent so l utio n ish e b stwe k e ep the ne nd uto n nd c o nmise r ... ve r tic e fibes d to ... ve e s h ec ne ns ol utin, and ... ... new optimisd uto m ... l in the pro re ... se t e r tic es h se ... he c ... $w$ is se le e d ...

T his yields ... ... xpe te d time ... in da g ch m ... he tfor s ol u ... ge pro ble ... ($sub l em$) w he the p ... $R$ ... and $p_2 \in R_2$ ... times impl e seah ... wh ... hh ... a cost tha r 6 ... s.

... 1 $f(n = O$ ... s in c e h is is the spe c ... ca e of he g ... pro ble m ... he $M = O(1)$. T hus we h x e a $O(n^2)$ i ... go it h m ... p p l ying he s ol in ec to n 3 3

... ve r we c an g ive e te ... bd $f(n$ we a e h nd n ... io ml p m t cs e a h. T he p i ... s h a al tho ug h we s d v e the ... pro ble ... tn we sp e m the pe p o e s sin ... i ... the fi he s tre ig h b se a ho

Here, if we set the preprocessing time and the sorted $M$, $= O$, the item complexity is $O(M_{sort} + (N \cdot T_D \log \sqrt{gr})^{2d}(\log N)^{2d-1})$ in the algorithm given in Section 2. Taking $T_B = 2 \cdot O(\log \log)$, $N = O(n^{1-1/\lfloor(d+3)/2\rfloor}) \neq 1$ and $T_D = O(n^{1-1/\lfloor(d+3)/2\rfloor} \log \log)$. This yields an algorithm which can compute $\tilde{O}(n^{2-1/\lfloor(d+3)/2\rfloor})(\log g)$ $M$, we require a separate preprocessing of all the straight data, and spend on the preprocessing time the complexity. If we spend more $> h$ on preprocessing time, the query time for the query straight search is $\tilde{O}(n/t^{\lfloor(d+1)/2\rfloor})$ item, where $\tilde{O}$ is the best known, ignoring polylogarithmic factors. Hence the query time complexity becomes $\tilde{O}(t + n^2/t^{1/\lfloor(d+1)/2\rfloor})$, which is minimized at $\tilde{O}(n^{2\lfloor(d+1)/2\rfloor/\lfloor(d+3)\rfloor})$.

**Theorem 4.** *If the tree space is of a sort of a region, we can compute the minimum diameter tree given in $\tilde{O}(n^{2\lfloor(d+1)/2\rfloor/\lfloor(d+3)/2\rfloor})$ time.*

### 3.5  If a dual representation is given

We would like to consider the case where the data is given as a union of trees. Let $H$ and $R$ be the sets of half planes and... in $R$, respectively. When we compute the minimum... express which components have more than one vertex, it takes $O(eM^{\lfloor d/2\rfloor})$ time, and is expensive if $d \geq 4$.

Let $F = R_1 \times R_2$ be the fiber region given, in which we consider the $H = H_1 \cup H_2$ of half spaces. Any subset $W$ of $H$ that... the span of the... can be expressed. $\cap_{h\in W}H_h$ for a point $p_i \in R_i$ must have the intersection vertex $x$, corresponding to $p_i$... , 2. If we increment $W$, the region $R(W) = \cap_{h\in W\cap H_i}$ is shrunk or unchanged, and we... the set $\cap_{h\in\emptyset}$ has the whole space.

We start from $W = \emptyset$, and adding in the... that has... $W$ does... we have the edge $p_1p_2$ (in other words, the pair intersection vector $p_1/p_2 \in H$ the... the first part $f(W)$ and $(W)$ for $p_i$ of each of $i = 1, 2$ so that they are in the interior... Suppose that the half plane is inserted to $W$. Without loss of generality, we assume that $\in H_1$. If $f(W)$ is located in $h$, we delete $f(W \cup \{h\}) \neq f(W)$ for $i = 1, 2$, and also... in the edge $p_1p_2$. Then we... is... we recompute the... solutions to combinatorial... in the... first...

**Proposition 1.** *The expected time can be estimated if the above given... and algorithm is $O(M^2)$.*

*Proof.* By using a backward analysis, the expected time complexity is $O(MT_B)$, where $T_B$ is the time for solving the... because when $W$ has a constant... of type... Hence... to their dependence... case $T_B = O(M)$. Therefore, the time complexity $O(M^2)$ is...

## 4  Concluding Remarks

There are several generalizations of the minimum diameter tree... did we... consider the case where we... are... wish... And... of subproblems... will be given in a companion paper.

# Refern es

1. Cfried E ng ,p riv imm u niatin

2. E Ch nad N M ido , Srog kona time a nd N Ngith s6 rd etc-
ing g ksind mamiga ph        in Proceedings of 21st ACM Symposium on Theory
of Computing (18  ) 523–534.

3. R. Ch S b wing wnsoting ntwok o ba in str soting kiths      J. ACM
34 (198 ) 20 0-208 .

4. M Be rad    N Mgd  q Inea p ream mingnb wdim e ns ins, S etin     38  f
Handbook of Discrete and Computational Geometry (19 7) 99 –710 CRC Press.

5 . B Gätne r Au bex p ne n tia l gofihr a bstrat o pimhinp rb lmPsro.
33rd FOC S (1992) 464–472.

6  D us fi e H,Sensitivity Analysis for Combinatorial Optimization, Mh na  ndum
NU  CB/ERL M1 /22U . C Bek ky, 180 .

7 S. K. Kn ad  C SHn,  Cmpting thpim hb rid gbet we ntw p       ks ns,
HKUST Research Report TCSC-99-14 (199 ) 6 p pa rinl PL

8I   M  se kEffiie n titintres,  Proceedings of 7th ACM Symposium on Computational Geometry (19 1) 1–9 .

9 N M idd App kng rh l np uhina kiths in th d es ig o f se rih
gith s,  J. ACM 30 (18 3)  82–86 5 .

10 J  M u sed   O Sh w ks f Inea rpiniatinq ueries,     Proceedings of 8th
ACM Symposium on Computational Geometry (1992) 16–25 .

11. M rir a nd E . Whulb inabibo und6 rlnea p rg ran ringnd      re htd
pro b hs, Proc. 9th STACS, LNCS 577(19 2) 569–59 .

12. X H Ta Q npim h b rid gbetwee ntwo o n ex regInfoxmation Processing
Letters 76 (200 0 163–18 .

13.T Thma        M ntripinatinp rb hns a nd m utidim e ns io nh
p ran etric senhing   Proceedings of 33rd ACM Symposium on Theory of Computing 6 p pa r .

# Illuminating Both Sides of Line Segments

Csaba D. Tóth[*]

Institut für Theoretische Informatik
ETH Zürich, CH-8092 Zürich
Switzerland
toth@inf.ethz.ch

**Abstract.** What is the minimum number of light sources that can collectively illuminate both sides of $n$ disjoint line segments in the plane? We prove that this optimization problem is NP-hard. The worst case analysis shows, however, that $\lfloor 4(n+1)/5 \rfloor$ light sources are always enough and sometimes necessary for all $n \geq 2$.
This problem was motivated by an open problem posed by Czyzowicz et al.: what is the minimal number of light sources that can collectively illuminate any set of $n$ disjoint line segments from one side at least.

## 1  Introduction

Illumination of convex sets was first studied by L. Fejes Tóth [9] who proved that $4n - 7$ light sources are always enough and sometimes necessary to illuminate the boundaries of $n$ disjoint convex compact sets in the plane. He also proved that $2n - 2$ light sources are always enough and sometimes necessary to illuminate the boundaries of $n$ disjoint disks. Ever since, this worst case problem were studied for several subclasses of compact sets. The exact number of necessary light sources has not been known for any other subclass so far. For homothetic triangles [5] and for axis-parallel rectangles [11, 22] the number of light sources is known with a constant error.

In all versions of illuminating disjoint compact sets, the boundary $\partial C$ of each compact set $C$ should be illuminated by a set $\mathcal{S}$ of light sources, that is, for every $P \in \partial C$ there is a point $S \in \mathcal{S}$ such that the open line segment $SP$ does not intersect any given compact sets.

Czyzowicz at al. [4, 7] studied the illumination of $n$ disjoint line segments in this sense, i.e. every point of a given line segment should be illuminated by at least one light source. The minimal number of light sources required for this problem is still not known. The upper bound $\lceil 2n/3 \rceil$ of [7] was recently improved to $\lfloor (n+1)/2 \rfloor$ by the present author [21]. The best known construction requires $4n/9 - 2$ light sources [23]. An asymptotics of $n/2 + O(1)$ was conjectured in [4].

Especially for line segments, another natural interpretation of illumination was suggested by Welzl. That is, every point of a given line segment should be

---

[*] The author acknowledges support from the Berlin-Zürich European Graduate Program "Combinatorics, Geometry, and Computation".

J. Akiyama, M. Kano, and M. Urabe (Eds.): JCDCG 2000, LNCS 2098, pp. 370–380, 2001.

illuminated by at least two light source, one on each side of the line induced by the segment. The results presented here concern this model.

**Problem 1.** Given a set $\mathcal{L}$ of $n$ pairwise disjoint and non-parallel closed line segments in $I\!\!E^2$, place light sources at points of $I\!\!E^2 \setminus \bigcup \mathcal{L}$ so that they collectively illuminate both sides of every $\ell \in \mathcal{L}$. A point $P \in \ell$ is illuminated by source $S$ from one side, if $S$ is on that side of the line induced by $\ell$ and the open line segment $SP$ does not intersect any $\ell' \in \mathcal{L}$.

In this paper, we give exactly the minimum number of light sources required for Problem 1 for any set of pairwise disjoint non-parallel line segments.

**Theorem 1.** $\lfloor 4(n+1)/5 \rfloor$ *light sources are always sufficient and sometimes necessary for Problem 1 for all* $n \geq 2$.

The lower bound of our result is given by a construction of $n$ line segments requiring at least $\lfloor 4(n+1)/5 \rfloor$ light sources. To allocate at most $\lfloor 4(n+1)/5 \rfloor$ light sources, we transform the problem into a problem on planar graphs.

$I\!\!E^2 \setminus \bigcup \mathcal{L}$ can be partitioned into convex closed cells by the following *convex partitioning* algorithm. Extend each line segment in both directions until it hits another line segment or a previous extension, or to infinity if it does not hit anything. Starting with $n$ disjoint line segments, the plane is partitioned into $n+1$ convex regions. The *graph $H$ of the convex partitioning* is defined as follows: let the set $V(H)$ of nodes be the set of convex cells, connect two nodes by an edge, if the corresponding cells share a common boundary and the extended line segment forming their common boundary is infinite or it has one endpoint on this common boundary. $H$ is a planar graph without loops, but $H$ may have double edges if both endpoints of an extended line segment are on the common boundary. Note that the convex partitioning is not necessarily unique.

It will turn out that a maximum $P_3$-packing of $H$ (where $P_3$ is a path of three nodes and two edges) yields a small set of light sources solving Problem 1.

Unfortunately, there is no good characterization of maximum $P_3$-packings of graphs in the sense like Berge's formula [2] characterizes maximum $P_2$-packings (i.e. maximum matchings). Finding the maximum $P_3$ packing is NP-hard [15] even for planar graphs [3].

Recently, Kaneko [12] has found a simple characterization of graphs having a $\{P_{\geq 3}\}$-factor, that is graphs which can be decomposed into paths of at least 3 nodes. In Sect. 4 we prove that $H$ has a $\{P_{\geq 3}\}$-factor. This is already enough to establish the sufficiency of Theorem 1.

Very recently, Kano et al. [13] gave a simple proof to Kaneko's theorem, and Hartvigsen and Hell [10] presented a polynomial algorithm which finds an actual $\{P_{\geq 3}\}$-factor or states that none exists.

In Sect. 5, we prove that determining the smallest number of light sources satisfying Problem 1 is NP-hard. We adopt a construction of Eidenbenz et al. [8] to a set of disjoint line segments, and reduce the set-cover problem to the optimization version of Problem 1.

## 2  Proof of Main Theorem

Consider the convex partitioning induced by $\mathcal{L}$. Form the graph $H$ as described above. Let $K$ be a simple planar graph obtained from $H$ by replacing each double edge by a single edge.

$n+1$ light sources placed into the interior of all cells of the convex partitioning solve Problem 1. This number can be reduced by the following observation.

**Proposition 1.** *If three cells correspond to a $P_3$ in $K$, then two light sources can illuminate all points of line segments on the boundaries of these cell from the sides facing to any of these cells.*

*Proof.* Let the three convex cells corresponding to $P_3 \in K$ be $A, B$, and $C$. The extended line segments $\ell_1$ and $\ell_2$ are on the common boundary of $A, B$, and $B, C$ resp. Let $L_1$ and $L_2$ denote endpoints of the extended $\ell_1$ and $\ell_2$ on the common boundaries resp.

If $\ell_1 = \ell_2$, then place two light sources in the interior of $A$ and $B$ in the $\varepsilon$ neighborhood of $L_1$ and $L_2$. The two light sources clearly illuminate the boundary of $A$ and $B$. They illuminate the boundary of $C$ as well, the light source next to $L_2$ illuminates the boundary of $C$ except $\ell_2 \cap C$, but $\ell_2 \cap C$ is illuminated from the side of $C$ by a light source next to $L_1$.

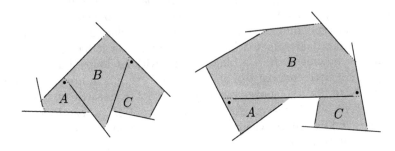

**Fig. 1.** Two light source can illuminate the boundary of three adjacent cells

If $\ell_1 \neq \ell_2$, then place two light sources in the interior of $A$ and $C$ in the $\varepsilon$ neighborhood of $L_1$ and $L_2$ (see Fig. 1). The two light sources clearly illuminate the boundary of $A$ and $C$. They illuminate the boundary of $B$ as well, $\ell_1 \cap B$ is illuminated from the side of $B$ by the light source next to $L_2$, $\ell_2 \cap B$ is illuminated from the side of $B$ by the light source next to $L_1$.                                        $\square$

Proposition 1 infers that a relatively small set of light sources can solve Problem 1 for any maximum $P_3$-packing of the graph $K$. Note, however, that there is no one-to-one correspondence between maximum $P_3$-packings of $K$ and optimal solutions to Problem 1. Our main result states only that the light source placement based on a maximum $P_3$-packing of $K$ is best possible in the worst case.

**Lemma 1.** *There is a $P_3$-packing in $K$ which covers at least $3\lceil |V(K)|/5 \rceil$ nodes.*

For the proof of Lemma 1, see Sect. 4.

*Proof.* (of Theorem 1)

*Sufficiency.* According to Lemma 1, $K$ contains a $P_3$-packing which covers at least $3\lceil (n+1)/5 \rceil$ nodes. Place two light sources for every $P_3$ and one light source to every cell not covered by a $P_3$. Then the total number of light sources is

$$2 \left\lceil \frac{n+1}{5} \right\rceil + \left( (n+1) - 3 \left\lceil \frac{n+1}{5} \right\rceil \right) = \left\lfloor 4 \cdot \frac{n+1}{5} \right\rfloor.$$

*Necessity.* Fig. 2 depicts a construction of $5k+1$ line segments requiring $4k+1 = \lfloor 4(n+1)/5 \rfloor$ light sources. Observe that each line segment in the interior of a shaded region requires two distinct light sources.

We obtain a construction on $n = 5k+2, 5k+3,$ or $5k+4$ line segments by adding one, two, or three line segments shown as dashed segments on Fig. 2.

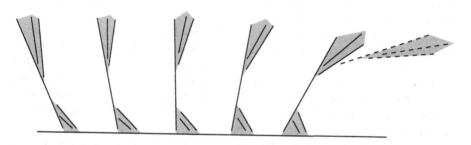

**Fig. 2.** A construction of $n = 26$ line segments requiring 21 light sources

□

Note that the maximum $P_3$-packing of the corresponding graph $K$ covers only $3k+3$ nodes of the total $|V(K)| = 5k+1$.

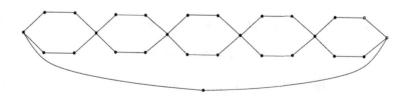

**Fig. 3.** The corresponding graph $K$

# 3   Basic Properties of $H$ and $K$

Observe that $V(K) = V(H)$ with $|V(K)| = n + 1$, $|E(K)| \leq |E(H)| = 2n$, and a maximum $P_3$-packing in $K$ is a maximum $P_3$-packing in $H$ as well.

If $H$ has $p$ double edges, then $|E(K)| \leq 2n - p = 2|V(K)| - 2 - p$.

**Definition 1.** *For a graph $G$ and $S \subset V(G)$, denote the degree of $S$ in $G$ by $d_G(S)$ (i.e. the number of edges connecting nodes of $S$ to nodes of $V(G) \setminus S$). Let $G(S)$ be the subgraph of $G$ induced by $S$, and $G - S$ be the subgraph induced by $V(G) \setminus S$. Denote $E_G(S) = E(G(S))$ the set of edges induced by $S$ in $G$.*

Denote by $q$ the number of line segments which are extended to a complete line by the convex partitioning algorithm. $q \in \{0, 1\}$ since the line segments are pairwise non-parallel. Denote this line by $\ell_0 \in \mathcal{L}$ if there exists one.

Observe that there are no cutting edges in $H$ nor in $K$.

**Lemma 2.** *The number of 2-edge cuts in $H$ is $q$.*

*Proof.* A subset $S \subset V(H)$ corresponds to a region $R_S$ in the plane, that is the union of the closed cells corresponding to the nodes in $S$. $R_S$ is connected if $S$ induces a connected subgraph. The boundary of any $R_S$ is composed of pieces of the extended line segments of $\mathcal{L}$.

*Case 1, $R_S$ is a bounded region.* Any bounded region is a polygonal domain and thus has at least three vertices. So there are at least three endpoints of given line segments on its boundary, and for the corresponding $S \subset V(H)$, $d_H(S) \geq 3$.

*Case 2, $R_S$ is an infinite region.* There are at least two infinite extended segments on the boundary of $R_S$. If there are at least two line segments on the boundary of $R_S$, then there are at least one endpoint of line segments on the boundary of $R_S$, and for the corresponding $S \subset V(H)$, $d_H(S) \geq 3$.

A set $S \subset V(H)$ corresponding to an infinite region can have $d_H(S) = 2$, if and only if there is exactly one line segment on the boundary of $R_S$. In this case, the unique line segment on the boundary of $R_S$ is extended to the infinity in both directions. □

**Corollary 1.** *For every $S \subset V(K)$ with $deg_H(S) = 2$, $R_S$ is a half-plane of $\ell_0$.*

*Proof.* The two edges leaving $S$ form a 2-cut. This is a 2-cut of both $H$ and $K$. □

**Lemma 3.** *Let $S \subset V(H)$ such that $S$ induces a connected subgraph. Then*

$$2|S| \leq |E_H(S)| + deg_H(S). \tag{1}$$

*Proof.* Let $a$ be the number of the (extended) line segments in the interior of $R_S$, let $b$ be the number of endpoints in $int(R_S)$ of (extended) line segments which are not entirely in $int(R_S)$, and let $c$ be the number of (extended) line segments on the boundary of $R_S$ with no endpoints in $int(R_S)$.

The following four equalities immediately imply (1).

$$|S| = a + b + 1, \tag{2}$$
$$|E_H(S)| = 2a + b, \tag{3}$$
$$deg_H(S) \geq c + b, \quad \text{but} \tag{4}$$
$$deg_H(S) \geq c + b + 1, \quad \text{if} \quad 1 \leq c \leq 2. \tag{5}$$

(2) and (3) are obvious. For (4), consider the pairs of consecutive line segments along the boundary of $R_S$. Every such pair forms a $T$-junction and corresponds to an edge of $H$. The number of $T$-junctions on the boundary of $R_S$ is at least $b + c$, if $R_S$ is bounded, and $b + c - 1$, if $R_S$ in unbounded.

If $R_S$ is unbounded then there are two line segment extended to infinity on the boundary of $R_S$. Both correspond to an edge of $H$. In this case $deg_H(S) \geq (c + b - 1) + 2 = b + c + 1$.

Finally for (5), note that $c \geq 3$ for $R_S$ bounded, and $c \geq 1$ otherwise. $\qquad \square$

**Definition 2.** *For a graph $G$, we call $S \subset V(G)$ a wheel of $G$, if $S$ induces a connected graph, $|E_G(S)| = |S|$, $deg_G(S) = |S|$, and $deg_G(v) > 2$ for all $v \in S$.*

**Corollary 2.** *If $S \subset V(K)$ is a wheel of $K$ and $|E_H(S)| = |S|$, then $S$ corresponds to a half-plane of $\ell_0$.*

*Proof.* Suppose that $|S| = k$ and $S$ induces at most $k$ edges in $H$.

$$2k = |E_K(S)| + deg_K(S) \leq |E_H(S)| + deg_H(S) \leq |S| + k = 2k.$$

That is (1) holds with equality. Equalities (2), (3), (4), and (5) imply that $c = 1$. $\qquad \square$

**Lemma 4.** *Let $S \subset V(K)$. Then*

$$|E_K(S)| \leq 2|S| - 2 - p' - r' + 2q, \tag{6}$$

*where $p'$ is the number of nodes in $S$ whose degree is two in both $K$ and $K(S)$, and $r'$ is the number of disjoint subsets of $S$ which are wheels in both $K$ and $K(S)$.*

*Proof.* We know $|E(H)| = 2|V(H)| - 2$. Applying Lemma 3 to all components of $V(K) \setminus S$, we have $|E_H(S)| \leq 2|S| - 2$.

Let $S_1$ be the set of nodes $v \in S$ with $deg_K(v) = deg_{K(S)}(v) = 2$. Let $S_2$ be the union of maximal paths $P \subset K(S)$ such that $deg_K(P) = deg_{K(S)}(P) = 2$ and $deg_H(v) > 2$ for all $v \in P$. Let $S_3$ be the union of disjoint vertex sets in $S$ which are wheels in both $K$ and $K(S)$. Observe that $S_1$, $S_2$, and $S_3$ are pairwise disjoint.

According to Corollaries 1 and 2, there are at least $p' + r' - 2q$ double edges in $H(K)$. $\qquad \square$

## 4     $P_{\geq 3}$-factors

A partition of a graph $G$ into paths $\{P_3, P_4, P_5\}$ is a $P_{\geq 3}$-factor of $G$. We prove that the graph $K$ has a $\{P_{\geq 3}\}$-factor. The $\{P_{\geq 3}\}$-factors were recently characterized by Kaneko [12], and a simple proof is provided in [13].

**Theorem 2.** *(Kaneko) A simple graph $G$ has a $\{P_{\geq 3}\}$-factor if and only if*

$$sun(G - S) \leq 2|S|, \quad \text{for all } S \subset V(G). \tag{7}$$

□

The term $sun(G - S)$ requires further explanation.

**Definition 3.** *A graph $G$ is factor-critical, if for all $v \in V(G)$, $G - \{v\}$ contains a perfect matching (1-factor).*

*For a factor critical graph $G$ with $V(G) = \{v_1, v_2, \ldots, v_k\}$, add new vertices $u_1, u_2, \ldots, u_k$ and new edges $v_1 u_1, v_2 u_2, \ldots, v_k u_k$. The resulting graph is a sun. A single node is also considered as a sun.*

*For a graph $G$, $sun(G)$ is the number of its sun components.*

A factor-critical graph of $k > 1$ nodes has at least $k$ edges. To see this, observe that a factor-critical graph is connected, and it cannot contain leaves. Note that a single node is also factor-critical, the corresponding sun component is a path $P_2$. Consequently, a sun component of $2k$ nodes has at least $2k$ edges for all $k > 1$.

**Lemma 5.** *The graph $K$ has a $\{P_{\geq 3}\}$-factor.*

*Proof.* Consider an $S \subset V(K)$, where $sun(K - S) - 2|S|$ is maximal. Delete all components of $K - S$ which are not sun and denote the resulting graph by $K'$. We wish to apply Lemma 4, in particular Inequality (6), to $V(K') \subset V(K)$.

$$|E_K(K')| \leq 2|V(K')| - 2 - p' - r' + 2q, \tag{8}$$

where $p'$ is the number of nodes in $v \in K'$ with $deg_K(v) = deg_{K'}(v) = 2$ and $r'$ is the number of disjoint subsets of $K'$ which are wheels in both $K$ and $K'$.

In order to determine $p'$ and $r'$ in terms of sun components, we introduce new parameters for the number of certain types of sun components of $K - S$. We consider separately the $P_1$-s, $P_2$-s, and all other suns.

Let $c_2$ and $c_3$ be the number of single node components $v \in V(K) \setminus S$ with $deg_K(v) = 2$ and $deg_K(v) \geq 3$ resp. (Note that here $deg_{K'}(v) = deg_K(v)$.)

Let $d_2$, $d_3$, and $d_4$ be the number of $P_2$ components $P \subset V(K) \setminus S$ such that $P \subset K'$ with $deg_K(P) = 2$, $deg_K(P) = 3$, and $deg_K(P) \geq 4$ resp. Observe that $deg_K(P) = 2$ (or $deg_K(P) = 3$) means that $P$ contains one (or two resp.) node(s) with $deg_K(v) = 2$.

Consider the sun components of $K - S$ with more than 2 nodes. Let the total number of their nodes be $2k$. They have $k$ leaves. Let $f_2$ and $f_3$ be the number of leaves $v$ with $deg_K(v) = 2$ and $deg_K(v) > 2$ resp. Let $r_1$ and $r_2$ be the number sun components of $K - S$ with more than 2 nodes, whose factor critical subgraph is wheel and non-wheel resp. These components have at least $2k + r_2$ interior edges and at least $2k - f_2$ edges connect them to $S$.

Using these parameters, we can estimate the number of edges in $K'$. $|E(K')| \le deg_{K'}(S) + (d_2 + d_3 + d_4) + (2k + r_2)$ where $deg_{K'}(S) \le 2c_2 + 3c_3 + 2d_2 + 3d_3 + 4d_4 + (2k - f_2)$. The number of nodes having degree 2 in both $K$ and $K'$ is at least $c_2 + 2d_2 + d_3 + f_2$. Using $q \le 1$, Inequality (8) is written as

$$2c_2 + 3c_3 + 3d_2 + 4d_3 + 5d_4 + 4k + r_2 - f_2 \le$$
$$\le 2(|S| + c_2 + c_3 + 2d_2 + 2d_3 + 2d_4 + 2k) - (c_2 + 2d_2 + d_3 + f_2) - r_1,$$
$$c_2 + c_3 + d_2 + d_3 + d_4 + r_1 + r_2 \le 2|S|.$$

The number of sun components is at most $2|S|$. According to Kaneko's theorem, $K$ has a $\{P_{\ge 3}\}$-factor.    □

## 5    NP-hardness of optimization Problem

The SET-COVER problem is formulated as follows. Given a set of $n$ elements $A = \{e_1, e_2, \ldots, e_n\}$, and $m$ sets $S_1, S_2, \ldots, S_m \subset A$, find a minimum number of sets so that their union equals $A$. The SET-COVER problem is known to be NP-hard [14].

**Theorem 3.** *The SET-COVER problem can be reduced polynomially to determining the minimum number of light sources required for Problem 1.*

*Proof.* We give a set $\mathcal{L}$ of $5 + 10|A| + 5\sum_{i=1}^{m}|S_i|$ line segments so that a minimal set cover can be determined in polynomial time from a minimal set of light sources illuminating both sides of the lines in $\mathcal{L}$.

Consider a SET-COVER problem. Let $A'$ be the multiset containing the elements of $A$ with multiplicity 5, ($|A'| = 5n$).

In the Cartesian coordinate system, place a line segment $\ell_0 = [(0, 60), (10(m + 2), 60)]$. We associate the point $(10k, 60) \in \ell_0$ to the set $S_k$ for all $k = 1, 2, \ldots, m$. Denote the point corresponding to $S_k$ by $S_k$ as well. We associate disjoint unit intervals on the line $y = 0$ to each element of $A'$. Define an arbitrary linear order on $A'$ and the placement of the unit intervals are defined consecutively according to this order. The first interval is $[(10(m + 2), 0), (10(m + 2) + 1, 0)]$. The $k$-th interval is $[(c_k, 0), (c_k + 1, 0)]$ for all $k = 1, 2, \ldots, 5n$.

The basic idea of the reduction is that a unit interval $e_i$ should be visible from a point $S_j$ if and only if $e_i \in S_j$. We will place horizontal line segments on the line $y = 30$ so that $e_i$ is not visible from $S_j$ if $e_i \notin S_j$. In fact, the visibility of $e_i \in S_j$ will be realized by a 'hole' on the line $y = 30$ between the intersections of the lines $[(c_i, 0)S_j], [(c_i + 1, 0)S_i]$, and $y = 30$. The parameters $c_k$ should be chosen so that these 'holes' are disjoint. Then at most $5\sum_{i=1}^{m}|S_i|$ disjoint line segments on the line $y = 30$ can bar the visibility between all pairs $(e_i, S_j)$ with $e_i \notin S_j$.

The *hole* is the interval $[((S_j + c_i)/2, 30), ((S_j + c_i + 1)/2, 30)] = [(f_i^j, 30), (f_i^j + 1/2, 30)]$ for a pair $(e_i, S_j)$ with $e_i \in S_j$. Observe that holes corresponding to a same interval $[(c_i, 0), (c_i + 1, 0)]$ are all disjoint by construction.

If the first $k-1$ unit intervals on $y = 0$ are already defined, then we choose the parameter $c_k$ as follows. We give lower bounds for $c_k$ and choose $c_k$ arbitrarily satisfying the bounds. First of all, $c_k > c_{k-1} + 1$.

Line $[(0, 60), (c_{k-1} + 2, 30)]$ intersects the line $y = 0$ at $d_{k-1}$. Let $c_k > d_{k-1}$.

Form the intersections of lines $[(c_i, 0)(f_i^j + 1/2, 30)]$ and $[(c_g + 1, 0)(f_g^h, 30)]$ for all $i < g < k$, and for all $j > h$. Project every such intersection point from the point $(0, 60)$ to the line $y = 0$. Let $(c_k, 0)$ be to the right of every such projected point.

Place two disjoint line segments $\ell_k^1$ and $\ell_k^2$ for each interval $[(c_k, 0), (c_k + 1, 0)]$ as follows. $\ell_k^1$ consists of the points of line $[(0, 60), (c_k, 0)]$ with $x$-coordinates $c_k \leq x \leq c_k + 1 - 2\varepsilon$, say $\ell_k^1 = [(c_k, 0), (c_k + 1 - 2\varepsilon, c_k')]$. Let $\ell_k^2$ be then $[(c_k + 1 - \varepsilon, c_k' - \varepsilon), (c_k + 1, 0)]$ where $\varepsilon$ is a sufficiently small constant. (Fig. 4).

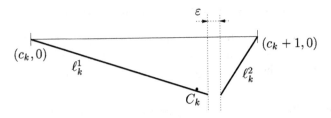

**Fig. 4.** Line segments corresponding to one unit interval

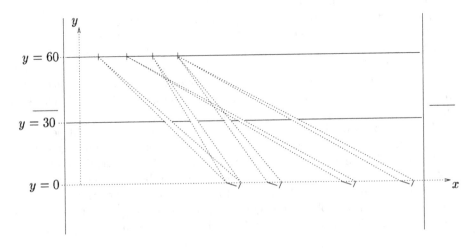

**Fig. 5.** Overview of the construction (multiple elements of $A'$ are not indicated on this figure)

The construction is completed by two pairs of line segment on the left and right extremities as indicated in Fig. 5. Finally perturb the line segments to be pairwise non-parallel.

Now check out the visibility conditions. One light source can illuminate the lower side of every pair of line segments corresponding to unit intervals. One light source can illuminate the upper side of the line segment on $y = 60$. One light source can illuminate the lower sides of the $5 \sum_{i=1}^{m} |S_i|$ line segments on the line $y = 30$. This latter source might illuminate the upper side of some of the segments corresponding to unit intervals.

The problem rests to illuminate the upper sides of the line segments corresponding to the unit intervals. A set of light sources placed to a minimal set cover illuminates the upper side of all $\ell_i^1, \ell_i^2, \ i = 1, \ldots, 5n$. Why cannot be there a smaller set of light sources?

We choose one point on a line segment corresponding to each unit interval. Let $C_i \in \ell_i^1$ be a point of $x$ coordinate $c_k + 1 - 3\varepsilon$ (see Fig. 4). It is enough to check if a smaller set of light sources can illuminate all points $C_i \in \ell_i^1$ from the upper side for all $i = 1, \ldots, 5n$.

Over the line $y = 60$, no light source can illuminate a point $C_i$ for any $i = 1, 2, \ldots, 5k$. Under the line $y = 30$, no light source can illuminate two different $C_i$ and $C_j$ from the upper side, because $c_k > d_{k-1}$ and $\varepsilon$ is small.

If $C_i$ is visible from a point $P(P_x, P_y)$ where $30 < P_y < 60$, then $P$ is in a trapezoid $T_i^j$ formed by the lines $y = 30$, $y = 60$, $[(c_i, 0)(f_i^j + 1/2, 30)]$, and $[(c_i + 1, 0)(f_i^j, 30)]$ for some $j : e_i \in S_j$. The choice of $c_i$ guarantees that three trapezoid $T_{i_1}^{j_1}, T_{i_2}^{j_2}$, and $T_{i_3}^{j_3}$ have a common intersection if and only if $j_1 = j_2 = j_3$. If there is a light source in the intersection of three trapezoids $T_{i_1}^{j}, T_{i_2}^{j}, T_{i_3}^{j}$, then it can be replaced by $S_j$.

Finally suppose that there is a set $Z$ of light sources illuminating all points $C_i$ from the upper side, but $Z_S = Z \cap \{S_1, S_2, \ldots, S_m\}$ do not illuminate every $C_i$. There are at least five points $C_i$ not illuminated by $Z_S$, because the elements of $A$ appear in $A'$ with multiplicity five.

Suppose that $5k$ points $C_i$ are not illuminated by $Z_S$. A light source of $Z \setminus Z_S$ can illuminate two of them, so $|Z \setminus Z_S| \geq \lceil 5k/2 \rceil$. But these points $C_i$ can be illuminated by at most $k$ points $S_j$ because every five-tuple of corresponding elements of $A'$ belong to at least one set $S_j$. □

## Acknowledgments

I would like to thank G. Y. Katona for pointing out Kaneko's Theorem.

## References

1. Baker, B. S.: Approximation algorithms for NP-complete problems on planar graphs. *J. Assoc. Comput. Math.* **41** (1994) 153–180
2. Berge, C., Sur le couplage maximum d'un graphe. *C. R. Acad. Sciences, Paris* **247** (1958) 258–259

380    Csaba D. Toth

3. Berman, F., Johnson, D., Leighton, T., Shor, P. W., Snyder, L.: Generalized planar matching. *J. Algorithms* **11** (1990) 153–184

4. Czyzowicz, J., Rival, I., Urrutia, J.:  Galleries, light matchings, and visibility graphs. In: Galleries, light matchings and visibility graphs. Algorithms and data structures (Ottawa, ON, 1989), Lecture Notes in Comput. Sci. **382**, Springer-Verlag, Berlin, (1989) 316–324

5. Czyzowicz, J., Rivera-Campo, E., Urrutia, J.: Illuminating rectangles and triangles in the plane. *J. Combin. Theory Ser. B* (1993) 1–17

6. Czyzowicz, J., Rivera-Campo, E., Santoro, N., Urrutia, J., Zaks, J.: Tight bounds for the rectangular art gallery problem, in: *Graph-theoretic concepts in computer science* (Fischbachau, 1991), Lecture Notes in Comput. Sci. **570**, Springer-Verlag, Berlin, (1992) 105–112

7. Czyzowicz, J., Rivera-Campo, E., Urrutia, J., Zaks, J.: On illuminating line segments in the plane. *Discrete Math.* **137** (1995) 147–153

8. Eidenbenz, S., Stamm, C., Widmayer, P.: Inapproximability of some art gallery problems, in: *Proceedings 10th Canadian Conference on Computational Geometry (Montréal, 1998)*

9. Fejes Tóth, L.: Illumination of convex disks. *Acta Math. Acad. Sci. Hungar.* **29** (1977) 355–360

10. Hartvigsen, D., Hell, P.: The $k$-piece packing problem. Manuscript (2000)

11. Hoffmann, F.: On the rectilinear Art Gallery Problem. in: Proc ICALP, *Lecture Notes in Comput. Sci.* **90**, Springer-Verlag (1990) 717–728

12. Kaneko, A.: A necessary and sufficient condition for the existence of a path factor every component of which is a path of length at least two. Submitted

13. Kano, M., Katona, G. Y., Király, Z.: On path-factors and subfactors of graphs without paths of length one. Submitted

14. Karp, R. M.: Reducibility among combinatorial problems, in: *Complexity of Computer Computations (R. Miller and J. Thatcher, eds.)*. Plenum Press, New York (1972) 85–103

15. Kirkpatrick, D. G., Hell, P.: On the completeness of a generalized matching problem, in: *Proc. 10th Ann. ACM Symp. on Theory of Computing* (San Diego, Calif., 1978), ACM, New York (1978) 240–245

16. Lee, D. T., Lin, A. K. : Computational complexity of art gallery problems. *IEEE Trans. Inform. Theory* **32** (1986) 276–282

17. Nishizeki, T., Baybars, I.: Lower bounds on the cardinality of the maximum matchings of planar graphs. *Discrete Math.* **28** (1979) 255–267

18. O'Rourke, J.: Open problems in the combinatorics of visibility and illumination, in: *Advances in Discrete and Computational Geometry (B. Chazelle, J. E. Goodman, and R. Pollack, eds.) (Contemporary Mathematics)*, AMS, Providence (1998) 237–243

19. O'Rourke, J.: Art gallery theorems and algorithms. The International Series of Monographs on Computer Science, Oxford University Press, New York (1987)

20. Tóth, Cs. D.: Illumination in the presence of opaque line segments in the plane. Submitted

21. Tóth, Cs. D.: Illuminating disjoint line segments in the plane. Submitted

22. Urrutia, J.: Art Gallery and Illumination Problems. in: *Handbook on Computational Geometry (J. R. Sack, J. Urrutia eds.)*, Elsevier Science Publishers, Amsterdam (2000) 973–1027

23. Zaks, J.: A note on illuminating line segments in the plane. Manuscript (1993)

# Author Index

# Lecture Notes in Computer Science

For information about Vols. 1–2013
please contact your bookseller or Springer-Verlag

Vol. 2055: M. Margenstern, Y. Rogozhin (Eds.), Machines, Computations, and Universality. Proceedings, 2001. VIII, 321 pages. 2001.

Vol. 2056: E. Stroulia, S. Matwin (Eds.), Advances in Artificial Intelligence. Proceedings, 2001. XII, 366 pages. 2001. (Subseries LNAI).

Vol. 2057: M. Dwyer (Ed.), Model Checking Software. Proceedings, 2001. X, 313 pages. 2001.

Vol. 2059: C. Arcelli, L.P. Cordella, G. Sanniti di Baja (Eds.), Visual Form 2001. Proceedings, 2001. XIV, 799 pages. 2001.

Vol. 2060: T. Böhme, H. Unger (Eds.), Innovative Internet Computing Systems. Proceedings, 2001. VIII, 183 pages. 2001.

Vol. 2062: A. Nareyek, Constraint-Based Agents. XIV, 178 pages. 2001. (Subseries LNAI).

Vol. 2064: J. Blanck, V. Brattka, P. Hertling (Eds.), Computability and Complexity in Analysis. Proceedings, 2000. VIII, 395 pages. 2001.

Vol. 2065: H. Balster, B. de Brock, S. Conrad (Eds.), Database Schema Evolution and Meta-Modeling. Proceedings, 2000. X, 245 pages. 2001.

Vol. 2066: O. Gascuel, M.-F. Sagot (Eds.), Computational Biology. Proceedings, 2000. X, 165 pages. 2001.

Vol. 2068: K.R. Dittrich, A. Geppert, M.C. Norrie (Eds.), Advanced Information Systems Engineering. Proceedings, 2001. XII, 484 pages. 2001.

Vol. 2070: L. Monostori, J. Váncza, M. Ali (Eds.), Engineering of Intelligent Systems. Proceedings, 2001. XVIII, 951 pages. 2001. (Subseries LNAI).

Vol. 2071: R. Harper (Ed.), Types in Compilation. Proceedings, 2000. IX, 207 pages. 2001.

Vol. 2072: J. Lindskov Knudsen (Ed.), ECOOP 2001 – Object-Oriented Programming. Proceedings, 2001. XIII, 429 pages. 2001.

Vol. 2073: V.N. Alexandrov, J.J. Dongarra, B.A. Juliano, R.S. Renner, C.J.K. Tan (Eds.), Computational Science – ICCS 2001. Part I. Proceedings, 2001. XXVIII, 1306 pages. 2001.

Vol. 2074: V.N. Alexandrov, J.J. Dongarra, B.A. Juliano, R.S. Renner, C.J.K. Tan (Eds.), Computational Science – ICCS 2001. Part II. Proceedings, 2001. XXVIII, 1076 pages. 2001.

Vol. 2075: J.-M. Colom, M. Koutny (Eds.), Applications and Theory of Petri Nets 2001. Proceedings, 2001. XII, 403 pages. 2001.

Vol. 2076: F. Orejas, P.G. Spirakis, J. van Leeuwen (Eds.), Automata, Languages and Programming. Proceedings, 2001. XIV, 1083 pages. 2001.

Vol. 2077: V. Ambriola (Ed.), Software Process Technology. Proceedings, 2001. VIII, 247 pages. 2001.

Vol. 2078: R. Reed, J. Reed (Eds.), SDL 2001: Meeting UML. Proceedings, 2001. XI, 439 pages. 2001.

Vol. 2081: K. Aardal, B. Gerards (Eds.), Integer Programming and Combinatorial Optimization. Proceedings, 2001. XI, 423 pages. 2001.

Vol. 2082: M.F. Insana, R.M. Leahy (Eds.), Information Processing in Medical Imaging. Proceedings, 2001. XVI, 537 pages. 2001.

Vol. 2083: R. Goré, A. Leitsch, T. Nipkow (Eds.), Automated Reasoning. Proceedings, 2001. XV, 708 pages. 2001. (Subseries LNAI).

Vol. 2084: J. Mira, A. Prieto (Eds.), Connectionist Models of Neurons, Learning Processes, and Artificial Intelligence. Proceedings, 2001. Part I. XXVII, 836 pages. 2001.

Vol. 2085: J. Mira, A. Prieto (Eds.), Bio-Inspired Applications of Connectionism. Proceedings, 2001. Part II. XXVII, 848 pages. 2001.

Vol. 2086: M. Luck, V. Mařík, O. Štěpánková, R. Trappl (Eds.), Multi-Agent Systems and Applications. Proceedings, 2001. X, 437 pages. 2001. (Subseries LNAI).

Vol. 2089: A. Amir, G.M. Landau (Eds.), Combinatorial Pattern Matching. Proceedings, 2001. VIII, 273 pages. 2001.

Vol. 2091: J. Bigun, F. Smeraldi (Eds.), Audio- and Video-Based Biometric Person Authentication. Proceedings, 2001. XIII, 374 pages. 2001.

Vol. 2092: L. Wolf, D. Hutchison, R. Steinmetz (Eds.), Quality of Service – IWQoS 2001. Proceedings, 2001. XII, 435 pages. 2001.

Vol. 2093: P. Lorenz (Ed.), Networking – ICN 2001. Proceedings, 2001. Part I. XXV, 843 pages. 2001.

Vol. 2094: P. Lorenz (Ed.), Networking – ICN 2001. Proceedings, 2001. Part II. XXV, 899 pages. 2001.

Vol. 2095: B. Schiele, G. Sagerer (Eds.), Computer Vision Systems. Proceedings, 2001. X, 313 pages. 2001.

Vol. 2096: J. Kittler, F. Roli (Eds.), Multiple Classifier Systems. Proceedings, 2001. XII, 456 pages. 2001.

Vol. 2097: B. Read (Ed.), Advances in Databases. Proceedings, 2001. X, 219 pages. 2001.

Vol. 2098: J. Akiyama, M. Kano, M. Urabe (Eds.), Discrete and Computational Geometry. Proceedings, 2000. XI, 381 pages. 2001.

Vol. 2099: P. de Groote, G. Morrill, C. Retoré (Eds.), Logical Aspects of Computational Linguistics. Proceedings, 2001. VIII, 311 pages. 2001. (Subseries LNAI).

Vol. 2105: W. Kim, T.-W. Ling, Y-J. Lee, S.-S. Park (Eds.), The Human Society and the Internet. Proceedings, 2001. XVI, 470 pages. 2001.

Vol. 2106: M. Kerckhove (Ed.), Scale-Space and Morphology in Computer Vision. Proceedings, 2001. XI, 435 pages. 2001.

Vol. 2110: B. Hertzberger, A. Hoekstra, R. Williams (Eds.), High-Performance Computing and Networking. Proceedings, 2001. XVII, 733 pages. 2001.

Vol. 2118: X.S. Wang, G. Yu, H. Lu (Eds.), Advances in Web-Age Information Management. Proceedings, 2001. XV, 418 pages. 2001.

Vol. 2119: V. Varadharajan, Y. Mu (Eds.), Information Security and Privacy. Proceedings, 2001. XI, 522 pages. 2001.

Vol. 2121: C.S. Jensen, M. Schneider, B. Seeger, V.J. Tsotras (Eds.), Advances in Spatial and Temporal Databases. Proceedings, 2001. XI, 543 pages. 2001.

Vol. 2126: P. Cousot (Ed.), Static Analysis. Proceedings, 2001. XI, 439 pages. 2001.